诺丁汉 100

——研究主导型设计室文化

Nottingham 100: Research-led Studio Culture

诺丁汉 100

——研究主导型设计室文化

Nottingham 100: Research-led Studio Culture

王琦　【英】达伦·迪恩 编著
Editors: Darren Deane, Wang Qi

中国建筑工业出版社
China Architecture & Building Press

图书在版编目(CIP)数据

诺丁汉100——研究主导型设计室文化／王琦，(英) 迪恩编著．
北京：中国建筑工业出版社，2013.12
ISBN 978-7-112-16240-6

Ⅰ．①诺… Ⅱ．①王… ②迪… Ⅲ．①建筑设计—英国—图
集 Ⅳ．①TU206

中国版本图书馆CIP数据核字（2013）第306742号

责任编辑：戚琳琳
责任校对：王雪竹 陈晶晶

诺丁汉100
——研究主导型设计室文化

Nottingham 100: Research-led Studio Culture
王琦【英】达伦·迪恩 编著
Editors: Darren Deane, Wang Qi
*
中国建筑工业出版社出版、发行（北京西郊百万庄）
各地新华书店、建筑书店经销
北京美光制版有限公司制版
北京方嘉彩色印刷有限责任公司印刷
*
开本：889×1194毫米 1/16 印张：15¼ 字数：480千字
2014年6月第一版 2014年6月第一次印刷
定价：158.00元
ISBN 978-7-112-16240-6
(24953)

ACKNOWLEDGEMENT - 致谢

Design teaching can thrive only under certain conditions, one of which might be described as an atmosphere of free thought. At the Department of Architecture and Built Environment, University of Nottingham, Professors Tim Heath, Brian Ford, Michael Stacey, along with Dr. Jonathan Hale, Dr. Laura Hanks, Dr Sergio Altomonte, and Mr David Short, have played an important role in creating a thoughtful curriculum. Additionally, nearly all full time staff supported this book in countless ways, whilst the extensive team of part-time tutors drawn from professional practice have ensured the currency and relevance of studio praxis. The students themselves, many of who could not be mentioned, have by far made the most significant contribution, as it is they who ultimately placed trust in guidance, and who extended advice much further than their teachers could ever imagine. In a project of this size and scope the list of contributors extends beyond those who are directly named in the volume itself. Any successful studio culture depends upon additional layers of support from individuals who, although not on the front line of teaching and research, make it possible for a department of architecture to function.

Without the careful translation of the following individuals the work contained in this volume would not have reached a Chinese audience in such high quality. Many thanks should give to Xiao Jing, Jia Min, Zheng Xiaofeng, and Gao Nan, who took the job of minor translation in different chapters. Special mention must go to Zhang Licheng and Ding Guanghui, who took on major tasks translating some of the more difficult sections first written in English.

With regards to the editors, Darren Deane would like to thank Adrian Ball of SatmokoBall Architects, London, for his long and valuable collaboration, along with Matt Mckenna who since 2008 has supported the Lateralisms series of studio production, and Tony Swannell for his boundless passion and energy. Finally, to Angela, Jude and Mimi for their patience. Wang Qi would like to thank Lei Yanhui for her attention to detail on page layout, proofreading, and overall invaluable encouragement, Wang Qimin for his initial conception of the project, and the devoted support of Qi Linlin from China Architecture and Building Press.

设计课程教学只能在具备某些必要条件的情况下繁荣发展，而其中一个条件或许就是那种鼓励自由思维的氛围。在诺丁汉大学建筑与建造环境学院里，蒂姆·希斯教授、布莱恩·福特教授、迈克尔·斯泰西教授，与乔纳森·赫尔博士、劳拉·汉克斯博士、塞尔吉奥·阿尔特蒙特博士，以及大卫·肖特先生等一道，孜孜不倦地鼓励并维护着各种思想百花齐放。此外，几乎全体教职员工也均对此书给予了各种各样的支持，同时一支来自设计实践行业的兼职教员团队也为建筑设计教学能够始终保持连贯性与相关性奉献颇丰。而学生们，尽管他们中的许多成员并没有在本书中被提及，却是做出了最为重要的贡献，正是他们将教导转化为无限的信任，将建议引申发展到即使老师自己也始料未及的深度。对于这样一本博采众长的集子而言，对它作出贡献的人们远远不止名列其中的那些。任何成功的建筑设计课程文化的形成，均需要来自不同个体的多层面支持，尽管他们未必站在教学与科研的第一线，但却对于整个建筑系的正常运转而言至关重要。

没有下列诸君缜密仔细的翻译，本书将无法以如此之高质量呈现在中国读者面前。肖靖、贾敏、郑小锋与郜楠帮助翻译了少部分文字，应予以诚挚的感谢。而特别的感谢则应给予张沥成和丁光辉，他们各自担负了主要翻译工作，将许多最为困难的部分译为流畅通顺的中文。

对于编者而言，达伦·迪恩希望特别感谢来自Satmoko Ball建筑师事务所的阿德里安·博尔所提供的长期且有价值的合作，迈特·麦肯纳自从2008年以来对“写实主义系列设计课程”所给予的支持，托尼·斯瓦内尔那无限的热情与能量，以及最后，感谢安吉拉、犹大和咪咪的耐心。王琦则希望特别感谢类延辉对页面设计的仔细校对、文字纠错，以及无价的全心支持，王奇民为此书的诞生提供了最初的概念想法，以及戚琳琳站在中国建筑工业出版社的角度对本书出版所给予的大力支持。

Darren Deane & Wang Qi
达伦·迪恩　王琦
2012/12

CONTENTS - 目录

FOREWORD - 序

ARCHITECTURE AT THE UNIVERSITY OF NOTTINGHAM
建筑学在诺丁汉大学

This magnificent book captures some of the recent highlights to emerge from the Department of Architecture & Built Environment at the University of Nottingham. The Department has an amazing track record and can trace its history back to 1843 with the first 'training' of architects in Nottingham and the establishment of the first department in 1865. In modern times, the Department has become one of the leading 'schools of architecture' across the world with students and alumni from most countries.

The breadth of expertise within the Department enables students to explore architecture through a plethora of approaches enabling them to develop a rigorous knowledge and understanding of the discipline and profession. Importantly, teaching is informed and enhanced by research being undertaken by academics and the projects in this book are organized by the four dominant themes that reflect their specialisms: architectural humanities, architectural environmental design, architecture and tectonics, and urban design. The department is also particularly proud of its 'live projects' that introduce students to specific challenges – such as the social architecture projects in South Africa and the Solar Decathlon project – and develop hands-on construction and management skills.

The chapters in this book offer a glimpse of the varied projects undertaken within the Department demonstrating the different scales of project, challenges being undertaken and issues being tackled. Most of all, it illustrates the intellectual and professional rigor with which our students address their work and the exemplary skills that have made of graduates highly sought after around the world. As the Department continues to flourish – and with the recent establishment of the course at our campus in Ningbo, China – there will be increasingly impressive student work and research emerging from the University of Nottingham.

在这本漂亮的集子中，收录了近几年来从诺丁汉大学建筑与建造环境学院所涌现出的一批优秀作品。学院本身拥有着令人自豪的历史轨迹，其历史最早可追溯到1843年为诺丁汉地区提供首次建筑师的培训，以及1865年学院正式成立。在当代，本学院业已名列世界知名建筑院校之一，而我们的学生与毕业生则遍布世界大多数国家，桃李满天下。

学院中各个专业间的广泛融合使学生们能够从不同的角度探讨建筑，进而既可发展出缜密综合的知识体系，又可了解不同的专业特点。更为重要的是，这里的教学系统通过教师所进行的学术研究而得到了支持与强化，正如在本书中所展示的那样，所有方案均可根据其专业特点而被细分入四个科研主题：建筑人文篇、建筑环境设计篇、建筑与建构篇，以及城市设计篇。除此之外，学院还为其拥有的“实践项目”而感到骄傲，诸如南非的社会建筑项目与太阳能十项全能项目等，均为学生们提供了特殊的挑战以及发展动手建造和项目管理技巧的宝贵机会。

本书中的各章节提供了一处小小窗口去了解学院中各种不同的设计方案，它们有着不同的尺度，迎接不同的挑战，以及试图去应对不同的问题。然而尤为重要的是，它不但通过方案体现了我们学生们的聪明才智与专业技能，而且展示了学生们杰出的设计技能——而正是这些技能使得我们的毕业生在世界各地供不应求。作为一所正在持续上升的学院，并且挟诺丁汉大学中国宁波校区最近设立建筑学专业的东风，我们相信在将来，必定会有更多引人注目的学生作品与研究成果从诺丁汉大学孕育而生。

Prof. Tim Heath
蒂姆·希斯 教授

Professor of Architecture & Urban Design
Vice Dean of Faculty of Engineering
建筑与城市设计教授
工程学院副院长

PREFACE - 前言

NOTTINGHAM – A WINDOW LOOKING INTO THE BRITISH SYSTEM OF ARCHITECTURAL EDUCATION

诺丁汉,一扇了解英国建筑教育体系的窗

In the UK the path to qualification as an architect is totally different to that followed in China. As we know to qualify as a Chinese level-A Architect, architectural students firstly need to complete the university degree; then based on the levels of their last academic degree, they are required to work additional years in practice before taking part in the National Architect Qualification Examination. This includes 9 modules that need to be passed in eight years – normally BArch students are asked to work in practice for three years and MArch students need to work for two. This system indicates that, to qualify as an architect in China one needs to spend a minimum nine years (5 years BArch + 3 years working experience + to pass all 9 modules in 1 year time if you are working hard enough) or up to maximum eighteen years (5 years BArch + 3 years March + 2 years working experience + to pass all 9 modules in 10 year time if you are not really lucky indeed) in the professional scope. By comparison, in the UK the route to Charted Architect status is much faster. All architectural departments validated and prescribed by the Royal Institute of British Architect (RIBA) and Architects Registration Board (ARB) can provide three-levels professional training courses: Part I, Part II and Part III. After successful completion of three years BArch course, students qualify as Part I Architects. Then after one year out in practice, if they choose to continue onto a two years Diploma course and pass it, the title of Part II Architect can be obtained. Finally after taking part in a one year part-time professional training course whilst working in practice, culminating in a series of complex course work, written examinations, and an interview, a title of Charted Architect is awarded. This shortest period in which this training can be completed is seven years in total and there is no additional specific registration examination involved. Indeed, to be qualified as a British architect, one simply needs to work hard in the university and to smoothly progress through the three-tiered training courses.

So clearly there is a significant difference in between the two qualification systems. Nevertheless, a true comparison cannot be based solely on the time it takes to professionally qualify and the style of progress. On the contrary, the most reliable criteria are embedded in the essence and strategy of architectural pedagogy. "Nottingham 100" provides such a small window offering an insight into architectural education in the UK. Although the aim is not to provide a comprehensive panorama, what follows is a snapshot of current trends – just like the old Chinese saying:"sensing the fall with one fallen leaf".

This collection contains 100 innovative student projects selected from the studio works completed in the Department of Architecture and Built Environment, University of Nottingham, between 2006 and 2012. Many were designed by year 5, year 6 and Masters students, with a small group of exemplary samples from year 2 and year 3 studios. As a 'Nottingham Book', it aims to reflect the full spectrum of architectural education within the Department and to illuminate the strong tradition and personality of a world-leading school of architecture. Furthermore, as a project archive representing one of the strongest validated departments in the UK, the book also demonstrates the principles and criteria advocated by the RIBA and ARB.

Nottingham's strategy of architectural education may be summed up in three key words. Then "humanities", "sustainability" and "freeness" could be the best extractions.

"Humanities" can be found in all Nottingham projects because it is the primary teaching principle. Crossing the different Years, the gauge of increasing difficulty in design is not measured by the complexity of functional arrangement or the size of constructional area. On the contrary, readers could easily find a large amount of projects no bigger than several hundred square meters and

在英国，想成为建筑师的年轻人要走一条与中国完全不同的资格认证途径。我们知道，为取得中国一级注册建筑师头衔，建筑学专业的学生们需要首先完成大学教育；然后再根据自己学术学位的高低而工作一定的时间——往往本科学位要工作三年，硕士学位要工作两年；最后在最多八年的时间里参加且通过九门国家一级注册建筑师资格考试。这就意味着想当一名中国建筑师，您需要经历短则9年（5年本科+3年工作+1年即通过全部9门考试）、长则18年（5年本科+3年硕士+2年工作+8年通过全部9门考试）的专业积累过程。而在英国，取得注册建筑师的头衔相比之下要快捷得多。通过英国皇家建筑学会（RIBA）与建筑师注册委员会（ARB）评估的建筑学院，均可以提供三阶段建筑师培养课程。这其中，完成三年本科学习的学生们将可获得建筑学学士学位（BArch）暨第一阶段建筑师头衔；此后，在工作实习一年后，他们可选择进而完成两年期的建筑学学位暨第二阶段建筑师头衔培训；最后，再在自己的工作岗位上接受为期一年的在职函授培训课程，并通过一系列复杂的课业，笔试与答辩，便可以成为一名英国注册建筑师了。这一过程总共加起来最短不过七年时间，而且学生并不需要参加额外的建筑师资格考试——想成为一名英国建筑师，只需要好好学习，顺利通过三阶段的培训就可以了。

不用说，从表面看来，英国的系统与中国的系统有着巨大的差异。然而，仅仅凭借着时间与考核方法的差别来判定孰优孰劣俨然有失严谨，而对建筑教育内涵与其指导思想的分析才可以提供最可靠的评价。本书正是为各位读者提供了这样一个了解英国建筑教育的小小窗口。从这100个方案中，纵然无法览其全貌，或许也可观其大意，见一叶而知秋至吧！

书中甄选了从2006年到2012年间，诺丁汉大学建筑与建造环境学院学生设计的100项独具创意，颇有特色的方案，这其中以五年级、六年级以及硕士研究生的设计方案为主，兼顾少数二年级与三年级的代表作品。作为一本诺丁汉大学的专辑，本书当然要在最大限度内体现出本学院的办学特色，并通过方案，将这所名扬全球的建筑院系的传统与性格阐述清楚。但除此之外，作为一家通过评估的传统强校的方案集，英国皇家建筑学会与建筑师注册委员会所倡导的建筑师培训宗旨自然也将在本书中得以体现。

如果试图用几个简单的词汇来总结诺丁汉大学的建筑教学理念，那么“人文精神”、“可持续发展”与“自由探索”可能是相对比较贴切的提炼。

“人文关怀”几乎印证在所有诺丁汉学子的方案之中，因为这是办学的首要理念。在诺丁汉的各年级中，建筑设计难度的递增并不是依照建筑功能的复杂程度与建筑面积的大小而定，与之恰恰相反，读者会轻而易举地在高年级甚至毕业班的设计方案

containing no more than a few simple functions, completed or even final year students. The first chapter (Architectural Humanities) and the fourth chapter (Urban Design) gathered the most of this kind, but this approach can be observed in other chapters, too. Among them, some chose historical buildings currently under threat and revitalized them through imaginative and sensitive reconstruction; some investigated neglected pockets of cities and brought them back to life through proper urban design; some explored hidden conflicts within multicultural society and reconciled harmony through cogent architectural intervention; and some sensed the loss of subtle functions within context and attempted to fill using contemporary re-programming of space. Indeed, it is not hard to see that, although the buildings may be small, their proposed social influences cover a vast area, and although the immediate internal function might be simple, their social functions are considerably rich and complicated.

In Nottingham, we emphasize that architecture, as a general type of artificial interference into society, regardless of size or complexity will bring significant and various effects to the surrounding built physical environment, demographic composition, cultural background and city operation. These effects may bring improvement or deterioration, prosperity or recession, revival or collapse in the built environment, but they are never useless or inactive. Undoubtedly, the positive impact is a natural expectation of everyone. However it can only be realized by manipulating meticulous analysis of local context, and establishing interwoven relationships with local context. Therefore, in Nottingham the Department emphasizes the importance of site analysis and concept generation throughout all studio works. The qualities have direct impact on a tutors' judgement. In brief, what we are looking for are projects that can illuminate the spirit of humanities for local communities, rather than beautiful and luxury "vases" – dazzling on the outside but hollow on the inside.

Based on the statistics of the UK Green Building Council, "globally, the built environment accounts for 40-50% of natural resource use, 20% of water use, 30-40% of energy use and around a third of CO2 emissions." This has led to more stringent targets to the extent that "in December 2006, the UK Government promised that all new homes would be 'zero carbon' from 2016."[1] This commitment suddenly raised wide-range echo throughout the field of architecture. Regardless of whether this target can be seriously achieved on time, it represents a positive attitude towards dealing with climate change. Subsequently, the entire British architectural profession adopted the new horizon of "Sustainable Architecture" as its guiding purpose, across the complimentary angles of planning, design, structure, construction and materiality. As a base for integrated research and education, the University of Nottingham has closely mirrored this green shift. In fact, the Department has committed itself to the research and application of sustainable building technologies since the 1990s. Its persistent development in this area has enabled it to gradually build up a high international academic reputation, and the Department has been recognized as a world-leading research centre on sustainable building technology. Here, not only has "sustainability" been defined as one of the key pedagogical strategies, but also many Department buildings have been converted into environmental friendly facilities that in turn can be used as testing facilities. Various passive sustainable technologies, such as solar panels, wind turbines, rain water collection and reuse, geothermal heat pumps, self-cleaning glazing, solar water heating system, natural ventilation, and light tube strategy, etc. have all been applied to old and new buildings for PhD students to test. This advantageous situation provides the students with a unique opportunity to experience the most advanced sustainable technologies whenever they are in the Department, and allows them to understand the strengths and weaknesses of technology in a direct way through their own feeling, to critically analyse their performance, and to inventively merge this knowledge into their own studio works.

The projects in the second chapter (Architectural Environmental Design) and the third chapter (Architecture and Tectonics) mainly embody this attitude. Within this section, the most striking are twelve Tall Building projects. Over 200 meters high, these skyscrapers are proposed in major metropolitan areas around the world. Although key issues like structural system and fire evacuation have been carefully discussed, the studio never detracts from their ultimate aim – the creation of self-sufficient, sustainable tall buildings. Besides applying various sustainable technologies to the buildings, these schemes also concerned with another kind of sustainability – social sustainable development. Therefore, these "Made in Nottingham" skyscrapers all become useful new members to different cities, who impose no burden on the cities' energy bill whilst enriching the local culture through their architectural language. The Environmental Design Studio, led by Professor Brian Ford, prefers to explore the essence of sustainability in a more traditional sense. Based on

1. New Build, from http://www.ukgbc.org/content/new-build [Access by 2012-11-27]

中，找到大量面积不过几百平方米，直接功能十分单一的方案。本书建筑人文篇与城市设计篇中集中了此类方案中的大多数，而在其余各篇章之中，该现象也是比比皆是。这其中，有的选择正在受到威胁的旧建筑遗存，通过建筑改造使其光辉重现；有的注意到城市中破落肮脏的死角，通过巧妙规划使其起死回生；有的探究现代多元社会中的潜在矛盾，通过建筑介入来重塑和谐；有的敏锐感知到某种微妙的功能缺失，通过建筑补丁来加以强化。不难发现，尽管建筑面积不大，但它们的社会影响面积却十分广阔；尽管直接功能简单，但它们起到的社会性功能却异常复杂。

在诺丁汉，我们强调建筑是对社会有着至关重要影响力的人工介入。一栋建筑的出现，无论其大小繁简，均会对周围的建造环境、社会组成、文化背景，以及城市运营产生各种复杂的影响——或提升、或降低、或繁荣、或衰败、或新生、或毁灭，却恰恰不会无为无用。积极的影响自然人人引领而望，然而得到这种结果却需要在设计上对周边文脉进行周密分析，并在建筑本体上建立起与周边环境密不可分的文脉衔贯。于是，在整个设计过程中，场地分析与概念提出就变得至关重要。它的成败与否直接关系到导师对方案的评判。简而言之，我们希望看到的是能够踏踏实实地为当地提供人文关怀的建筑，而不是漂亮而昂贵的花瓶，尽管夺目却无甚内涵。

根据英国绿色建筑委员会发布的统计结果，"在全球范围内，建筑环境消耗了40%-50%的自然能源，这其中包括20%的水资源，30%-40%的能源以及大约三分之一的二氧化碳排放"。为此，在2006年12月英国政府做出了一项极富挑战的承诺——"即从2016年开始，所有的新建住宅均必须达到零碳排放标准"。[1] 这一誓言随即在整个建筑界内激起了广泛的反响，且先不论届时其目标是否真的可以在严格意义上得以实现，它至少表明了一种积极的，应对全球环境持续恶化之事实的态度。而整个业界也随之在规划、设计、结构、构造、材料等专业领域全面开启了对"生态可持续建筑"的探讨。作为科研与教学单位，诺丁汉大学自然没有落后。事实上，自从1990年代以来，诺丁汉的建筑与建造环境学院就致力于生态可持续技术的研究与应用。如今，该学科发展已在国际学术界中极负盛誉，成为全球领先的生态可持续建筑技术研究中心。在这里，不但"可持续发展"成为贯穿教学体系始终的核心战略之一，而且学院也身体力行，将很多教学建筑进行了节能生态化改造。大量的被动式生态建筑技术，诸如太阳能发电板、风力发电、雨水收集、地源热泵、自净化玻璃、太阳能热水、自然通风、光管自然采光等等，均被广泛应用在新老教学建筑之上，供博士研究生们测试研究。这一得天独厚的条件使得诺丁汉的学生可以随时体验到世界上最先进的生态建筑技术，以自身的感知来理解其好坏优劣，并进而进行批判性分析，然后再融会贯通入自己的设计方案之中。

本书的建筑环境设计篇与建筑与建构篇中，所收录的方案多是这一指导思想的具体表现。这其中最为醒目的当属那12个超高层设计方案。这一栋栋超过200米高的摩天大楼均计划位于世界各地的大都市中，虽然师生在诸如结构体系与消防疏散等重要方面均进行了细致的探讨，但设计组的终极目标却始终聚焦于如何使这些大厦实现自给自足的可持续性发展。事实上，除了各种巧妙融入建筑本体的生态技术之外，年轻的建筑师还进而对另一种可持续发展——社会可持续发展——给予了充分的关注。于是，这些"诺丁汉制造"的摩天大厦成了各个城市中有用的新成员，不但不会在能源消耗方面为城市增加负担，而且还以独特的建筑语言丰富文化，回馈社会。布莱恩·福特教授领衔的环境设计组则更多地在传统意义上试图探讨生态可持续的本质。他们基于英格兰中东部，结合当地条件，在小型公共建筑领域进行了卓有成效的尝试。迈克尔·斯泰斯教授领导的五年级设计组——零碳建

the context of East Midlands in England, they have completed successful small scales public buildings. Professor Michael Stacey, along with his year 5 studio – Zero Carbon Architecture Research Studio (ZCARS) and year 6 studio – Making Architecture Research Studio (MARS), discuss the applications of sustainable concepts from the angle of architectural tectonics, which results in a series of successful schemes such as Nottingham Meadows Sustainable Housing Project, the Solar Hydrogen Centre on Trent River and the Water Squares in Liverpool with tidal energy, etc.

Architecture is a professional field where you can extend imagination into any possibility. An architect without free-thinking is like a soulless person. "Freeness" is an indispensable element in the strategy of architectural pedagogy of Nottingham. Yes, we highlight the importance of humanities thinking and social responsibility, and we encourage the exploration of sustainability. However, all of these cannot be alive without respecting the students' individual developments. Here, design studios don't provide the students with any explicit design specification but just detailed project briefs. What included in a brief are explanations of project aim, definition of optional sites, introduction of local context, useful background theories, and studio schedules, but parametric data like how many square meters the whole project is, how many different functions need to be included, and what size of every single functional room should be, etc. is strategically excluded. In general, individual student is free to analyse the project and to draft design generators by themselves. The result is a high degree of diversity within design approaches, even within one studio. During this process, every student will becomes the true "author", who will "write" a unique architectural story in their own architectural language. Furthermore, for the students with special talent we will provide them enough room to harvest it. For instance, the exploration of surreal art within architectural scope is strongly encouraged, and even the practice of pure fine art is developed in some schemes. The 100th project of this book – Stories at Nottingham – is one such example, which interprets the built environment through graphic art. The copyright of these two paintings has also been officially purchased by the University and these beautiful works have become a permanent collection displayed in the Department of Architecture and Built Environment.

It was on the basis of the strategic principles of "humanities", "sustainability" and "freeness" that the Department achieved its current reputation. And by examining the profile of this typical British school of architecture, the readers can obtain an understanding of the distinctiveness of the British training system. Through the seven years study, students are shaped as British Charted Architects with qualities of social responsibility, contextualism, environmental awareness and the spirit of innovation. These qualities of architects can, for such a professional occupation that can bring significant physical influence to society, be more important than other skills.

However, for the honorific Chinese readers, the social soils where different educational systems growing up should be noted, by the way blind acceptance from externality should be discreetly avoided. The British architects training system is decided by the current situation of British post-industrial society. Indeed, all British cities and towns have completed their infrastructural development some time ago, which resulted in little change to the current built environment and fewer opportunities for designing large scale projects. Over time industrial estates declined and stagnated becoming places of social deprivation in need of urgent improvement. Additionally, because British communities hold relatively more conservative opinions supported by a reverence for tradition, which has resulted in the economic climate being less robustly equipped for change, it is quite rare to see local governments approving large scale projects. As a result, the architects who acquire the capability for interpreting social context and the reconfiguration of existing building find it easier to survive. What is more, the geography of the United Kingdom is not irrelevant as it is not located on a seismic belt. The territory is defined more by flat moors and plains than steep mountains and valleys. There are no major rivers with significant vertical drop. This relatively stable geological characteristic naturally creates a more flexible atmosphere for architectural practice, which relies less on anti-quake structure design and disaster resilience design. In general, all these external conditions are key factors determining the British architectural training system. In contrast, China is currently at a stage of economic prosperity, which naturally increases the number of large-scale projects resulting in a decreased time-span for design. Furthermore, China is located on seismic belt; its geological situation is much more complex with large areas of mountains and limited areas of plains; many major rivers feature with high vertical drop; and all these characteristics result in the frequent occurrence of various natural disasters. Based on this intricate situation, architectural design must strictly follow the national standards to satisfy the requirement of health and safety.

筑和六年级设计组——材料研究工坊从建构技术角度探讨了可持续理念的应用。诺丁汉梅多斯区的生态住宅，特伦特河上的太阳能制氢中心，以及尝试运用潮汐能的利物浦水之广场等方案均十分难能可贵。

建筑学是一个驾乘着想象之车翱翔九天的专业。没有了自由发挥空间的建筑师就如同人失去了魂魄。“自由”在诺丁汉的建筑教育宗旨中不可或缺。是的，我们强调人文关怀思想与社会责任感，我们鼓励对生态可持续元素的探讨，但是这一切都必须基于对学生个性的足够尊重。在这里，几乎所有设计组均不会为学生提供明确的设计任务书，取而代之的则是一份非常详细的设计大纲。这里面包括对方案主旨的阐述、可选择场地的界定、场地周边文脉的介绍、有用的背景理论简介，以及课程进度安排等，但独独不会告诉学生建筑总面积是多少平方米、要包括多少不同的功能，以及每项功能的具体面积有多大，等等。总而言之，每个学生均有权利去对该项目进行自主分析，并最终根据结果来自己确定设计任务书的内容。于是，同一个设计组中往往会产生截然不同的设计方向。而在此过程中，每个学生都变成了真正意义上的“作者”，运用自己的建筑语言来“书写”独特的建筑故事。此外，对于在某个方面天资秉异的学子，我们也为他们提供足够的发展空间。例如，我们鼓励超现实主义艺术在建筑领域的应用，甚至绘画艺术创作也可以发展成为独树一帜的设计方案。本书收录的第100个方案——诺丁汉故事——正是以平面艺术手段阐述建筑环境的佳作。这两幅作品的版权也已被大学正式购买，从而成为建筑与建造环境学院的永久收藏。

通过坚持以“人文关怀”、“可持续发展”与“自由探索”为其办学之根本，学院逐渐成就了今天的声誉。而从这所典型英国院校的轮廓中，读者或许可以理解英国建筑师培养体系的独到之处。通过这七年的学习，年轻的建筑学子们被培养成为具有社会责任感、尊重文脉、关心环境且崇尚创新与探索的英国建筑师。而这些品质，对于建筑师这一能对社会产生明显物质性影响的职业而言，似乎比其他方面更加重要。

然而，对于尊敬的中国的读者，我们也应该正视不同教育体系所生根发芽的社会土壤，不能对外部事物一味地盲目接受。首先，英国建筑师培养计划是基于英国社会的后工业发展现状而定的。如今，英国所有城镇早已完成了基础建设，改动很少，大型方案凤毛麟角；而与此同时，那些衰败的旧厂房厂区和破败的市中心却逐渐成为滋生犯罪的温床，迫切需要进行改造。其次，英国社会又是一个相对保守的群体，加之经济环境多年来也不甚强劲，于是地方政府很少会做出大拆大建的决定。因此，具有很强社会文脉阅读能力与更新改造能力的建筑师往往更容易生存。最后，英国并非位于地震带，全境多平原，少山地，没有什么高落差的大川大河，地质环境相对稳定，这在结构抗震与抗灾防灾方面为建筑设计提供了相对较大的冗余度，也自然有利于建筑师的自由发挥。总的来说，这些外部条件都是催生英国建筑师培养体系的重要元素。而与之完全相反，中国现在正处在经济高速发展阶段，大型项目自然层出不穷，设计周期也必然一再压缩。另外，中国处于地震带之上，地质条件复杂，山地多，平原少，大川大河垂直落差显著，各种自然灾害发生频繁，建筑设计必然要严格地遵循国家规范，保证其坚固安全。综合这些因素，就不难发现中国建筑师的确需要拥有在短期内处理复杂大型项目的能力，要对规范标准了然于胸，另外丰富的工作经验也非常重要。依此公论，现有的中国建筑师培养体系可以中肯地说适应了社会的要求。

综上所述，我谨希望读者们能够客观看到两国建筑师培养体系的差别，并能够以批判性的眼光来审视来自诺丁汉的100个方

Summarizing these factors together, it is not difficult to figure out that Chinese architects indeed need come terms with large scale projects designed in a limited time, being extremely familiar with building standards, and obtaining rich and varied working experience. On this point, the current architects training system in China could pertinently adapt the social requirement.

Summing up the contents of this volume, it is hoped that readers will begin to understand the difference between the two training systems from a more objective point of view, and to read the 100 projects from Nottingham with a more critical judgement. On a personal note, what is collected in this book are 100 high-quality architectural design projects, but I can never affirm them as "perfect" because architectural design is an extremely hard field to assess. The benevolent see benevolence and the wise see wisdom! One can only comment but probably never can determine the final quality! So, only if this book inspires or helps a little, that will for me bring the greatest encouragement and joy!

案。公允而谈，这里仅集中了诺丁汉大学100个优秀的建筑设计方案，但我决不能断言它们是最好的100个，这是因为本身建筑设计就是一个极难评判的领域，仁者见仁智者见智，可以评说，却不可定性！于是，如果本书或能启发灵感，或能取长补短，则甚幸，甚慰！

王琦 博士
2012年11月27日凌晨
于诺丁汉阿滕伯勒小村

Dr. Wang Qi
Dawn, 27th Nov. 2012
Attenborough, Nottingham

PEDAGOGICAL METHODOLOGY - 教学方法论

FRAGMENT ON METHOD: PRESERVING THE MIDDLE GROUND
方法的碎片：保留中间阶段

If the design process has a beginning, middle, and an end, where is the act of design located? The answer seems obvious enough – throughout - for design is a threefold combination of investigation, development and resolution. Design is often used to describe the general overall 'flow' between the three realms. But perhaps the truth is more complex, for each stage is characterised by different intentions and modes of thought. There is also a difference between 'flow' and 'structured movement'. What is often glossed over in the flowing definition of design is the fine-grain creative practice, or particular interactions between modalities of thought, that enables the continual progression of the process. Without such middle ground operations thought can soon become discontinuous (how many times has insightful research and investigation failed to live up to expectations?). In a data-led culture where increasing emphasis is placed upon preliminary research, it is no accident that mapping techniques have become so predominant. However this is often at the expense of the middle ground, which as a result tends towards instrumental automation and scripting of form. The automation of creativity is perhaps inevitable as it is often developmental time that is squeezed out by tight budgets. And yet without a well-rounded experience of how the middle ground of project-work operates, design cannot be mastered and the sequence of perception-intuition-intention remains under-articulated. One of the primary roles of architectural education is therefore to cultivate this gestational ground, and for students of architecture to learn how it can be abbreviated and preserved in professional practice. Much of the material collected in this volume has benefitted from a reconditioning of the developmental phase of project-work, so what follows is an account of its structured movement.

如果设计的流程由开始、中间和结束组成，那么设计行为本身应位于何处？答案似乎很明显——在整个过程中——因为设计是一个集调查，发展和解决问题为一体的三重组合。设计经常被用来描述这三个阶段之间的总体“流动”。但也许事实的真相更加复杂，因为每个阶段都有不同的意图和思维模式。而“流程”与“结构型推进”也有不同。在强调设计的“流程”属性时，那种细腻的，富有创造性的实践或者不同思维方式之间的碰撞通常被掩盖了，而它们恰恰能够使设计过程不断向前发展。于是，如果没有这种在中间阶段的推进，思想很快就会变得不够连续（想想有多少次有见地的研究和调查辜负了期望？）。在以数据为主导的文化里，初步研究得到越来越多的重视，因此整体计划技术自然而然地变得如此突出。然而，这往往是以牺牲设计的中间进程为代价，其结果趋向于工具性的自动操作和形式的脚本化。创意的自动化也许是不可避免的，因为对设计的深化时间往往受到紧张预算的影响。假如没有一个成熟的，在中间过程操作项目的经验，设计便不可能被掌控，而且感知——直觉——意图的过程也无法被明确表达。因此，建筑学教育的主要任务之一便是培养这种孕育性的中间阶段，并让建筑学学生了解如何在职业实践中将其提炼且加以应用。本书收集的许多素材受益于对设计作业深化阶段的重新整理，因此，接下则来是关于教学大纲的结构说明。

Making Room for Thought

The creation of a room remains a guiding intention in architecture. Given how difficult it is to specify their edges with any precision, the conventionality of something like a room makes it more, not less profound as a design task. Despite its bounded nature, the boundaries of a room 'move' up and down, side to side, and back and forth in order to enrich and situate our experience of inhabitation. Interiority and exteriority can be challenged without destroying the identity and integrity of a room. Like clearings in a thicket, they are spatial conditions allowing phenomena to emerge into nearness, express and swell in presence, and then contract and withdraw. In essence the room exists somewhere between metaphor and thing. Continually made, unmade and remade in accordance with everyday life, a room is a fundamentally unstable, manifold reality, as opposed to a box. It is a hinge within a threefold spatiality whereby the background space of the world is accommodated and measured by placed rooms in the context of which human activity takes hold. David Leatherbarrow explains this threefold spatiality in the following way:

"When the building is freed from technological and aesthetic intentionalities, we discover its lateral connections...performance in architecture unfolds within a milieu that is not of the building's making. A name for this milieu is topography, indicating neither the built nor the un-built world, but both. Three characteristics of topography sustain the building's performativity: its wide extensity, its mosaic heterogeneity, and its capacity to disclose previously latent potentials. There is always more to topography than what might be viewed at any given moment. Excess is implied in its ambience, for what constitutes the margins of perceptual concentration always exceeds the expectations of that focus." [1]

If, through their creation and performance, rooms condition the slow acts of secular revelation, then

为思考提供一个房间

设计一个房间仍然是建筑中的一个首要意图。考虑到精确限定房间的边界是多么的困难，即使一个通常的房间作为设计任务而言也是极其深奥的。尽管其有一定的范围，房间的边界依然可以上下、左右、前后来回“移动”，目的是丰富我们的居住体验。在不破坏一个房间的可识别性和整体性的同时，其内在和外在属性亦可改变。就像位于丛林中的空旷地，它们的空间状态允许现象就近涌现，表达存在，膨胀，然后收缩并退出。在本质上，房间存在于象征性和物质性之间。在日常生活中，一个不断经历变化，不变，和再变的房间从根本上来说是一个不稳定的，丰富多样的实体，而不简单是一个盒子。通过将其纳入人们的活动之中，房间就变成了一个三重空间里的中枢，容纳并测量着背景世界。大卫·莱瑟巴罗这样解释三重空间性：

“当建筑脱离技术和美学的意图时，我们发现它的横向联系……发生在建筑中的各种行为超出了房屋建造本身的预料。这个范围的名称就是建筑拓扑学，它意味着既不是建成的世界也不是未建成的世界，而是两者的叠加。建筑拓扑学的三个特点，即广泛的延展性，拼图般的异质性，以及揭露先前隐藏潜能的能力，支撑着建筑物的表现。当然除了观察的那一时刻外，还会有更多的变化可能。过剩总是隐含在它的周围环境中，因为浓缩的感知边缘总会超出对重点的预判。”[1]

1. David Leatherbarrow, 'Architecture's unscripted performance', in Kolarevic, B. and Malkawi, A.M., eds., Performative architecture: beyond instrumentality (New York: Spon, 2005), pp.5-19

the following question arises: are the conditions of architectural design analogous to the threefold spatial structure of a room? In other words, is the primary task of design one of (re)articulating and reconditioning the 'project-work' - a realm which has an interiority all of its own, both real and virtual – that is capable of articulating the slow, lateral emergence of architecture? In the words of Theodor Adorno, every project-work is a potential space or force field. Viewed in this way, project-work, along with the its space of production, can be defined as a constellation of potentiality, the purpose of which is to set out the connective structure for gathering and aligning human, inhuman and cosmic action.

Descending-Ascending Thought

Combining scales of operation, degrees of embodiment, and networks of equipment, project-work has a generational order of its own, all of which is housed and registered in the studio environment. Dwelling within a manifold of materialities, thought is transmitted through media of variable density and quickness, some of which have greater living presence than bricks and mortar – art, language, philosophy, and narrative. But in the same way that actual buildings must speak through less-animate materials, design is drawn down into fabric that delays verbal communication. 'Fabricated thought' is pulled backwards through a phenomenal world of dim materials and flattened figures forcing its utterances to become more archaic, embodied, and less explicit. At this level the voices expressed are more akin to the murmuring background surfaces of a room that witness everyday life, which only occasionally reach into the foreground. For the same reason that the room has a lateral or distended spatiality, so does the process which generated it. Lateral can refer to many things including peripheral, overlooked and edge phenomena but here it is used to designate the confluence of manifold realities within non-linear project-work.

In the reconditioned atmosphere of project-work, things come into being through artifice. It is a performance space in its own right, where cultural material and content interact. Animated from within as though it were a second nature, an immanent structure instils project-work with agency and instinct. In an academic context, project-work exists to intensify and thicken this field and as such, act as a compressed generating matrix analogous to the creative potential of nature. The fictional writer Italo Calvino once declared "my working method has more often than not involved the subtraction of weight. I have tried to remove weight, sometimes from people, sometimes from heavenly bodies, sometimes from cities." (Calvino, 1996) In architecture the opposite also holds. Forced to start from nowhere, the first act is to assemble a layered field of data similar to Robert Smithson's 'Heap of Language'. Eugene Dupréel's philosophical study of slowness, density and consolidation in *Théorie de la consolidation* (1931), was opposed to the simple generative concept:

"Life has not moved from an original nucleus towards an indeterminate development: it seems to have resulted from an advance from the external to the internal, from a state of dispersal to a final state of continuity. It has never been like a beginning from which a consequence results, but it was from the first like a frame that is filled, or like an order that has gained in consistency through, if we may be permitted to use the expression, a kind of gradual stuffing... Life is certainly growth, but all growth that is in extension, like fabric that stretches or individuals that proliferate, is only a particular case; life is essentially growth through density, an intensive progress." [2]

Simple driving concepts are emancipated from the slow accretion of an idea cultivated in the middle ground ecology of project-work. In architectural education the 'straight jacket' of the first formal concept needs radical and urgent retroversion.[3] As Simon Unwin put it, nothing will come out of nothing. Project-work has the capacity to invert ex-nihilo beginnings into the proper conditions of possibility of design. Single grand concepts, posited at the start and obeyed at all costs, are suspended and transformed into ideas and intentions embedded with their own particular scaffold, or preparatory architecture, that has weight.[4] Within a reconditioned project-space, architectural thought can move within different processual orbits simultaneously, thus helping to loosen the determining grip of one all-encompassing technique. Another way of putting this might be that heterogeneous materials and slower processes can interfere with and enrich design practice.

如果说通过房间自身的创造和表现，它们容纳了日常的普通行为，那么接下来的问题是：建筑设计本身是否就是对房间三重空间结构的体现？换句话说，是否设计的首要任务就是对项目工作进行再深化，再调整，从而制定一个包含全部自我，集真实和虚拟于一体的领域，再进而保证之后那缓慢，按部就班的建造过程得以顺利进行吗？用西奥多·阿多诺的话来说，每个设计作业都是一个有潜能的空间或力场。从这个角度来说，设计作业以及它创造的空间，可以被定义为一个由许多潜能之星组成的星座。而在它那星辰构络之间，人类的，非人类的，以及宇宙活动均可被妥善安排与放置。

思想的回落与升华

对各种不同尺度的操作，对想法的具体表达以及对工具的使用这三个维度拥有一定的共性结构，而这种共性结构恰恰在设计室环境中均可以找到。寄居在物质表现之中，思想通过各种不同密度，不同速率的媒介传播，其中有一些存在得甚至比砖头和砂浆更长久，比如，艺术、语言、哲学和叙事。但是，实际的建筑必须通过不那么有生机的材料说话，而设计则需要向那延迟言语交流的建筑构造妥协。“构造的思想”在这个由冰冷材料和二维图形构成的现实世界里，不得不以一种更加档案化的，具体化的，和不明确的方式表达自己的声音。在这个层面上，要表达的声音更类似于房间里的背景声，沙沙潺潺似有似无地作响，只是偶尔才变得引人注意。出于同样的原因，房间有一个横向性的或扩张性的空间感，产生它的过程同样如此。横向可以指很多事情，包括外围的，被忽视的和边缘的现象，但在这里它是用来指一个非线性设计作业中各种各样实体的汇集。

在一个经过调整的方案设计氛围中，事情的形成有赖于各种策略。在这个说到底是一个表演性空间，在其中文化素材和内容相互互动。在这种类似第二自然的环境中受到鼓舞，一种内在的结构赋予方案设计以力量和本能。在学术语境中，设方案计的存在可以对其进行强化，进而作为一种浓缩的基质激发无穷的创造潜力，就如同大自然那样。小说家伊塔洛·卡尔维诺曾经宣称：“我的工作方法往往涉及减去重量。我努力消除重量，有时是消除人的重量，有时是消除神灵的重量，有时是消除城市的重量”（卡尔维诺，1996）。在建筑中刚好相反。设计可以说从零开始，其首要行为就是组合一份包含各种有关数据的多层资料库，类似罗伯特·史密森的“语言堆。”尤金·杜普瑞尔在其1931年《固结理论》一书中关于缓慢、密度、固结的哲学研究与这个简单生成概念相反：

“生命不是一个原始的细胞核向某一个不确定的方向演化：它似乎遵循某种从外部到内部，从分散的状态到最终连续状态的演变规律。它从来不是有个开头其结果就尾随而至，而是像一个被填充的框架，或者说像一个在这个规律中被遵循的秩序，假如我们乐意使用这样一种表述方式，则这是一种逐步填充……生命当然在成长，但这种扩展性的成长，就如被撑开的面料，或繁衍的个体，都仅仅是一个个案；生命本质上是一个增加密度的成长过程，一个连续不断的进程。”[2]

对设计起到推动作用的简单概念来自对想法的反复推敲，并在方案设计的中间阶段——一种有益的类生态环境——得到培育。在建筑教育中，第一个正式概念往往需要激进且紧迫地返工。[3] 正如西蒙·昂温说的那样，“无物生于无中”。设计作业可以把空无的初始阶段转化为可以用于开展设计的适当条件。不

2. As quoted in Gaston Bachelard, Dialectic of Duration, translated by Mary McAllester Jones (Manchester: Clinamen Press, 2000), pp 95-6.
3. See Jonathan Crary's reading of Olafur Eliasson: "Eliasson's work must be seen as part of a counter tradition of machinic production in which the dominant contemporary values of storage, speed, productivity, uniformity are discarded in favour of techniques for the creation of singular and non-recordable phenomena...[that reverse] an escalating mechanization and most often an impoverishment of human perception."
Jonathan Crary, Olafur Eliasson: Visionary Events. http://www.olafureliasson.net/publications/download_texts/Eliasson_Visionary_Events.pdf, Accessed 1st December 2012, 13.45.
4. In Greek scaffold translates as katabasis - that which lies beneath.

The middle ground is where project-work takes on thickness. Thickness is the direct result of the forced compression – simultaneity - of many planes of thought into one composite space. Project-work has the capacity to bifurcate concept, bequeath infinite potential to finite thought, and transform it into a convivial architectural topic. Laterally orchestrated creative practice also exceeds graphics. The process, its crystallisation, and subsequent presence, are one. Embodying and thickening abstract diagrams (swollen thought) transform through projection into a worldy body. Vice-versa, when flattened into plan corporeality is pressed into a rarefied intermediate state (inter-reality) between the ideal and the real where it becomes subject to proportional measure. The multiple flattening and elevating of minor events into major situations is the traditional procedure for progressing matter (procession) and the realisation of meaningful, lyrical form.

A persistent tension running through the intermediate stage of architectural thought, something that is often left unexplored, is the manner in which it is stretched (distended) across a three-fold spectrum of practice – the abstraction of language and conception, concrete materiality, and embodied spatial practice. This is a unique position, or threshold, in the overall design process.[5] Such ambiguity allows the project-work to enter into dialogue with culture through combinations of textual practice, the practices of inhabitation, and technological expertise. The American architect Louis Kahn once recalled how he failed physics at school because he couldn't come to terms with its abstract, numerical communication methods, but had he been asked to "draw physics" as he put it, his access to the phenomena would've improved. There is an important point to this: architects are privileged with the role of translating disembodied conceptual knowledge into tangible, public phenomena. Despite our growing ability to talk about architecture using increasingly more sophisticated language, combined with the enhancement of expertise offered by digital technology, in the end architecture is essentially interplay between the refined and the unrefined. Its creative potential lies in how phenomena shuttle (translate) across a spectrum between the material and the de-materialized.

Thought in Motion

Meaning, ideas and creative thought are carried by and communicated through things. Such re-embodiment is a sacrifice of pure thought. This descent allows the world to grip and takes hold of ideas, which can once again gain traction. The truth of thought is tested by movement in and through an inter-reality of matter-mind. Here thought becomes chimerical and no less noble, as the most elevated mind can soon become impoverished through solipsism. The paradox is that architecture is an impure discipline governed by ontological interplay. The philosophy of group-work is another medium of mobile thought that expropriates the ego. Within architectural education subjective authorship and the signature artifact still seem to carry the greatest significance, often at the expense of other voices that need to be registered in project-work. Solo and group performance need to be reconciled. The search for an enriched understanding of synesthetic, multidimensional space is therefore an ongoing, iterative task in the academy. A common mistake is to view project-work as a mapping of an introverted, mental landscape. But design is not the externalised psychology of the genius. Gianni Vattimo contrasts this strong trait latent within modern aesthetics with a deliberately "weak ontology": "The occurrence of Being is...an unnoticed and marginal background event." He goes on to say that which "endures" does so "not because of its force...but because of its weakness."[6] One way of interpreting weakness is to view project-work as a gathering and amplification of distributed utterances - voices spoken by the site, the whispers of history, the contingent life of materials and their temporal behaviour, not to mention the voice of the individual architect whose task it is to make sense and orchestrate the chorus. Despite our modernity the structure of the architectural imagination may still be said to resemble the creative exchange and distributed intelligence that once took place in the ancient polis, in the midst of which the individual voice and its lyrical intent becomes distinct.

The problem of integration is a further manifestation of mobile, collaborative thought, one that seeks to address the atomisation of project-work, along with the lifeworld, into competing specialisms. If architecture, as a reunification of thought and existence, provides the primary synthetic layer of culture, the question remains whether integrative, environmental science, which is itself an extension of modern instrumentality, can achieve such reconciliation. One route out of this situation could

惜一切代价而坚持下来的初始概念，会被搁置，被转化为各种想法和意图，进而嵌入它们自己特定的支架或预型建筑之中，而这些表征均具有重要之处。[4] 在一个经过调整的项目中，建筑思想可以同时在不同的思考轨道上运动，从而尽可能地摆脱一个所谓的全能方法的决定性桎梏。另一种有益的方式或许是尝试多种材料和放慢节奏，它们均可以促进和丰富设计实践。

设计的中间阶段是让作业在深化过程中呈现“厚度”。厚度，就是指一种强迫压缩的结果，同时，不同层次的思想可被集中呈现在一个复合空间之中。通过细分概念，把无限的潜力变成明确的想法，设计作业就会成为一个有意思的建筑主题。这种在横向上精心编构起的创造性实践超越了图形位界。设计的构思，实施以及建成则是一个一体化的过程。在这个过程之中，表现与厚化抽象图示从一个映像转化为一个存于世间的实体（思想膨胀）。反之亦然，当一个具体建筑呈现在二维的图纸上，有形的东西就被变成一个服从于比例法则，介于理想与真实之间的稀薄产物（内部真实）。这种将许多小事件的多重平面化与立体化交织在一个整体约束之下的过程，则恰恰是如何将设计推进并产生富有意义的、诗意的形式的传统真谛所在。

尽管时常被忽略，一种持久的思考张力应贯穿于对建筑思考的中段。这是一种对实践的三重维度的伸展——对语言和概念，对具体的材料以及真实空间的抽象化概括。这在整个设计过程是一个独特的位置，或阶段。[5] 这种模糊性使得方案设计作业可以通过整合文字表达，对栖居场所的设计以及表达的技巧和知识，与文化进行对话。美国建筑师路易斯·康曾经回忆起他在校时物理科目没有考试及格，是因为他不擅长那种抽象的数字表达方式，但是当有人让他“画物理”时，他却可以胜任。这一点很重要：建筑师很荣幸的角色就是能把无形的概念知识转化为有形的、公共的现象。尽管我们越来越善于使用日益复杂的语言来谈论建筑，再加上数字化技术所带来的丰富的专业知识，但是在本质上建筑仍是被细化和未被细化的结合体。其创造潜力体现在现象如何在物质实体和非物质概念这两个领域间穿梭交互之上。

思想的活动

意义，想法和创造性思想蕴含在事物当中，并通过事物来沟通。思想的具体化是以牺牲其纯粹性为代价的。这种过程可以让现实世界得以把握思想，进而产生联系。思想的真谛存在于介于物质和精神的内在真实状态并在活动当中得到检验。这里思想也容易变成妄想且高高在上，因为如果过于自我，即使最有活力的头脑也可能瞬间变得贫瘠。吊诡的是，建筑作为一个不纯粹的学科依然受到本体性的制约。作业中的团队合作为思想的交锋提供了土壤也可以抑制过强的个人表现。在建筑教育中，设计者的署名权似乎仍是最重要的，但这往往以牺牲作品中其他必要的有意思的他人见解为代价。个体和群体的表现需要协调整合起来。另外，对复杂多维空间的深入理解依然是大学教育中一个中心任务。一个常见的错误就是把设计作业看作一个封闭的内心反映。但设计也不是天才心理的外在表达。詹尼·瓦蒂莫把这种蕴含在现代美学中的强烈特质与刻意的“弱本体论”进行了对比：“事物的发生……是一个被忽视和边缘化的背景事件。”他接着说，之所以能够“持久”的东西，“不是因为它的强力……而是由于它的脆弱。”[6] 一种解释这种脆弱的方式就是把设计作业看成许多分散声音的聚集和放大——包括现场的故事，历史的诉说，以及材料的生命和其特定时间的表现，更不必说那个负责让这些声音完美合唱的建筑师的操作。尽管我们身处现代时期，建筑创作的深层结构仍然可以说与古代城邦中那种创造性的交流、多元的

5. The idea of distension is derived from St Augustine's notion of the threefold present, or the idea that human intentions are composed from memory, attention, and anticipation.
6. Gianni Vattimo, The End of Modernity: Nihilism and Hermeneutics in Postmodern Culture (Baltimore: John Hopkins University Press), p. 86.

lie in the concept of intermundia or metakosmia, an Epicurean concept translated as 'interworld'. Referred to by Tennyson as the 'lucid interspace of world and world', it is a notion that appears attuned with current visions of multidisciplinary overlap. Characterised by motion and potentiality, intermundia suggests a de-specialised and de-tooled, spatio-material totality prior to division and separation. De-tooling is a provisional stage of integration that avoids the pitfalls of speculative collaboration - throwing expert tutors together during the early stages of the design process. Offering multidisplinary teaching at a formative stage of a project is of course part of the solution, but there is more to it than simply juxtaposing art and science in a shared space. A deeper question needs to be asked: what exactly happens to each discourse during the encounter, i.e. the middle ground, as thought moves back and forth between art and science? The environmental consultant Max Fordham and the architect Peter Clegg in a joint lecture held at the University of Nottingham in January 2009 gave a first-hand insight into this mobility:

"Listening to Peter [Clegg] reminded me of the different things between being an engineer and being an architect. When you are an engineer you can have a very narrow vision which most engineers do, and in a way that narrowness of vision gives you the freedom to have really crazy ideas. When you look at how a plan developed with Peter doing the architecture you can see that there is a lot more to it than my simplistic little lighting idea. And that is very important. People who benefit from the architecture draw from a whole lot of values and emotions that have to be satisfied in the building. So it is no good just having a simple view."[7]

Sustaining Thought

The point here is that the purpose of architectural development is to reverse engineer science and reconnect it with its phenomenal origins in the lifeworld. Art and science were at some point in their history kindred spirits, and contemporary design needs a strategy for blending archaic residues of experience and meaning with exact measure and technology. The current understanding of 'environment' is a reduced version of a pre-enlightenment spatial condition which, prior to its conversion into the void of modernity, was perceived to be alive in a semi-materialised state. Constructing a well-rounded environment therefore requires a lateral, enriched account of space as a stratified reality where the distinction between thing and world although less clear, is more revelatory and truthful about the human condition. To that end the goal of a culturally-attuned, environmentally-aware, design education, ought to be one of grasping the primordial spatial intermundia that spans, envelopes, layers and sustains life.

Architecture always already exceeds instrumental framing by any single technocratic field. Rooms orientate inhabitation with respect to cultural and natural horizons, and as such are synthetic assemblages of cultural-tectonic intentions. Existential space is defined as non-neutral or qualitative; it replaces dimensions and ergonomics with human relational categories such as foreground, middle-ground and background; nearness and remoteness; generic horizon and particularised rooms. When we inhabit space we often speak in terms of reach, horizontality and verticality, centre and periphery, above and below, seen and unseen, stillness and movement, active and passive, thickened and rarefied. It is a process that relies upon the renewal of a particular type of discourse for interpreting space. A shared design vocabulary helps to build understanding. Hence, largely assisted by the writings of David Leatherbarrow and the phenomenological tradition, we describe rooms in terms of 'recession, spectrum of articulation, sedimentation, extended temporality, crystallisation (of the vicinity), lingering, sectional strata, horizon, framing, focus and dissolution (of objects), lateral drift, prominence and retreat, commingling of the near and far, thick space, concentration and decompression, fluctuation and zones of attention.'[8] This language is a linguistic representation of pre-scientific intuitions of space, and is based on our primordial encounter with the world. Earlier Heidegger had similarly defined the relationship between a thing and its spatial sphere in terms of 'gathering of the fourfold, staying, dwelling together, enfolding, mutual belonging, revealed, bringing near, mirroring or interplay of spatial regions, expropriation (dissolution of the edge of discrete objects), worlding, nestling, inconspicuously compliant thing, stepping back (recession), and shining forth.'[9] Even though this discourse is untimely, it is inconceivable how architects can operate without a fundamental understanding of this influential tradition. This has to be the aim of design education today: to develop ways of seeing buildings as things or concentrated realities (contra objects) in

智慧没有太大差异，因为两者当中类似的就是个人声音和意图的表达。

整合是对动态思想收集的深入推进，它强调将设计作业结合生活世界并转化成富有竞争力的独特方案。假如说建筑，作为思想和存在的统一体，是文化的主要组成部分，那么整体的环境科学——作为现代社会工具理性的一个延伸——能否促进这种整合依然是一个未知的问题。一个叫作intermundia或metakosmia的伊壁鸠鲁哲学概念，或许能为摆脱这种局面提供一条可能的出路。这个可翻译为“interworld”的概念就是丁尼生所指的世界和世界之间的明了间隙，与当今多学科交叉重叠的想法相吻合。以运动和潜力为特征，intermundia建议去专业化的、去工具化的、空间物质整体性优先于分工与分离性的想法。去工具化是整合过程的一个临时阶段，目的是为了避免投机性的人际合作的陷阱，也就是说在设计过程的早期阶段将多名专业老师集中在一起指导。在设计的成形阶段，为学生提供多专业交叉教学当然是解决方案的一部分，但除了这种简单地把艺术和科学并列在一起的方式，它还意味着很多其他层面。一个更深层次的问题是：当思想在艺术和科学之间来回游走时，不同学科相遇时，比如在设计的中间阶段，到底会发生什么？对于这种流动性，环境咨询顾问马克斯·福特汉姆和建筑师彼得·克莱格在2009年1月在英国诺丁汉大学举行的联合演讲给出了第一手洞见：

“彼得（克莱格）的话让我想起了工程师和建筑师的差异。当你是一个工程师，你可以像大多数同行那样，有一个非常狭窄的视野，在这个范围内，你可以有很疯狂的想法。但当我看到彼得设计一个建筑平面时，我就明白了，在我那简单的，关于照明设计的小点子之外还有很多重要的事情。从建筑中受到启发，你就知道其中还有很多因素要考虑。因此，只有一个简单的想法是不够的。”[7]

保持思考

这里要指出的是，建筑发展的目的并不仅仅是为了展示工程科学，而是把工程科学和其在生活世界中的最初意图联系起来。历史上，艺术与科学在某种程度上是类似的，现代设计需要一个能够把古老的前人流传下来的经验和意义，与当今的方法和技术相互结合的策略。我们目前对 “环境”的理解是一个着重在前启蒙时代空间环境里的简化版，在其经历现代性的虚无状态之前，它是一个鲜活的半物化的存在。因此，构建一个完美的环境需要对空间进行全面而丰富的诠释，而在这个多层次的释义里，事物与世界的区别，尽管依然不甚明显，却需要更具有启发性地，更真实地揭示出人类的生存状况。为了实现这一目的，一个强调文化自觉与环保意识的设计教学目标就应该尽量把握那本初的，空间的intermundia，因为它可以跨越、包容、层切并延续生命。

建筑总是超越任何单一技术领域所限定的工具性框架。从文化的和自然的属性来说，房间作为一个集文化建构意图于一体的有机组合，是用于栖居的。真实的空间是有其属性的，而非中性的；它取代了尺度维度和一些与人有着密切的关联人体工程学名词，比如前景、中景和背景，亲近与偏远，普通用途和特殊用途等。在我们所居住的空间里，常常讲到远近，水平和垂直，中心和外围，上面和下面，看得见和看不见，静止和运动，主动和被动，厚重和稀薄。这个解释空间的过程依赖于特定类型的话语。共同认可的设计语汇有助于增进彼此之间的了解。因此，在很大程度上借助于莱瑟巴罗的著作和现象学传统，我们这样描述房间的空间意向：诸如空间的退让，明确的空间表达，沉淀，丰富的时间感，（对周围环境的）结晶，不断的迂回反复，剖面地层，

7. Unpublished lecture held as part of the 'Making Architecture' series organised by Michael Stacey and Darren Deane during 2008-9.
8. David Leatherbarrow, "Table Talk", in Architecture Oriented Otherwise (New York: Princeton Architectural Press, 2009), pp. 119-141.
9. Martin Heidegger, "The Thing", in Poetry Language Thought, translated by Albert Hofstadter (New York: Harper and Row, 1971), pp. 163-86.

relation to a wider horizon of reference. This will in turn enhance understanding of the 'sustainable metabolism' between things.

Meaningful Thought

Meaning is carried in the articulation of human situations. Architecture in turn operates as a framework for human action. The renewal of interest in architectural meaning amongst students of architecture is of paramount importance. Even though today architects are aided by increasingly sophisticated software, cad/cam interfaces and predictive environmental modelling, and have become experts at explaining in exact terms the performance of materiality and atmosphere, this has come at a cost: a growing inability to elucidate the meaning of the work; its situation in cultural history, or even account for how the design process generates and sustains meaning. Our discomfort when faced with the question, 'so, what does it all mean', or 'what contribution does it make', partly derives from the poor reputation meaning inherited from postmodernism. The suppression of the problem can also be linked to recent anti-representational tactics promoted by North American pragmatists who claim that "practices imply a shift to performance, paying attention to consequences and effects. Not what a building, a text or a drawing means, but what it can do: how it operates in – and on – the world."[10] Process, performance and the actative now appear to carry greater value than theories of representation, a work's content, or slower acts of interpretation. Our ability to restate the problem has to be recovered.

The production of meaning in architecture is founded on a dialogical intertwining of everyday rhythmic repetition and routine, with contemporary residues of myth, symbol and representational form. Architecture has the broad capacity to orchestrate and reconcile these two orders of time governing representation and performance of space. The nature of meaning can be revealed by something as simple as a fountain. Spanning the scale of furniture (1:1) and urban block (1:200), this humble piece of equipment brings the horizon of the countryside into contact with the representative order of the town, uniting material and environmental processes with iconography and inhabitation, all within one highly charged setting. A civic-scale room can also articulate the relationship between memory and ordinary habit; relationships that unite historic interior fresco with everyday performance in a single spatial encounter. This is an example of how architecture, in projecting a single image onto the life of the town and yielding through appropriation, leads us to dismiss as unhelpful the straightforward opposition between representational form and spatial practice. The production of meaning always involves a double movement between, on the one hand, the conviviality of room that suspends for contemplation the life and history of a place, and on the other, a wider urban field consisting of individual moments of social appropriation. Meaning is an ongoing event conditioned by the intersection between idealised space, and its deformation within a manifold, democratic reality. This became manifest in cosmological drawings that document in semi-abstract terms the latent meanings and relational field into which things, including buildings and bodies, are inextricably woven.

(Syn)thetic Thought

Despite our post-enlightenment condition, the one characteristic that architecture still holds in common with ancient thought is its ability to draw together, within a single horizon of understanding, a multidisciplinary field of knowledge. This volume captures a sample of project-work from the Part I and II courses at the University of Nottingham, UK, which involved iterative combinations of writing, making, and poetic, sustainable programming. The architect has always been a polymath guided by synthetic ideals that condense the richness of our world into an integral, convivial figure. Process and progressive synthesis amount to the same thing, with the designer acting as an opportunist who steers and listens to the research material at his or her disposal. A research project thus always has the capacity to 'project itself'. As Louis Kahn put it, 'it is always the hope on the part of the designer that the building, in a way, sort of makes itself, rather than be composed with devices that tend to please the eye'.[11] The final year (from where much of the work contained in this volume is drawn) is an opportunity to deal with a complex body of research that unfolds across a structured middle ground. The sixth year thesis begins with the construction of a charged constellation of research out of which arise cultural topics relevant to our time. A thesis cannot be posited as a single concept. It is made from intersecting strands of intelligence that crystallise into a culturally relevant final

地平线，限定空间的框架，（将对象）集中和分散，横向移动，突出和隐退，结合远景和近景，空间的丰富感，浓缩和释放空间，将注意力集中和打乱。[8] 这些词汇是透过语言学对空间认知的前科学阶段表达，是建立在我们对世界最本初的认知基础上。与此类似的是，海德格尔在早期也这样定义过事物与其空间的关系：四重的集合，停留，聚居，拥抱，相互隶属，揭示，拉近，空间区域的反射或相互作用，征用（离散物体边缘的消解），世界化，偎依着，不引人注目的抱怨，退步（衰退）和灯塔效应。[9] 虽然这些话显得不合时宜，但如果说建筑师没有从根本上理解这个有影响力的传统，很难想象他将如何去实践。因此这一点有必要成为今天设计教学的目标：在更广阔的视野内尝试把建筑看作事物或者浓缩的现实（而不是一些简单物体）。这将反过来提高对事物之间“可持续性新陈代谢”关系的理解。

有意义的思想

意义蕴含在对人类处境的清晰表达中。反过来建筑为人类行动提供一个框架。学生恢复对建筑意义的兴趣是非常重要的。尽管今天的建筑师获益于日益复杂的软件，例如，电脑辅助设计（CAD）和电脑辅助制造（CAM）以及用于模拟环境的模型，并已成为可以用精确的语汇来解释材料和环境性能的专家，但是代价却是：越来越无力去阐明作品的意义，以及其在文化历史中的地位，甚至说明设计过程是如何产生和维持意义的。那么，当我们面对诸如“这一切都意味着什么?它的贡献体现在哪里？”等这类问题时，那种潜在的不自在感可能部分源于声誉不佳的后现代主义对解释意义的影响。对意义的抑制也和北美一些实用主义者所主张的反对再现的策略有关，他们声称，实践意味着关注建筑的表现，后果以及影响——这不关乎建筑物，文本或图纸意味着什么，而关键在于它们可以做什么：即建筑如何在物理大环境上（或内）起到作用。[10] 现在，过程，表现以及激活作用看起来比再现理论，作品的内容，以及缓慢的阐释更有意义。我们应该重新正视这一问题。

建筑所衍生出的意义来自于日常生活中的反复使用与一些神话的片段，象征以及再现性形式的相互对话和交织。建筑具有协调和整合那分别掌管着空间表现和空间性能两种时间秩序的能力。一个简单的喷泉就可以揭示出意义的本质。这个不太起眼的但多用途的装置横跨家具（1:1）和城市街区（1:200）两种尺度，将乡野景致的魅力带入城市的秩序性表征之中，在一件被高度掌控的设施中，将物质与环境之形成过程和图示与居住结合为一体。一个宜人尺度的房间也可以阐明记忆和日常行为之间的关系；在一个单一的空间里，历史悠久的室内壁画可以和日常生活紧密共存。这个例子说明了建筑如何通过占有一块土地且进行开发建设而在城市生活中投影下一个单一印象，并促使我们直截了当地摈弃了具象形式与空间实践之间那毫无益处的对立。意义的产生总是会涉及双重的互动，一方面，互动存在于那房间所代表的，暂停了对生活沉思的欢乐与场地的深沉历史之间；另一方面，存在于其他更广泛的，包括社会专有的个人时刻的城市领域之中。意义是一种不断发展的事件，本身则受到理想化的空间，以及它在多元化，民主化现实中的畸变所影响。这一点清晰地体现在融合着宇宙哲学的图纸上，因为这里以一种半抽象的方式记录了潜在的意义以及事物之间的联系，包括建筑和人体的千丝万缕的关联。

（综合）决定的思想

尽管我们身处后启蒙时代，但建筑学一个自始至终始终保持着的特点就是在一个单一理解范围内综合多学科的知识。本书

10. Stan Allen with Diana Agrest, Practice: Architecture, Technique and Representation (Amsterdam: G+B Arts, 2000), p. xxv.
11. Louis Kahn, 'The Architect and the Building', in What Will Be Has Always Been: The Word of Louis I. Kahn (New York: Rizzoli, 1990), p. 5.

idea or grounded topic. Architects don't stick to concepts; they arrive at, or are propelled towards a synthetic idea as the fulfilment of research.

Project-work is another name for that structured movement; one might even refer to it as a topography of thought. The outcome stands at the intersection of x-y-z axes representing historical continuity, environmental responsibility, and humanised technology. This intersection marks the site of architectural meaning. What counts is to make sense of thick spatial conditions by forming questions that frame a topic; translating these questions into intentions; embodying intentions as spatial settings, and finally transforming the topic into a culturally relevant made work or room. One definition of a thesis could therefore be to make sense of the world. 'Thesis' derives from the Greek word thetos or tihenai, meaning 'placed' or 'position'. Some positions can be speculative – hypothetic – whilst others are synthetic. An architectural thesis situates and orientates design with respect to a set of historical conditions and possibilities. The notion of a 'thesis' is also linked to thematic construal, i.e. the act of drawing out and crystallising ideas over the time span of a project. In this respect a thesis is a disclosure of truth and is immanently moved by recursive oscillation between design and interpretation, much of which is made to happen in the middle ground.

精选了英国诺丁汉大学建筑学专业从第一阶段建筑师教育到第二阶段建筑师教育的一些学生作业，综合涵盖了联系写作、制作、诗意与可持续环境的设计方案。一直以来，建筑师作为一个具有综合知识的博学者始终致力于把我们丰富的世界浓缩成一个整体的、愉悦的图形。过程和进步性综合意味着是同样的事情，就是设计师作为一个机会主义者根据自己的需要“凝视，倾听”研究素材。因此一个研究项目，总是可以去“展示自己。”正如路易斯·康所说的那样，“设计师总是希望建筑能以某种方式自我逐渐形成，而不是将一批取悦眼睛的装置组合在一起。”[11] 最后一学年（本书中的许多作业是这一学年做的）为学生提供了一个机会，可以让他们在一个理构建起来的中期阶段中逐步展开复杂的研究课题。第六年的毕业论文始于一个严谨而丰富的研究建构，学生从中提炼出与当代有关的文化主题。一篇论文不能被假定为一个单一的概念。它是由相互交叉缠绕的智慧组成，最后凝练成一个与文化相关的想法或明确的主题。建筑师们不必拘泥于概念；他们主动或被动地构建一个能体现研究成果的综合构想。

方案设计作业是这种结构化行动的另一个名字，人们也可以把它看作思想拓扑学。它的成果建立在XYZ三轴的交点之上，代表着历史的延续性、环境的责任，以及人性化的技术。这个交汇点也是建筑意义的所在之处。设计过程中最重要的是通过提出架构起主题的问题来确定一个拥有厚度的空间条件；将这些问题翻译成意图；再把这些意图具体化为空间设置，并最终把主题变为一个与文化相关的作品或房间。因此，六年级设计论文的一种定义就是要把世界变得有意义。“论文”这个词来自于希腊字thetos或tihenai，意为“被放置的”或“观点所在”。有些观点是投机性的，假设性的，而有些是综合性的。通过对一系列的历史条件和可能性的尊重，一篇建筑设计论文定位且引导着设计方向。“论文”的概念也与专题研究有关，即在一个项目的设计周期内想法的提出和表达的呈现。在这一点上，一篇论文也是一个对真相的揭示，并在设计和阐释之间来回摇摆，而这其中大部分过程则是发生在教学的中期阶段。

Dr. Darren Deane
达伦·迪恩 博士

Chinese Translators: Ding GuangHui, Jia Min
中文翻译：丁光辉　贾敏

LIST OF PROJECTS - 方案名录

建筑人文篇 ARCHITECTURAL HUMANITIES

建筑环境设计篇 ARCHITECTURAL ENVIRONMENTAL DESIGN

建筑与建构篇 ARCHITECTURE AND TECTONICS

城市设计篇 URBAN DESIGN

实践设计项目篇 LIVE PROJECTS

—— ARCHITECTURAL HUMANITIES
—— 建筑人文篇

The projects presented here under the banner of Architectural Humanities are united by a number of common themes. Predominant among them is an interest in the role of history, memory and human behaviour in the creation of meaningful places. As researchers in architectural humanities (architectural history, theory, culture and design) we also share an interest in the 'reading' of history through the built remains of past cultures and situations, an interest in supporting those processes – both natural and cultural – that respond dynamically to changing conditions and allow places to transform themselves gradually over time in response to new requirements.

A number of the projects here could be described – to adapt a term from Kenneth Frampton – as explorations in 'Critical Conservationism', dealing creatively and carefully with redundant buildings, landscapes or fragments of urban infrastructure. A key aspect of this approach to renewal is to offer new life to old places while maintaining the histories that are embedded in their physical fabrics – the slowly accumulated traces of previous occupation that give many historic structures their almost magical sense of significance, of personality, and 'aliveness'. One group of projects deals specifically with the creation of new exhibition spaces within existing twentieth-century buildings in Nottingham, and another group consists of final year students' thesis projects with individual programmes responding to the particularities of their contexts.

This quality of aliveness is also pursued in a more literal way through a number of projects that deal with the dramatic possibilities offered by new technological developments such as 'protocell' structures. By suggesting an architecture that can literally grow – and heal – itself like a living organism, this work also overlaps with projects based in landscape locations which are subject to other kinds of dynamic and transformational forces. This work demonstrates that natural processes such as coastal erosion and deposition, seasonal flooding and other fluctuating climatic phenomena, demand an architecture capable of responding dynamically to these ultimately uncontrollable factors. Many of these projects also engage with significant social programmes, such as public places of gathering for celebration, commemoration or other civic rituals.

The political possibilities implied by these proposals for public places also reminds us of the importance of what Kenneth Frampton called (after the German philosopher Hannah Arendt) the 'space of appearance' – that realm of social interaction through which we fulfil ourselves both as individuals and as a members of a social group. Another interest that much of the work suggests in materiality, construction tectonics and sensory perception, is not unrelated to this social dimension – if we think of the overarching importance of the relationship between architecture and the body. The 'architecture of embodiment' is perhaps the key underlying interest that links all of the work presented – reminding us of the fact that architecture is ultimately a three-dimensional art of objects arranged in space and experienced in time, 'speaking' to us directly through the language of the senses.

聚集在“建筑人文篇”旗下的项目大多围绕着一些共同的主题。在它们之中，我们最感兴趣的是历史，记忆以及在创造富有意义的场所时人类行为举动。作为建筑人文学（建筑历史，理论，文化和设计）研究者，我们不但通过凝结在建筑遗存中的那些逝去的文化与场景来品味历史，而且乐于支持这些遗存的变化过程，不管是从自然的方面还是从文化的方面，我们希望这种变化能够使建筑遗存对不断变化的外界条件做出积极的反映，并使得其场所能够通过逐步改变自身来适应新的要求。

该章节里的一些项目可以被描述为符合肯尼思•弗兰姆普敦思想的，对于“批判性保护主义”的探索——即创造性地、细致地处理那些废弃不用的建筑、景观场地，或是城市基础建设中的“碎片”。这个更新过程中的一个关键方面是如何赋予老场所新的生命，并同时保留它们肌理中所蕴含的历史印记。这是一种由一栋建筑曾经拥有过的各种不同功能缓慢积累起来的历史痕迹，赋予了许多历史建筑魔力，使它富有意义，获得人格，甚至拥有生命。这其中，一组项目专注于如何将诺丁汉现有的20世纪旧建筑改造为新的展览空间，而另一组项目则包含了毕业班的学生作品，他们必须根据各自项目的文脉特殊性来独自完成设计。

生命的力量也在一些项目上通过一种更为直接的方式来加以探索，它们主要回应了新技术发展所带来的戏剧化效果，如‘原细胞’结构。这种理念认为建筑自身可以实际生长，甚至恢复，就像一个活的有机体。而在一些项目中，建筑所在的地区正遭受着各种活跃的，持续变化的外力影响，该理念就有了一展身手的舞台。这些作品展示了——诸如海岸侵蚀与沉积，季节性的洪水以及其他变化不定的气候现象等等——这些自然过程均要求建筑能够对根本无法控制的因素做出积极的回应。还有很多项目也与重要的社会现象紧密相连，例如对集会性公共场所的重视，或为庆祝，或为纪念，或为其他城市仪式典礼等。

这些针对公共空间的方案设计暗示了其所隐含的政治可能性，同时也提醒了我们回忆起肯尼思•弗兰姆普敦所提出的（受到德国哲学家汉娜•阿伦特影响）“（公共）参与空间”的重要性。这是一种特殊的社会互动领域，作为单个个体和社会集体中的一员，身处其中，我们便可以使自己过得充实而满意。本篇中的另外一个兴趣点主要集中在材料，建构技术与感官知觉方面。假如我们仔细考虑存在于建筑与人体之间的重要关系，就会发现这一点也与先前提出的社会维度相关。或许，“建筑的具体体现”就是那个能够将所有入选作品整合在一起的潜在性关键兴趣点。它不断提醒着我们一个事实，即建筑终归还是一个物质化的三维艺术，被组织在空间之中，接受岁月的洗礼，并通过感知的语言与我们直接对话。

—— **Dr. Jonathan Hale**
乔纳森 · 赫尔 博士

The Head of the Architectural Humanities Research Group
建筑人文研究组主任

001

DESIGNER-设计人: Taylan Tahir

Museum of Eccentric British Genius, Nottingham

Does size matter? "Sometimes the smallest things take up the most room in your heart". I responded to the brief to create a 'Museum of Eccentric British Genius', housing five chosen items and one of my own. A key driver in the design of this project was the journey through the building and the order in which visitors would appreciate each object. Objects were initially ordered in physical size however as visitors pass through the exhibitions they begin to understand that as the objects get smaller they actually relate to them on a more personal level. Tea, personal transport and music have a much greater cultural importance and a larger impact in society than a Harrier jet. The concrete monolithic external appearance hides the definitive slope that cuts through the museum from front to back, linking levels and exhibitions and providing circulation and viewing points.

英国奇珍异宝博物馆，诺丁汉

尺寸很重要么？"有时，最小的东西也能占据你整个心扉"。受这句话的启发，我设计了一个"英国奇珍异宝博物馆"，收藏了5件固定的展品，以及一件我本人挑选的展品。该项目的主导思想是整个参观流线，以及如何让参观者在穿越整个建筑的同时欣赏到每一个展品。这些展品将事先根据大小排列好，但是当参观者开始游览时，他们会发现展品的尺寸越来越小，却越来越接近人们的普通生活。与一架海鸥式战斗机相比，茶、私人交通工具与音乐在人们的文化社会层面要显得更加重要。隐藏在巨石般的混凝土外表下，一条明显的斜坡由前至后贯穿了整个博物馆，将不同高度的平台和展览相互连接，为游客提供灵活流线以及可以驻足的观景平台。

▼ - Ground Floor and First Floor 一，二层平面图

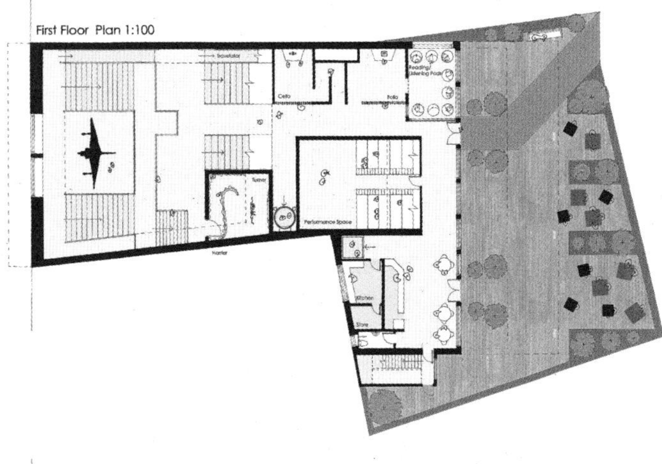

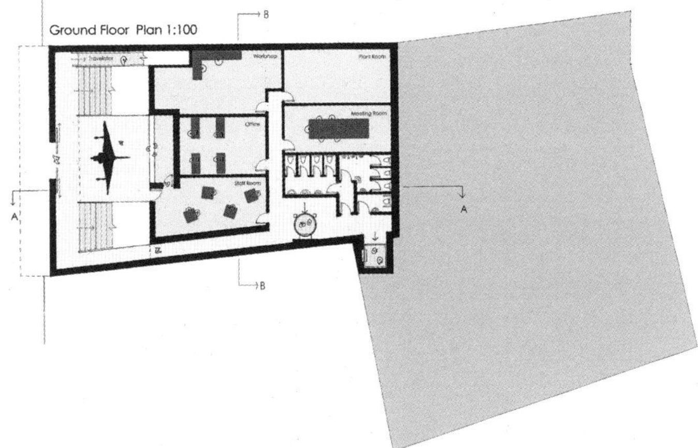

▼ - Second Floor and Third Floor 三，四层平面图

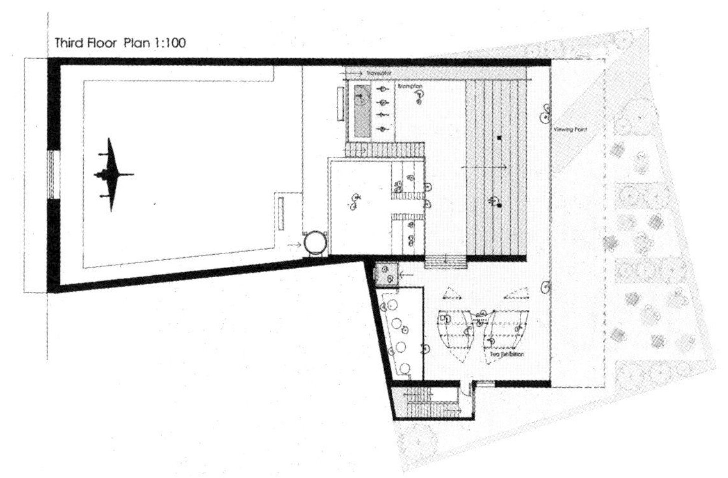

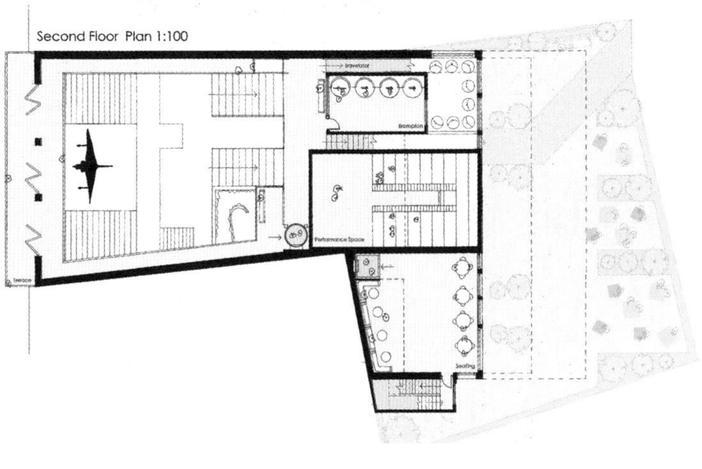

▼ - Concept Sketches 概念草图

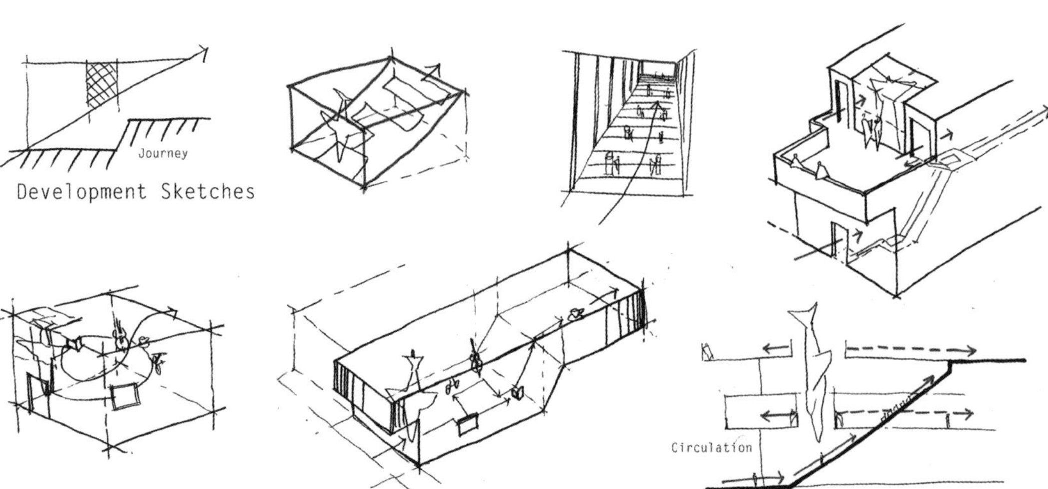

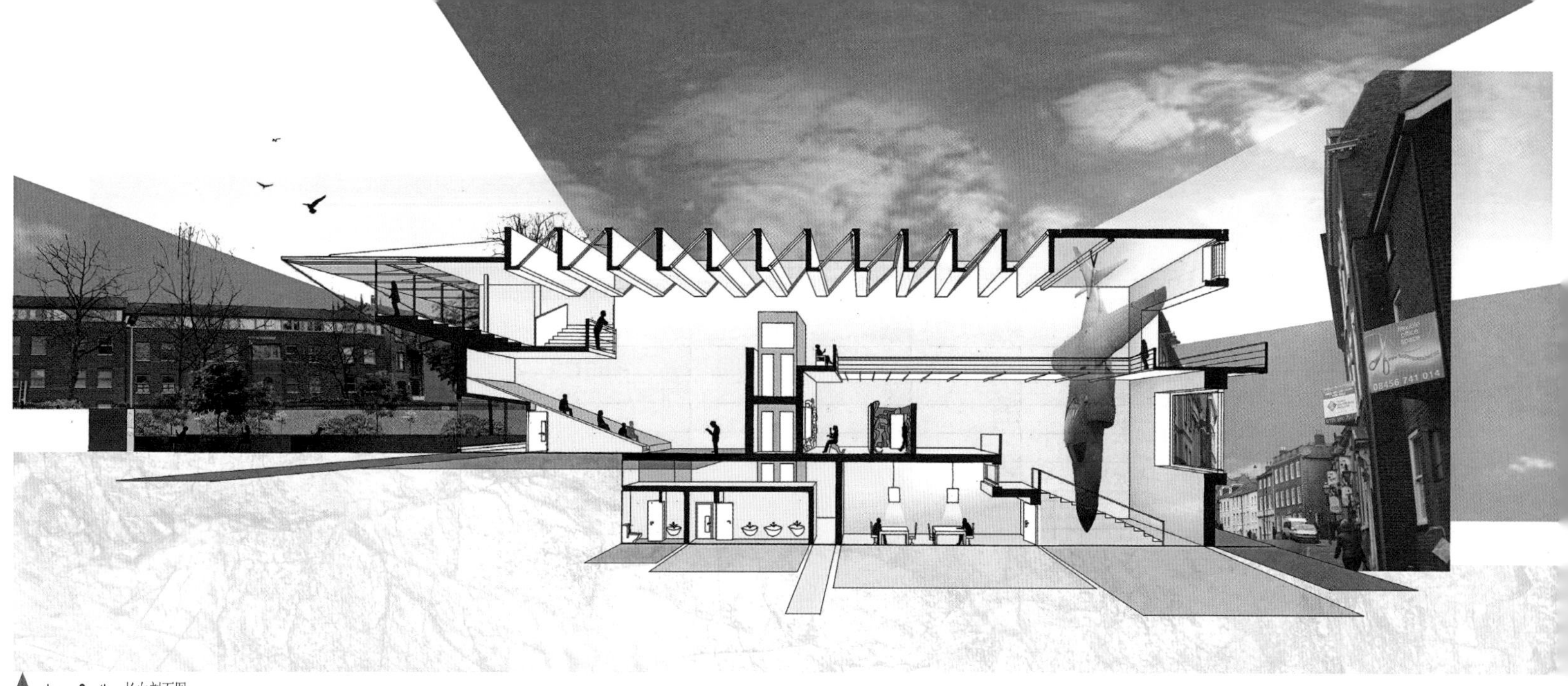

- Long Section 长向剖面图

View from above of the museum and surrounding buildings

Internal Perspective

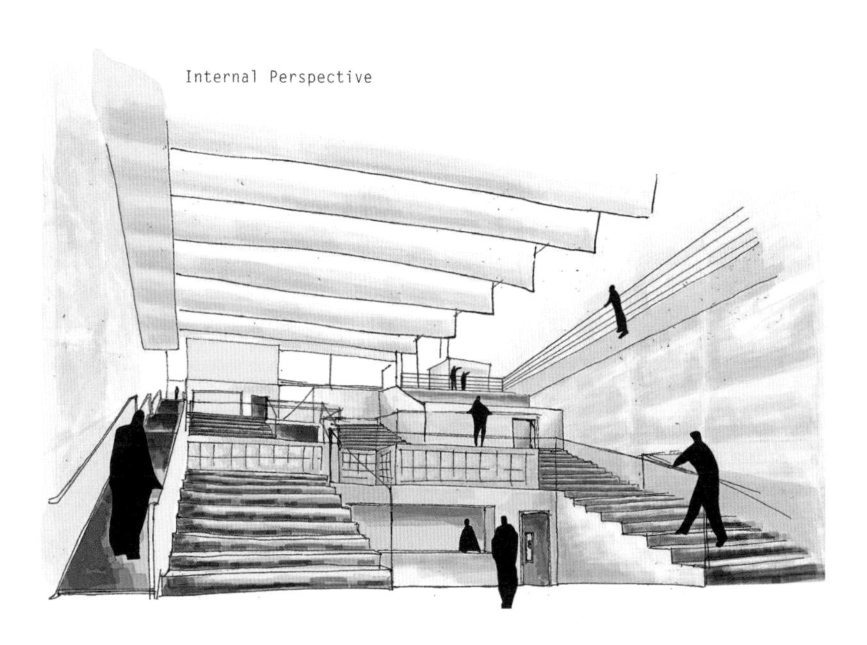

Hierarchy of size and cultural importance

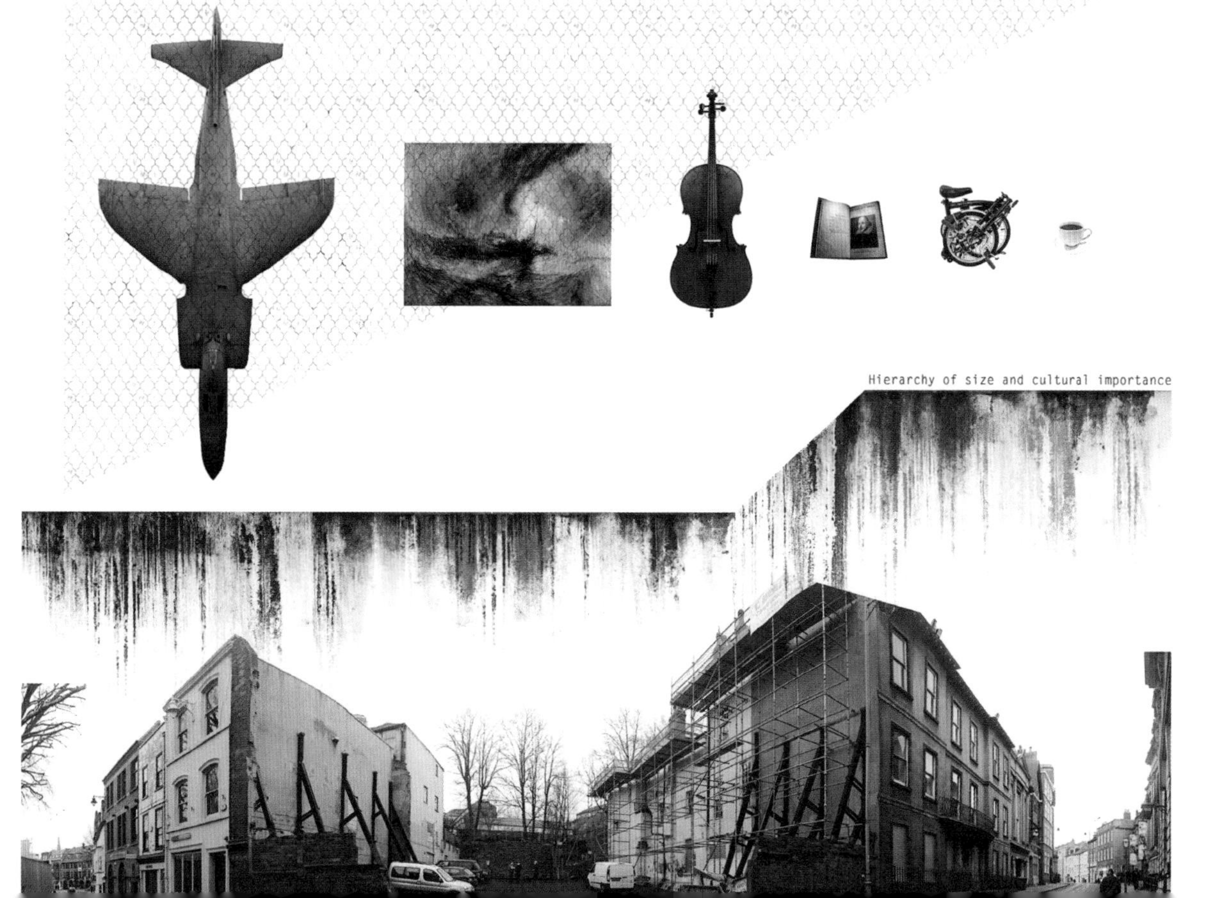

- Site Bird View 基地鸟瞰图

- Interior Perspective 内部透视图

- Exhibition Narrative 展陈叙述概念设计

002

DESIGNER-设计人: Benjamin Ferns

Devoured Entropy

St. Mary's Church is an entropic product, dealing with a series of interrelating networks in a dynamic, fluctuating and self-augmenting system. Mechanisms maintain the coastal balance, whilst the architectures of Happisburgh are continually consumed through a cycle of 200 years. Elements from previous systems become reclaimed devices, suspended for eternity in the sunken courtyard of the time arena, a metamorphosis of energy. The architecture seeks to establish a new legacy of St. Mary, one of saviour and recollection, sacrificing the graves of HMS Invincible to the storms, in exchange for the salt lines that inform the scavenging mechanisms of the impending loss. A system of petrification and archive, these lands may be re-augmented but they will never be forgotten.

噬熵

圣玛丽教堂是一个具有高熵值的人工物，在一种动态的、波动的以及自我膨胀的系统内形成了一系列相互关联的网络。虽然借助机械装置，我们可以维持海岸线的平衡，但黑斯堡建筑却在200年中不停地被侵蚀。来自原有系统的元素变为了再生装置，在时光之环的下沉庭院中追求不朽，进而成为一种能量的转化。这个建筑试图谱写圣玛丽教堂的新历史，成为一处救世堂与一部回忆录。通过牺牲战舰的墓场（纪念地），将在英国皇家海军历史上战功彪炳的7艘无敌战舰奉献给狂涛骇浪，却换来了机器人可以平静地在岸线峭壁上的盐层上采集海盐结晶，使得这一随着海浪冲刷日益消失的资源得到利用。随着类似石化与归档过程的建筑实现，这些地块也许会再次膨胀，但它们将永被铭记。

- Ground Floor Plan 底层平面图

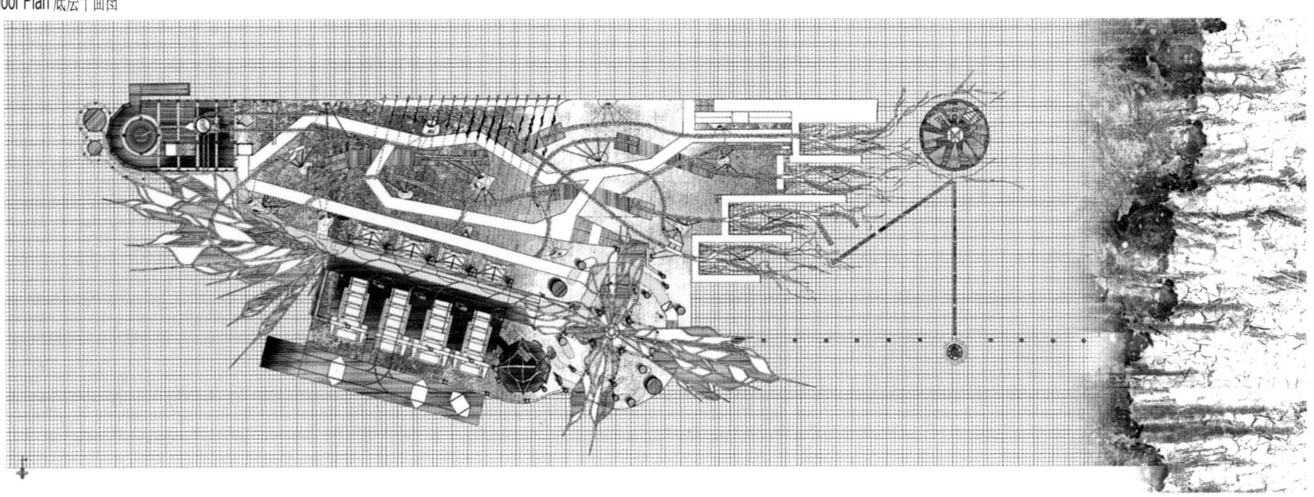

- Salt and Machine Detail 采盐机器人细节设计

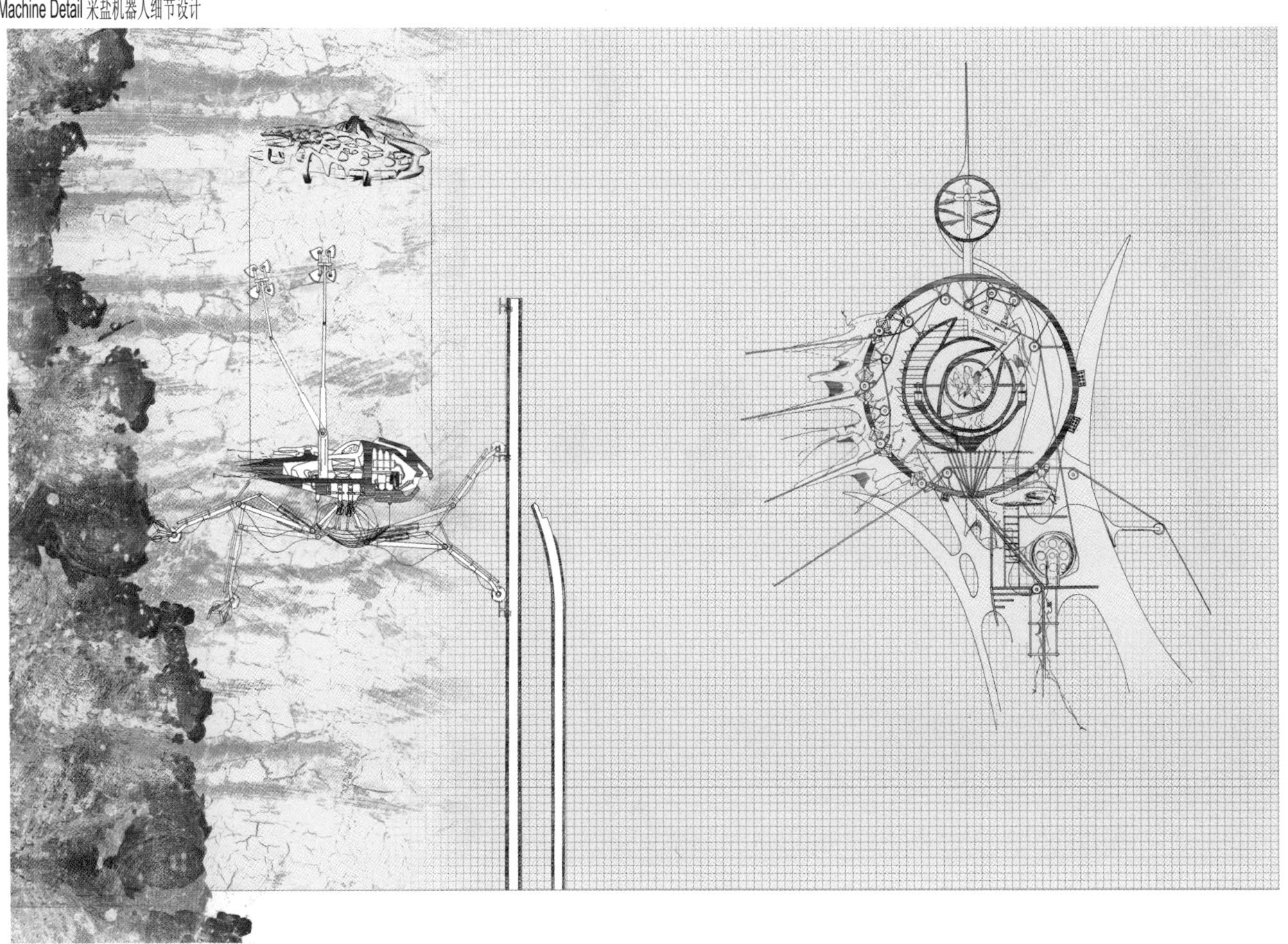

▲ - The Nave 中厅

▲ - The Hanging Cemetery of Lost Sailors 死去水手的悬挂公墓

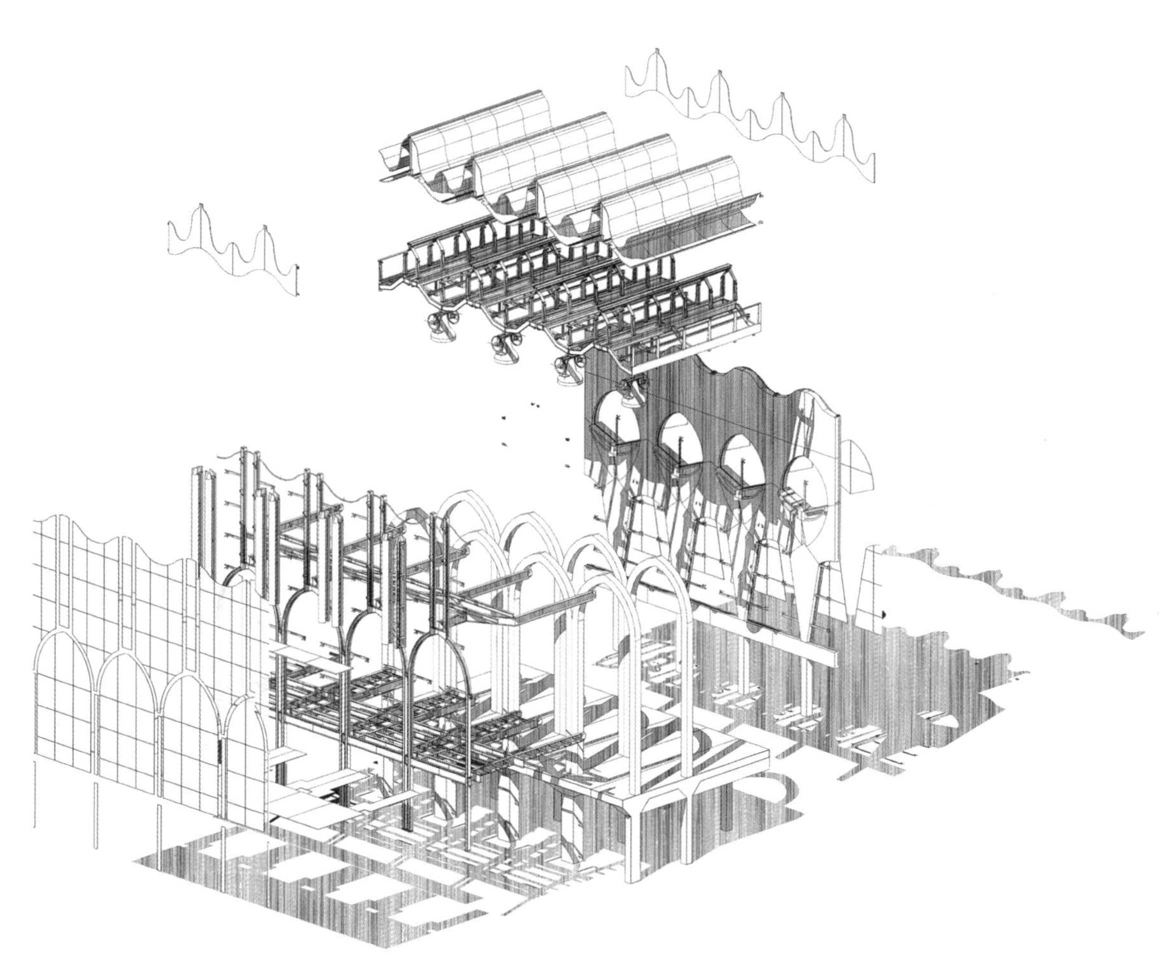

▲ - Construction Elements 构造元素分析

▼ - Cliff Perspective 剖切过悬崖的透视图

003

DESIGNER-设计人: Helen Battison

Mincemeat

In around fifty years the village of Covehithe situated on the eroding Suffolk coastline will be lost to the sea and forgotten. The place embodies its dismal future. It has a powerful feeling of melancholy and dislocation; the community is shrinking, the shore is littered with remnants of previous dwelling and the impending doom of the historic church, four hundred metres away from the soft sand and clay cliffs, plays on the mind. The Institute for the Study of Pottery would breathe life back into Covehithe. Archaeologist and Potter would work together utilising the surrounding ground. They would collect pottery shards from all over East Anglia and distribute newly crafted pots made from Covehithe clay; poignantly affirming man's relationship with the land in a place where land is to be lost to the sea. The institute would consolidate over time as the landscape diminishes. It would transform from a place of culture to a place for the study of culture, housing and embodying the things that gave Covehithe meaning, allowing it to be remembered.

肉馅酱

再过大概50年，位于萨福克郡那饱经侵蚀的海岸线上的科夫海斯村就会被波涛所吞没，被人们所忘记。这个地方处处体现出阴暗的前景，充满了绝望的忧伤与背井离乡的凄凉；社区在萎缩，海岸上四散着以前住家丢弃的杂物，而在离由松软的沙土和黏土构成的悬崖岸线400米远处，一座历史悠久的教堂正直面着这即将发生的厄运。而这座陶器研究中心将为科夫海斯带来新生。考古学家与陶艺匠人将一道开发利用周围的土地。他们将在整个东盎格利亚范围内收集陶器碎片，售散由科夫海斯泥土新烧制而成的陶壶；深刻地探讨在一片即将被海水淹没的土地上，人与大地间的关系。久而久之，研究中心还将随着大地的不断减少而不断巩固自身，从一处文化场所转变为一处研究文化的场所，容纳并展示那些赋予科夫海斯意义的事物，使它们被人永记于心。

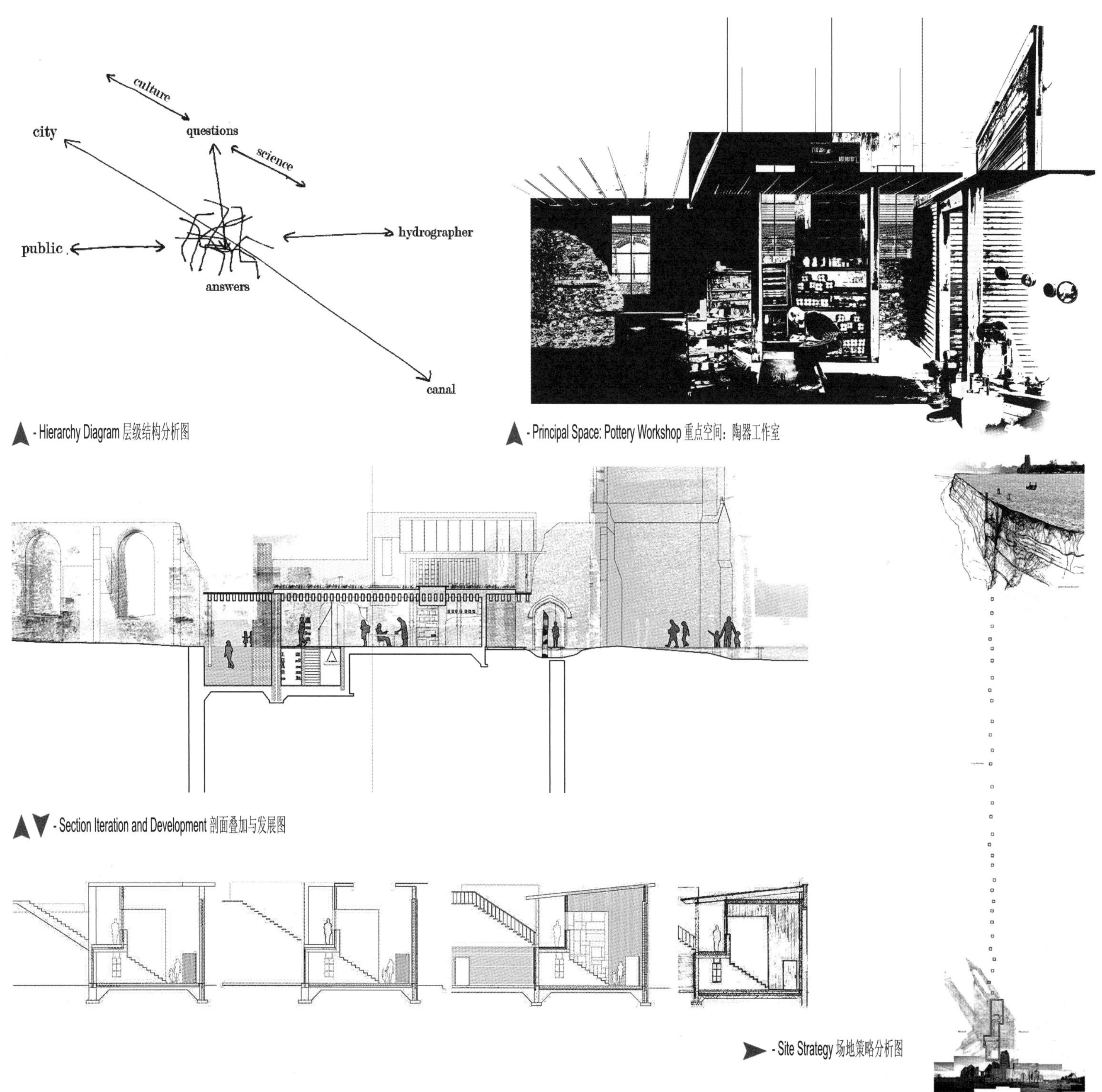

- Hierarchy Diagram 层级结构分析图

- Principal Space: Pottery Workshop 重点空间：陶器工作室

- Section Iteration and Development 剖面叠加与发展图

- Site Strategy 场地策略分析图

▼ - Plan 平面图

▲ - Final Model 最终模型

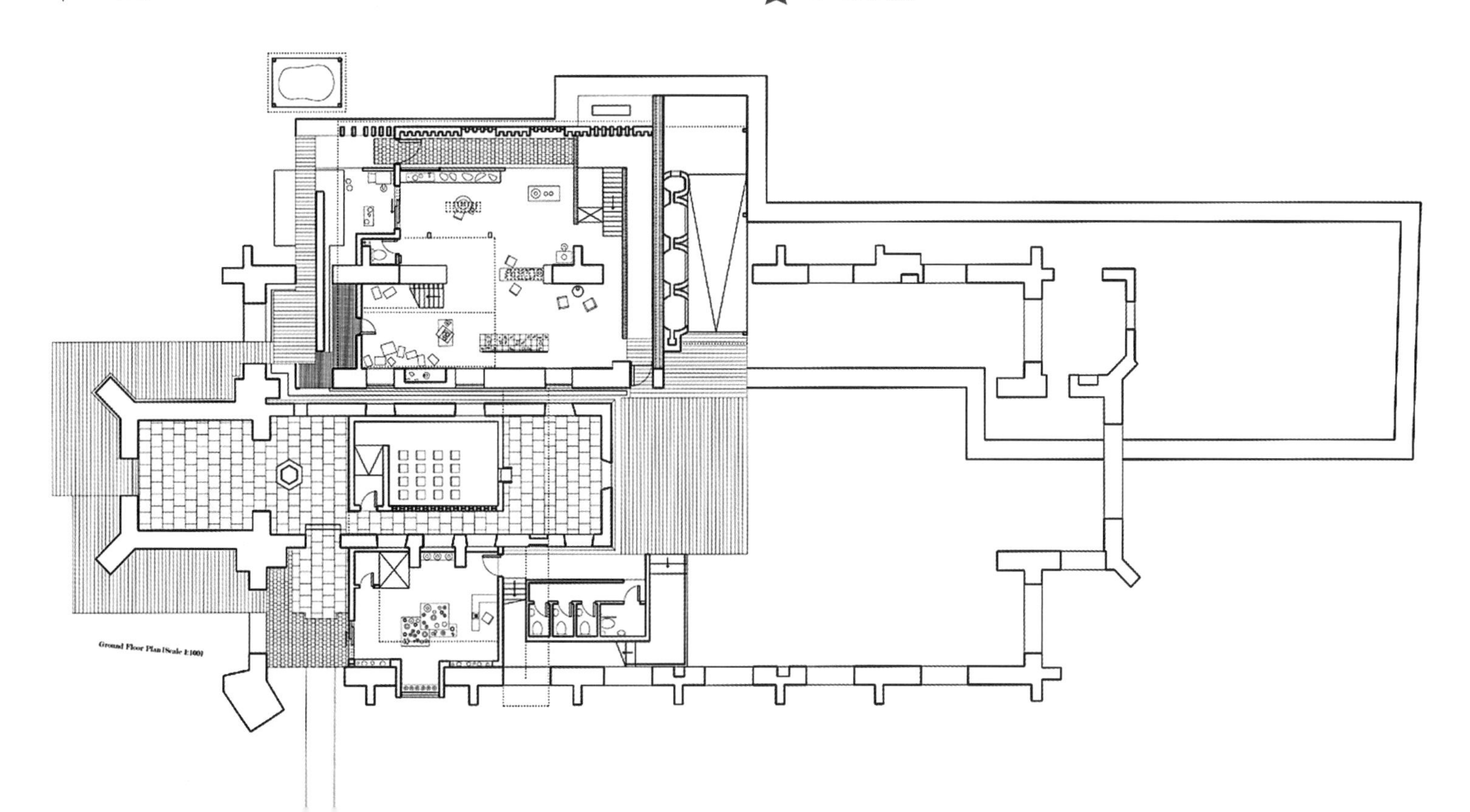

004

DESIGNER-设计人: Martin Punch

The Book and The Building

The project is based on information archaeologists extract and learn from the historic land. It is concerned with the piecing together of materials and knowledge of the landscape and the translation of this process into the gathering of information and construction of a library. The architecture provides a supportive focus and identity to a slowly de-contextualising, fragmented place; a place which suspends the mind in a threshold between permanence and change. The project tries to embrace Victor Hugo's metaphoric language of the book and the temporality associated with a fluctuating landscape through the process of translating an archaeologist's fragment into the bounded artefact of a book. In this project the dialectical interplay between cultural tensions enriches lateral spatial connections. When private investigation, public activity, individual speculation and sacred knowledge meet, there a richer poetic condition emerges.

书与建筑

该方案的发展基于考古学家对历史区域的提炼与研究。它关注对大地环境知识与材料的拼补过程，更关注这一过程转变为信息收集与一所图书馆的建立。这一建筑为一处正在缓慢丧失文脉，逐渐破碎掉的，挣扎在永恒与改变之间的地区提供了支持性焦点与可识别性。通过学习将考古学家的片段性成果转变为一本完整的著作的过程，该方案试图去拥抱维克多•雨果书中那充满隐喻的语言以及与起伏不定的大地景观相关的暂时性隐喻。在本方案中，存在于文化张力之间的辩证性相互影响丰富了横向空间联系。当私人调研、公众活动、个人思考与神圣的知识相遇时，一种极富诗意的情形随即浮现。

▼ - 3-D Diagram of Massing 三维体块研究

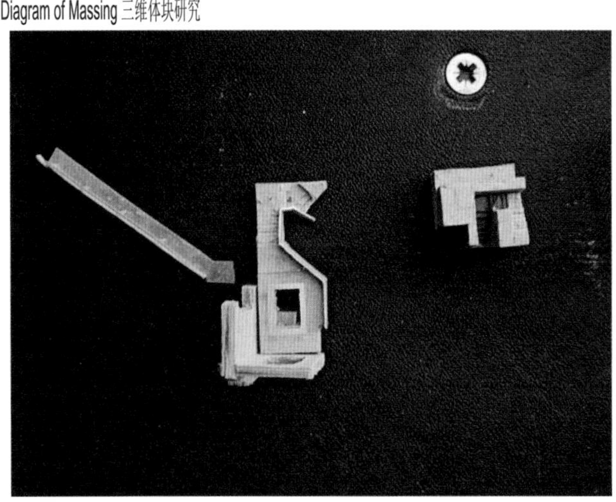

▼ - Ground Floor Plan 底层平面图

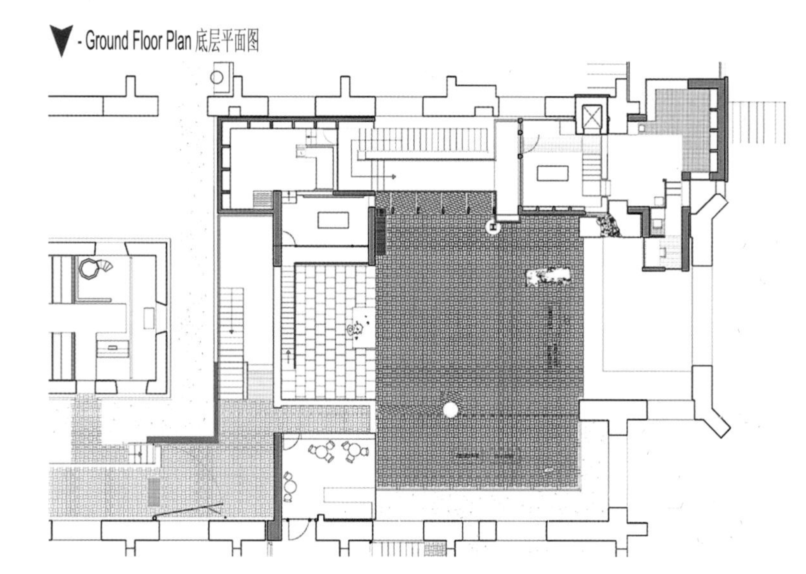

▲ - Relationship Between Courtyard and Lateral Field 庭院与建筑外围元素关系

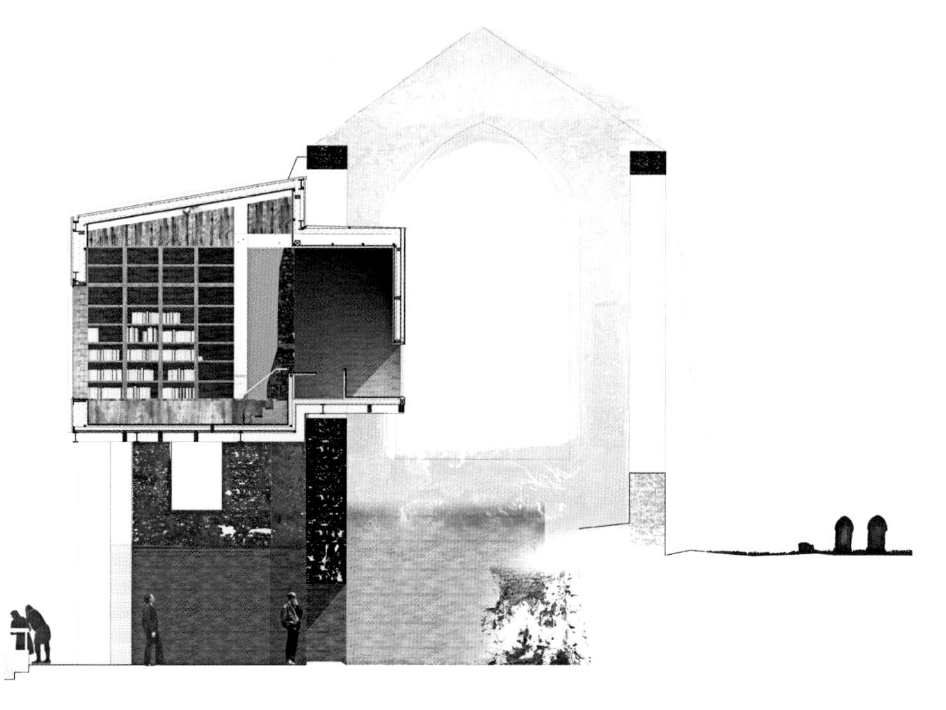

- Section 剖面图

- Model Fragment: Library 局部模型：图书馆

005

DESIGNER-设计人: Thomas Wells

House for an Ecologist

The project for Chiswell on the Isle of Portland reconnects the community with a valuable resource – the Sea. A place for ecologists who work with local fishermen teaching them about sustainable resource management, thus assisting them reinvigorate their industry for the future. The building houses a local marine ecologist (from the Marine Conservation Society) who also monitors fish stock and informs the public about their surrounding coastal waters and marine wildlife. There will be a dialog of information exchange as the fishermen add their extensive local knowledge to the expertise of the ecologist. This scheme choreographs a series of crossing paths and shared territories, creating a middle ground between responsibility & reward, brought about by better management of the Island's surrounding waters. The facility enables a lateral dissemination of ideas and local knowledge, forming a constant flow between Public, Ecologist and Fishermen.

生态学者之家

位于波特兰岛奇斯韦尔项目重新建立了当地社区与大海这一无尽宝藏之间的联系。在这儿，生态学家们可以与当地渔民一起工作，传授可持续的经营理念，并协助他们重振渔业的未来。建筑本身计划为一名来自海洋保护协会的生态学家提供住所，他在这里肩负着监视鱼类资源，向公众通告沿岸海域水文变化与野生动物状况的责任。同时，当地的渔民也可将自己多年积累的实践经验告诉生态专家，从而实现一种信息交互的对话。因此，如同设计了一套优美的舞步，这一建筑在责任与回报之间创造了一块中间地带，实现了一系列交融与共享，进而激励了对海岛周围水域的更好管理。而从一个侧面，该项目也达成了对不同思想与地方知识的传播，从而在公众、生态学者与渔民间形成了持久的互动与联络。

▼ - User Group: Pragmatic Field Collage 使用者概念分析图 - 不同功能领域组合

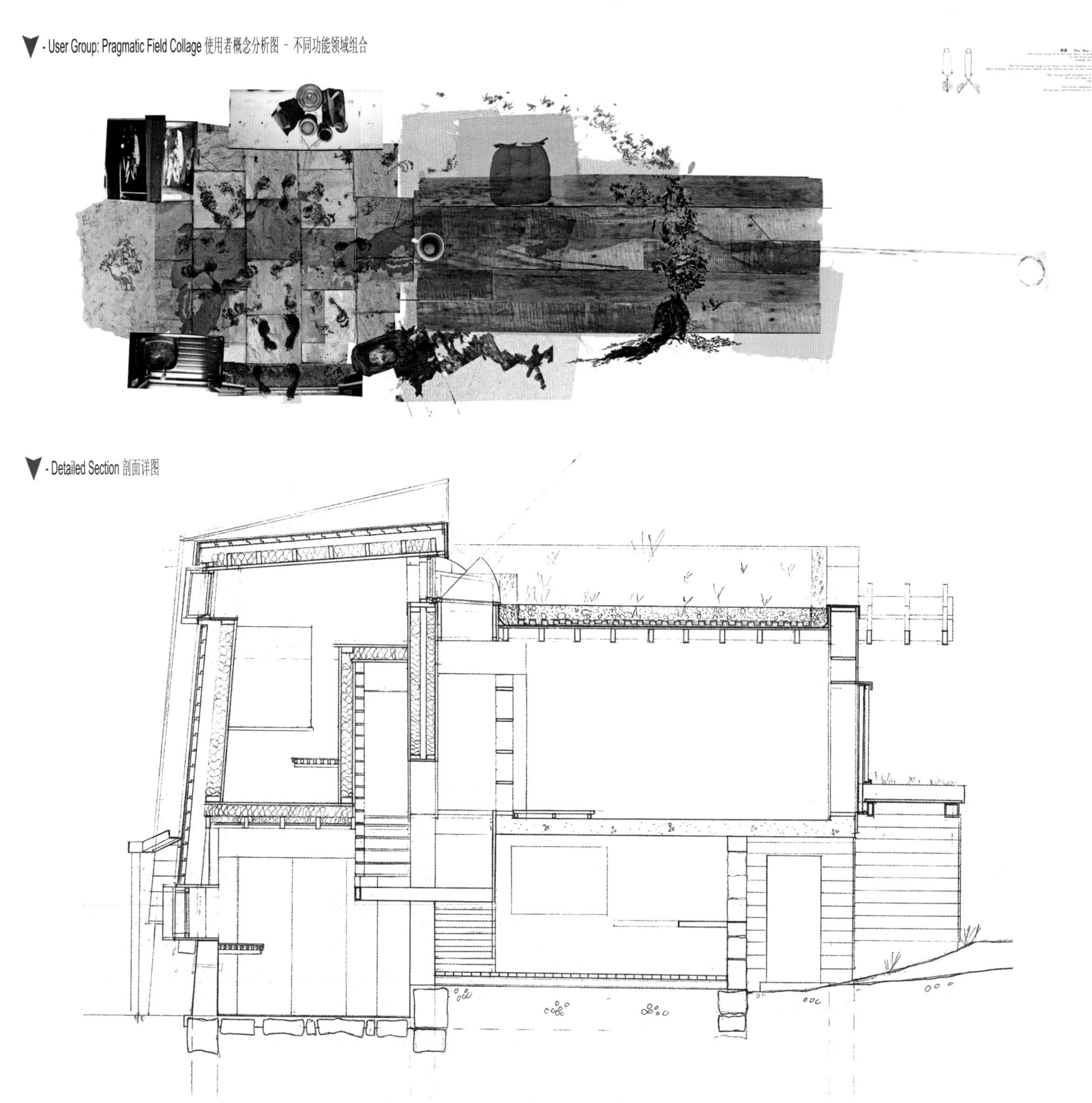

▼ - Detailed Section 剖面详图

- Tectonic Axonometric 轴测图

- Tectonic Maquette 概念模型

006

DESIGNER-设计人: Michael Ramwell

Institute for a Changing Landscape

The research institution is dedicated to furthering knowledge and education of the naturally occurring land processes of Portland – a place subjected to regular landslides caused by oversaturation within the rock strata that makes up this limestone rich Island. The institute seeks to gather interest, stimulate wider education into the process of morphology, field study, and to encourage visits by local residents. The scheme is situated in a fragile position along the Jurassic coast. It is positioned around a fissure (a scar caused by land slippage) – into which the entrance is constructed. The fissure sits just off a coastal path where ramblers/ public walkers trek across the coastal line. The institute is positioned as part of a fragment away from the urban realm, and not a fragment lost to sea - it is an element that lies between. The institute provides a place of retreat for public walkers to take shelter, or to enjoy the certain platform within the landscape.

大地景观变化研究中心

这一研究中心致力于将文化与教育引入波特兰内一处自然发生地质变化的区域，在这里，由于构成这一石灰岩小岛的岩层含水过于饱和，滑坡时常发生。研究中心希望能够在地质形态学、地理研究方面激发公众兴趣，促进广泛的教育，并鼓励当地人前来参观。该方案坐落在沿侏罗纪海岸线的一处脆弱的地点。它围绕着一处地缝（由于滑坡而形成的大地伤疤）修建——地缝中即是建筑的入口。这条地缝恰好位于漫步者小道/公共步道切向海岸线，与沿海岸小路相交的地点。而坐落于此的研究中心既是从城市机理中分离出来的一个片段，又不是丢失在大海中的一片碎片——却是位于两者之间的一个元素。研究中心为在此漫步的公众提供了一处屋檐，既可以来此遮风避雨，也可以享受这么一个嵌入大地景观中的平台。

▼ - Being a Geologist: Categorising Earth 作为一名地质学家，首先要学会为岩石分门别类

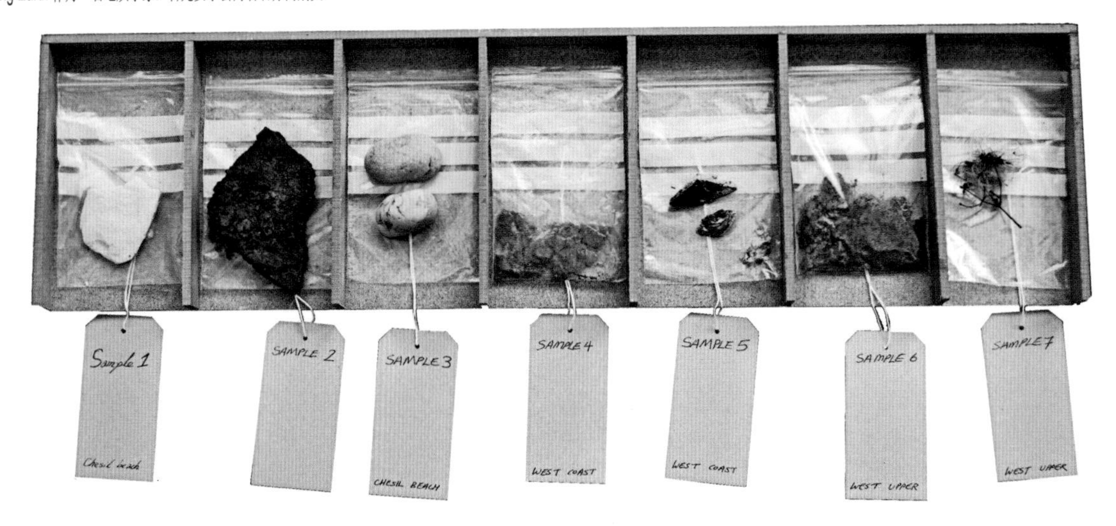

▲ - Relationship to Coastal Edge 建筑与海岸线的关系

➤ - Staircase Study 楼梯系统研究

▲▼- Interior View 内景透视图

▼- Physical Delamination Translated into Model 实体模型分层研究

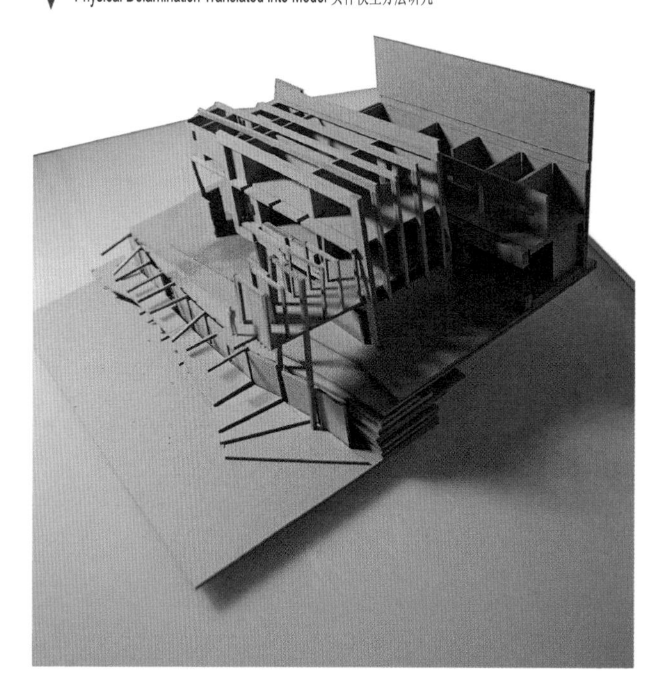

007

DESIGNER-设计人: Laura Mitchell

Microbrewery: Between Lithic and Atmospheric

"Ambient landscape is not what surrounds and supplements the building, but what enters into, continues through, emanates from, and enlivens it." (David Leatherbarrow) The raw materiality of Portland has been overly exploited, and within 30 to 40 years its natural reserves of stone will become scarce. The post-industrial landscape has been left scarred resulting in scattered communities clustering around the edges of quarries. The connectivity to the landscape needs to be re-established, infused by a traditional, cultural identity. The programmatic intention of the proposal is to develop a microbrewery that will restore this fragmentation. The infusion of processes from brewing to beekeeping will coexist to produce a honey beer that is cultivated from the land. Craft beer is considered to be a social lubricant in many societies, connecting people across local and regional boundaries. Implying or facilitating relationships between the community of beer drinkers, tourists and locals creates opportunities or nodes for interaction.

现酿现卖：岩石与大气

"周边环境并非仅指的是那些存在于建筑周围的，辅助性的东西，而是一种特殊的东西，它可以进入建筑，通过建筑而延伸，从建筑中向外辐射，且使建筑富有生气"（大卫•莱瑟巴罗）。波特兰地区的原料已经被过度开采了，在未来30到40年里，这里的自然石料资源就会枯竭。可后工业时代的大地环境已经被破坏得千疮百孔，结果形成了一串串散布在采石场的边缘的社区。与大地环境的联系需要通过传统的，当地文化的可识别性来重新建立，而该方案则通过设计一处现酿现卖啤酒厂来整合这些碎片。在激励实现这一整合过程中，酿酒业与养蜂业将会相互依存去生产从当地土壤中孕育出来的蜂蜜啤酒。当然，精酿啤酒被很多社区视为一种社会润滑剂，可突破界线，将当地与周边地区联系起来。因此，这种暗示或促进喝啤酒的人们，旅游者与当地人之间联系的状况的确可以创造更多的互动交集与良机。

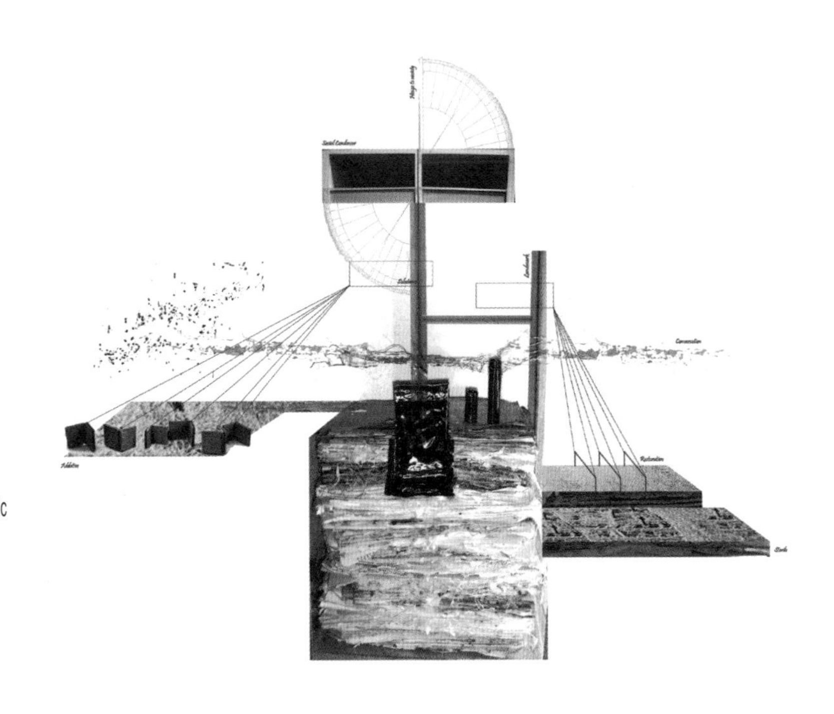

- Poetic Environmental Study: Lithic to Atmospheric
诗意环境研究：从乱石满地到意境深远

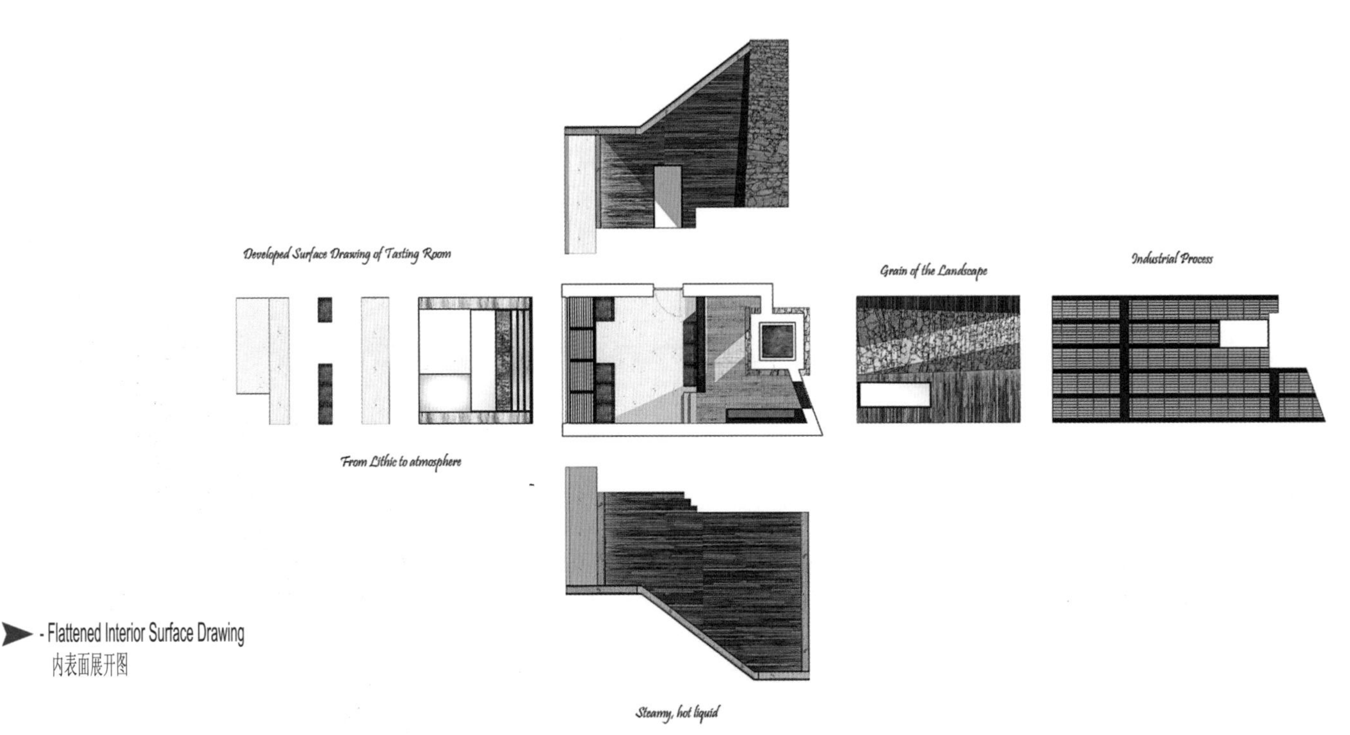

- Flattened Interior Surface Drawing
内表面展开图

- Model Study, Tasting Room 模型，品酒室

- Section 剖面图

- Relation between Architecture and Agriculture 建筑与农业的关系 - 平面图

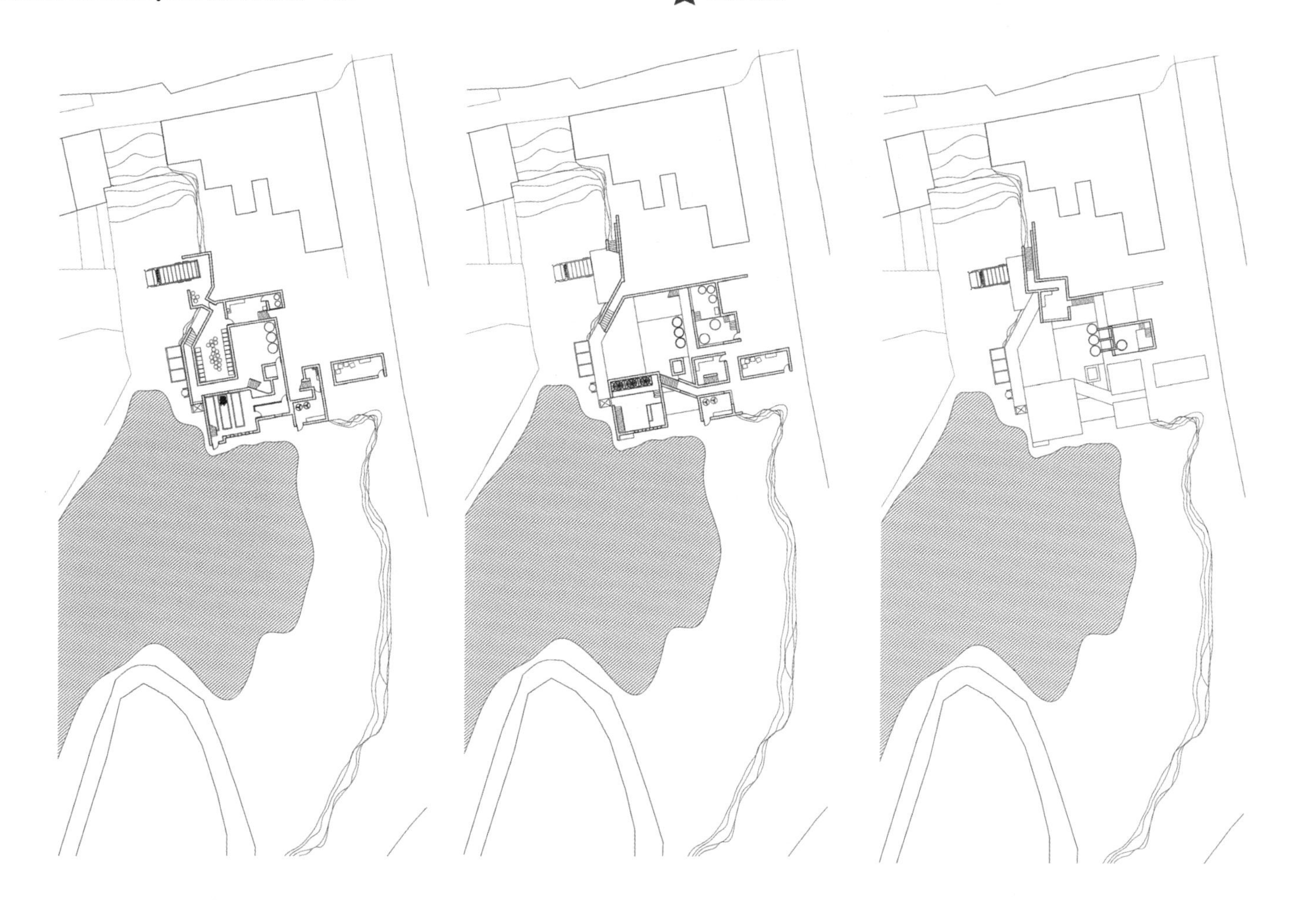

008

DESIGNER-设计人: Daniel Ladyman PRIZE-获奖: Runner-Up, Philip Webb Award 2010

Oral History Museum

"Oral history preceded the written word, oral history is having people tell their own stories and bringing it forth. That's what history is about: the oral history of the unknowns that make the wheel go round." - (Studs Terkel) Portland, 'the peninsula carved by time out of a single stone', is saturated with dying culture, both written and oral. The semi-deconstruction and remaking of Rufus castle, a 15th-century ruin, the proposal does not frame a historical fortification but reconfigures and generates a new typology. By capturing the initial dissemination of active spoken knowledge, the preservation of oral history provides an invaluable learning resource for local inhabitants and distant academic scholarship. The reactivation of a historically sedimented site of ruins positions the project in suspended animation between past, present and future. The formal transformation of a defensive medieval structure into a 21st-century accessible typology governed the development of the architecture.

口语历史博物馆

“口头语言的历史要早于书写语言，口语使得人们可以讲述他们自己的故事并将其实现。这即是历史的意义所在：对未知事物的诉说推动着历史车轮前进。”——（斯塔兹•特科尔）波特兰，“一处由时间在一整块巨石上凿出来的半岛”，充斥着死气沉沉的，口述的或书写的文化。那引人注目却半壁颓塌的卢夫斯城堡，建于15世纪却几近废墟，而本项目并非意在框构出一处历史军事要塞，而是旨在重塑并产生一种新的类型。通过对口述知识之本初传播状态的捕捉与收集，我们旨在保护口头语言的历史，而这种保护行为为当地人和身处远方的学者提供了极有价值的学习资源。方案激活了积淀着厚重历史的废墟，同时也将自身置于过去，现在，未来之间的动态悬浮状态之中。这是对一处中世纪防御建筑的正式改造，将它带入21世纪并转变为可使用一类新的建筑类型。而正是这种行为在主导着建筑的发展。

▼ - Recomposition of Old and New Structure 新旧结构体组合

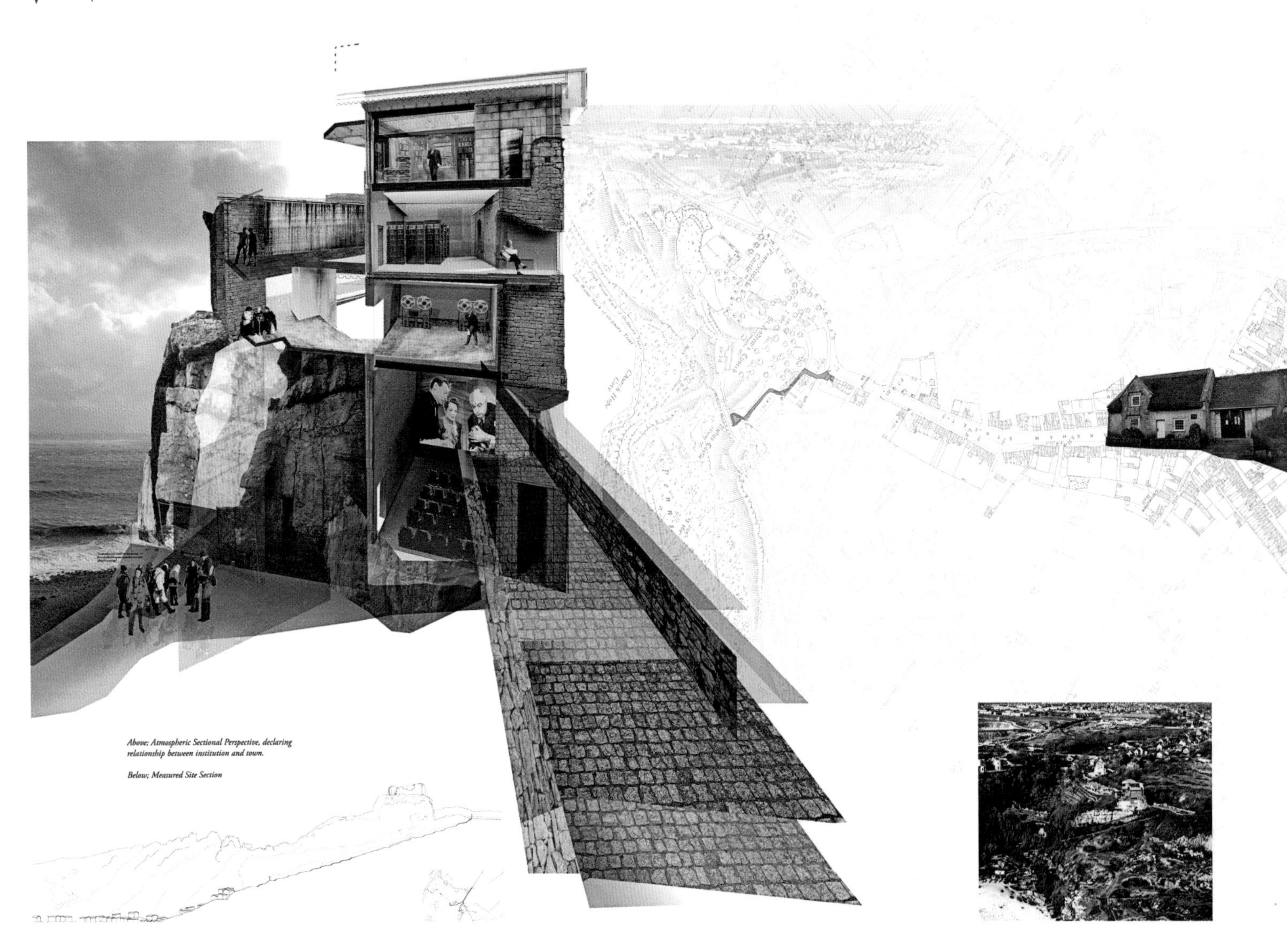

Above; Atmospheric Sectional Perspective, declaring relationship between institution and town.
Below; Measured Site Section

- Context, Section, Axonometric View and Models 文脉研究，剖面图，轴测图以及实体模型

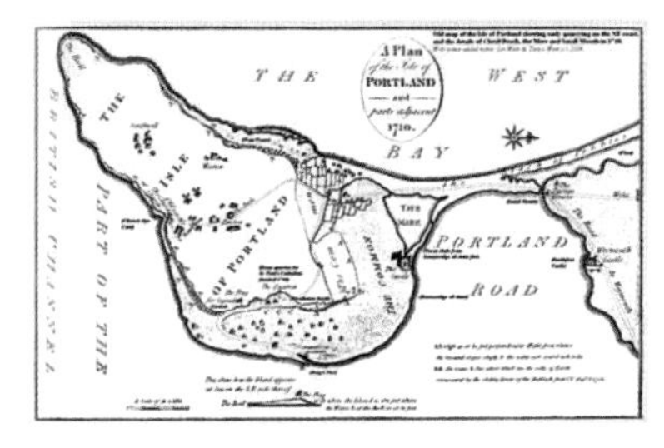

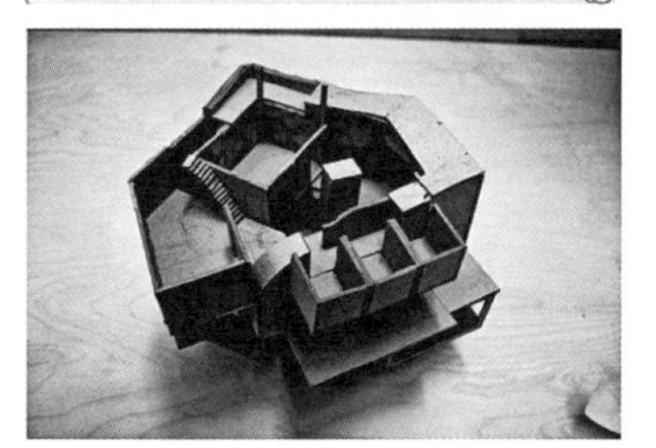

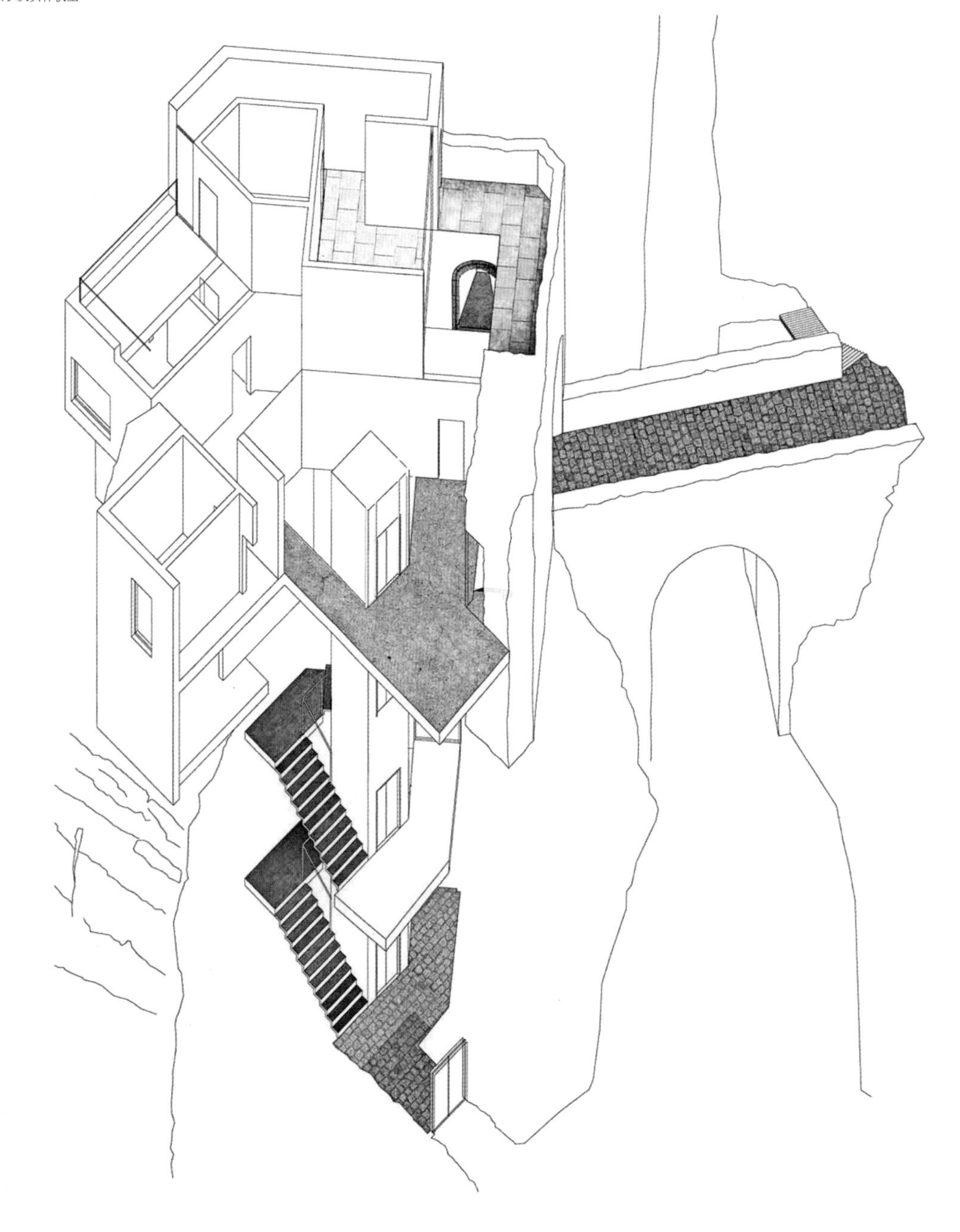

Above Right; Axonometric - Entrance Sequence

Left; Developing Articulation through Models

Below Right; Tectonic Section

Below; External Perspective showing relationship between existing fabric and the new

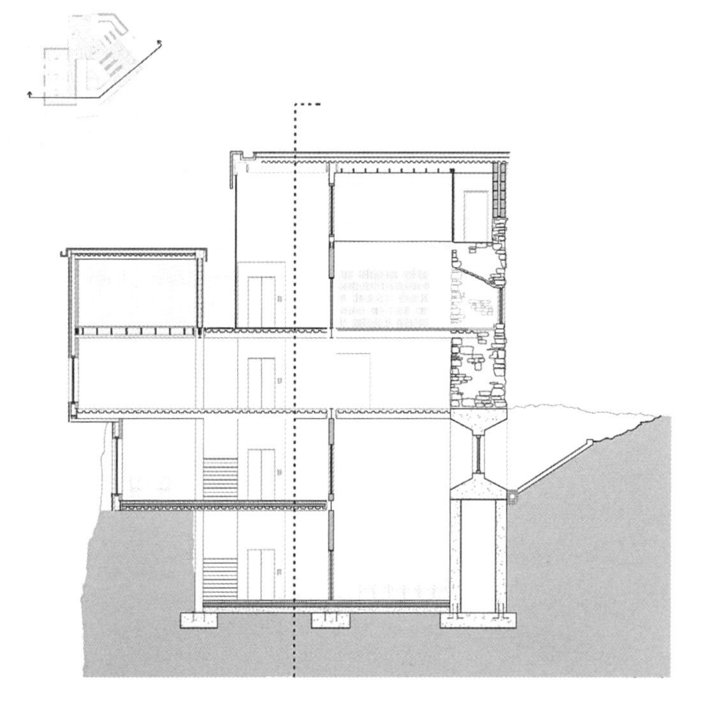

009

DESIGNER-设计人: Christopher Ansell

Remediating the Prison

Despite being an omnipresent sear upon the horizon with a scale alien to the rest of the island, the young offenders prison on Portland acts as a void through the heart of the community - repelling and incarcerating at one and the same time. There are moments, however, when this ribbon of wall that surrounds the prison ruptures, offering views into the heart of the institution. One of these blisters is exploited to incorporate a centre where families can come and see offenders, while giving offenders opportunities for the future beyond the wall. This institution has a dual-axis approach and acts as a neutral environment between two extreme worlds. It is neither this side, nor that, of the wall - it becomes a part of the wall, a part of the fabric of the community, a piece of the prison reaching out over the wall to manifest itself beyond. Much of the focus of this analysis has been in the form of looking at thresholds, and the impact that these have upon the existential experience of the occupant and the operation of the institution. An institution reaches out and influences the wider pragmatic field surrounding it while at the same time appearing impenetrable.

治疗监狱

尽管有着与岛上其他建筑环境迥乎不同的尺度，如同在天际线中一处目光无法避开的焦痕，波特兰少年监狱依然在当地社区的心底中仿佛一处空白——似被驱逐在外，像是幽禁自身。然而，那堵如带子般环绕着监狱的高墙却在某些地方被“撑裂开”，为人们提供了一睹其核心情景的机会。我希望在这些如“水泡”般的“裂缝”中选择一处进行改造，将其整合为一处犯人家属探视中心，从而给这些犯人提供越过高墙阻隔，展望未来的机会。这里联系里外两条轴线，在两种截然不同的环境间提供了一处中和的环境。它不属于墙的任何一边，它本身就是墙的一部分，是当地社区文脉中的一部分，是监狱的一部分，却探出了高墙，对外展示着自己。该方案的绝大部分分析都集中在处理隔阂，以及其对服刑者造成的在存在体验方面的影响。而这样一个机构恰可以提供对外接触，在广泛的相关实用领域内产生影响且同时表现得无懈可击。

- Boundary Study “边界”概念研究

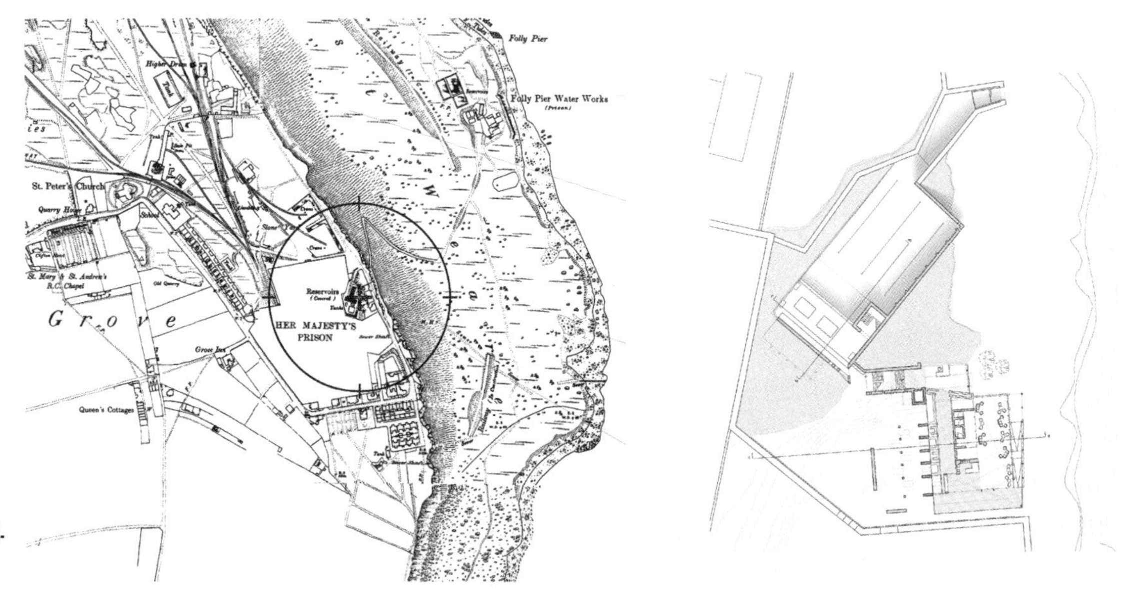

- Relationship to Prison Wall, and Edge-land Condition 与狱墙关系研究，边界地带环境研究

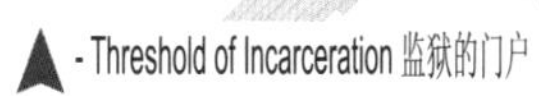
- Threshold of Incarceration 监狱的门户

- Entrance 入口

010

DESIGNER-设计人: Gareth Marriott

Landscape Reminiscence: Lighthouse for Military Trauma

The Kings Lynn WWII Coastal Defence Battery in the reclaimed Fenland Marshes was in action from only 1940-1943. A history now left derelict. Needs deteriorate much more hastily than the structures put in place to facilitate them. This project explores the landscapes and people that have been forgotten, neglected and abandoned. The design explores military vernacular through carved form and facade fortification detailing. Drawing also on specific light qualities, through principles of Phototherapy, Vision Therapy, Sleep Therapy, Art/Sculpture etc. to benefit military veterans left behind by their communities, returning to find conflict with physical, sensory, or mental imbalance. The Memorial Pathway that brings together these scattered relics, continues across the landscape to an active bombing range. And through bio-stimulation gradually regenerates the "poppy-lands" that once washed the fields here crimson. Flowers further used in Phytoremediation, removing explosive residues and heavy-metals from the soil, in a long-term commitment to the area.

景观回忆录：战争纪念灯塔

利用芬兰德沼泽地建设起来的金斯林二战海岸堡垒，仅仅在1940年到1943年间投入了使用。现如今，它的历史正在一点点被遗忘。当地对这栋建筑的需求正在急剧退化消失，其速度甚至比支持着这些功能的结构毁灭得还快。这个设计试图将那些被遗忘，被忽视，被废弃的景观和历史人物重新呈现给大众。通过雕刻形式和防御工事的立面细部，设计试图来表现军事传统建筑的风格。光线效果也受到了特殊的关注，比如光线疗法、视觉疗法、睡眠疗法、艺术／雕塑等，从肉体上、感官上，甚至是不平衡的心理感受上，来帮助那些留在当地社区的战后老兵找回当年血战沙场的记忆。一条纪念小路将那些散落的历史遗迹都串联了起来，穿过整个景观区，引导入一处还在使用的投弹场。通过生物激活的技术，我们可以使曾经将此处染为一片深红色的罂粟花田重新逐渐兴盛起来。（注：在英国，人们用点点罂粟花来象征战争中牺牲士兵的血迹，因而变成了战争纪念日的标志，与毒品无关。）此外，罂粟花还可以进一步用作植物修复，移除土壤里的爆炸残留和重金属，长期为这个地区服务。

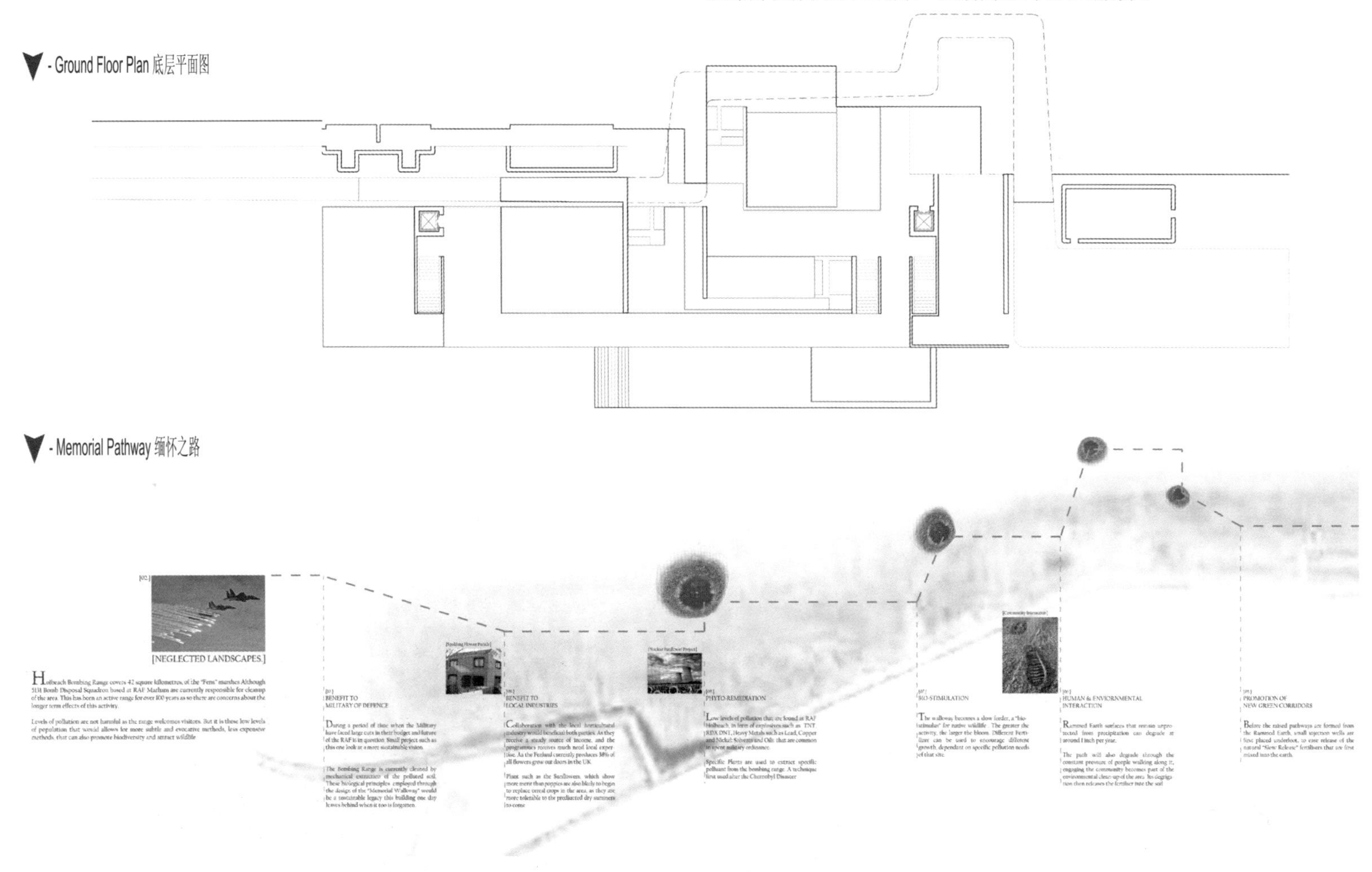

- Concept Drawings

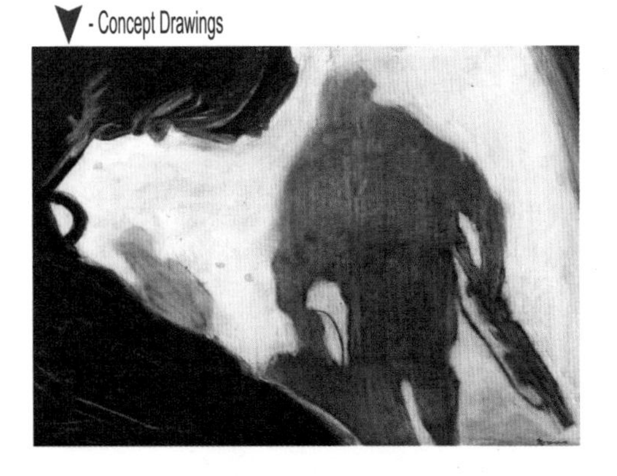

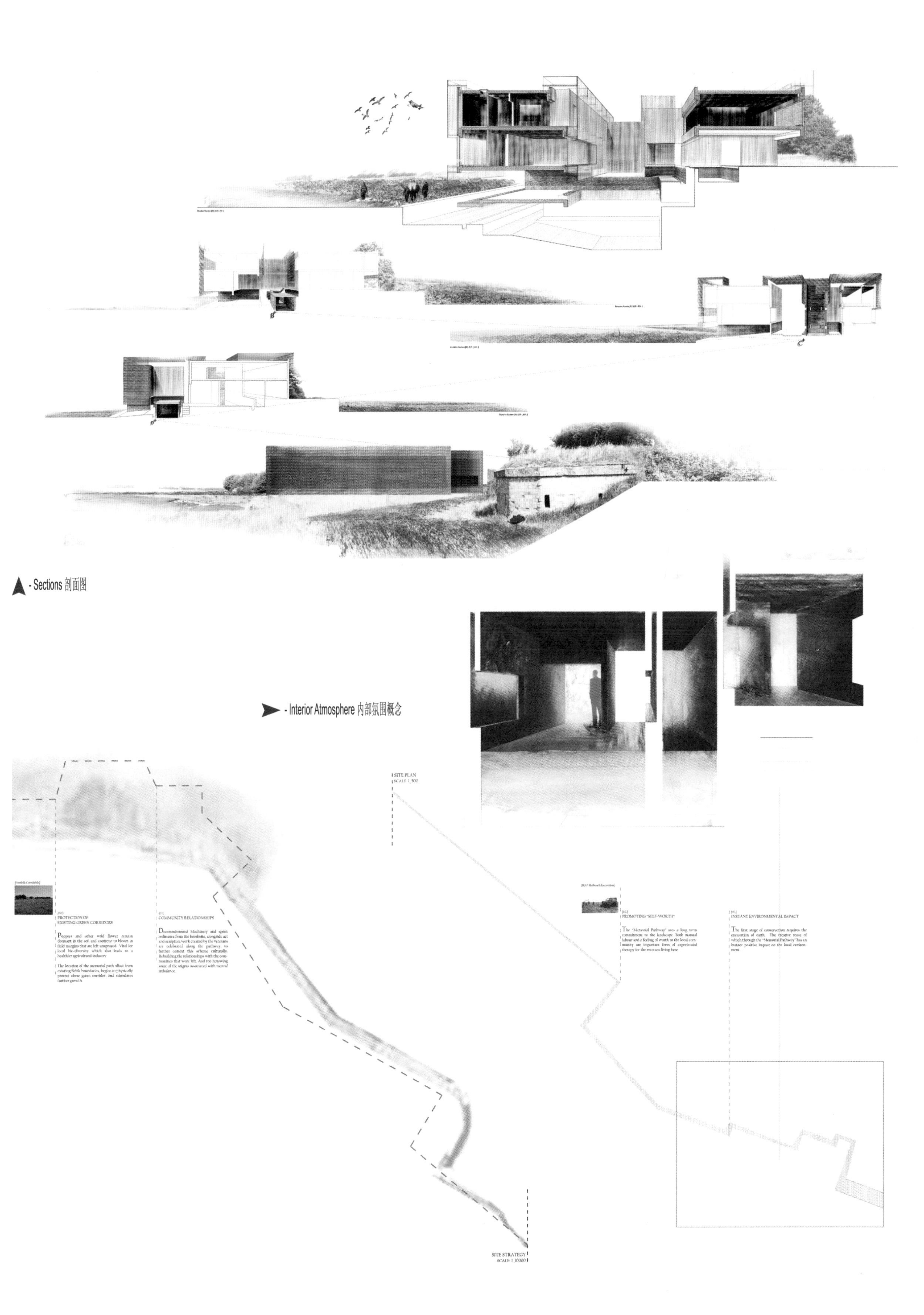
- Sections 剖面图
- Interior Atmosphere 内部氛围概念
SITE PLAN
SCALE 1_500
PROTECTION OF
EXISTING GREEN CORRIDORS
Poppies and other wild flower remain dormant in the soil and continue to bloom in field margins that are left unsprayed. Vital for local bio-diversity which also leads to a healthier agricultural industry
The location of the memorial path offset from existing fields boundaries, begins to physically protect these green corridor, and stimulates further growth.
COMMUNITY RELATIONSHIPS
Decommissioned Machinery and spent ordinance from the bombsite, alongside art and sculpture work created by the veterans are celebrated along the pathway, to further cement this scheme culturally. Rebuilding the relationships with the communities that were left. And too removing some of the stigma associated with mental imbalance
PROMOTING "SELF-WORTH"
The "Memorial Pathway" sees a long term commitment to the landscape. Both manual labour and a feeling of worth to the local community are important form of experiential therapy for the veterans living here
INSTANT ENVIRONMENTAL IMPACT
The first stage of construction requires the excavation of earth. The creative reuse of which through the "Memorial Pathway" has an instant positive impact on the local environment.
SITE STRATEGY
SCALE 1_30000

011

DESIGNER-设计人: Wang MangYuan

Secrets of The Magdalen Laundries

This project is about the memory of Magdalen Laundries, to commemorate those victims who had been incarcerated into the Cork Magdalen Asylum. This Project is a combination of a chapel, a public baths, a memorial space and a Magdalen Laundries Archive. The baths connected to the chapel, is a modern way to reinterpret the lost positive meaning of Magdalen Laundries which considered sins could be washed away along with the stains. Therefore the baths provides a place for people whose guilty in mind could be released and filtered through a series of spiritual bathing process. The memorial space mainly focuses on the inner worlds of those victims, as Fenster (2000) pointed out, "they (victims) lived in a private world of desire, longing, and unreachable fulfilment, forced into a mundane ritual of service without pleasure of amenities, their vitality and eros were bound by the superficial morality of the church." This space tries to explore their incarcerated individual identities through a subtle architectural way.

从良妓女洗衣店的秘密

这个项目为我们描述了从良妓女洗衣店的往事，记录了那些被迫监禁在科克从良妓女庇护所的受害者，她们在那里度过了漫长的岁月，有的甚至是一生。整个建筑包括了一个小教堂、一个公共浴室、一个纪念性场所和一个从良妓女洗衣店的档案馆。将浴室与小教堂相连是一种现代的手法，隐喻了罪过也能像污渍一样被冲洗干净，这也诠释了从良妓女洗衣店那消失的积极含义。因此，在浴室里，人们可以通过一系列富有精神性的洗浴仪式，冲刷耻辱，放松和净化心灵。而纪念性场所则主要关注受害者的内心世界，正如芬斯特（2000）所指出的："她们（受害者）生活在一个充满欲望、思念和永远无法满足的私人世界中，被迫从事世俗低贱的服务，没有任何的快乐，她们的活力和情爱被肤浅的教会准则所束缚。"这个空间试图通过微妙的建筑手法来释放她们心中被禁闭的自身认同感。

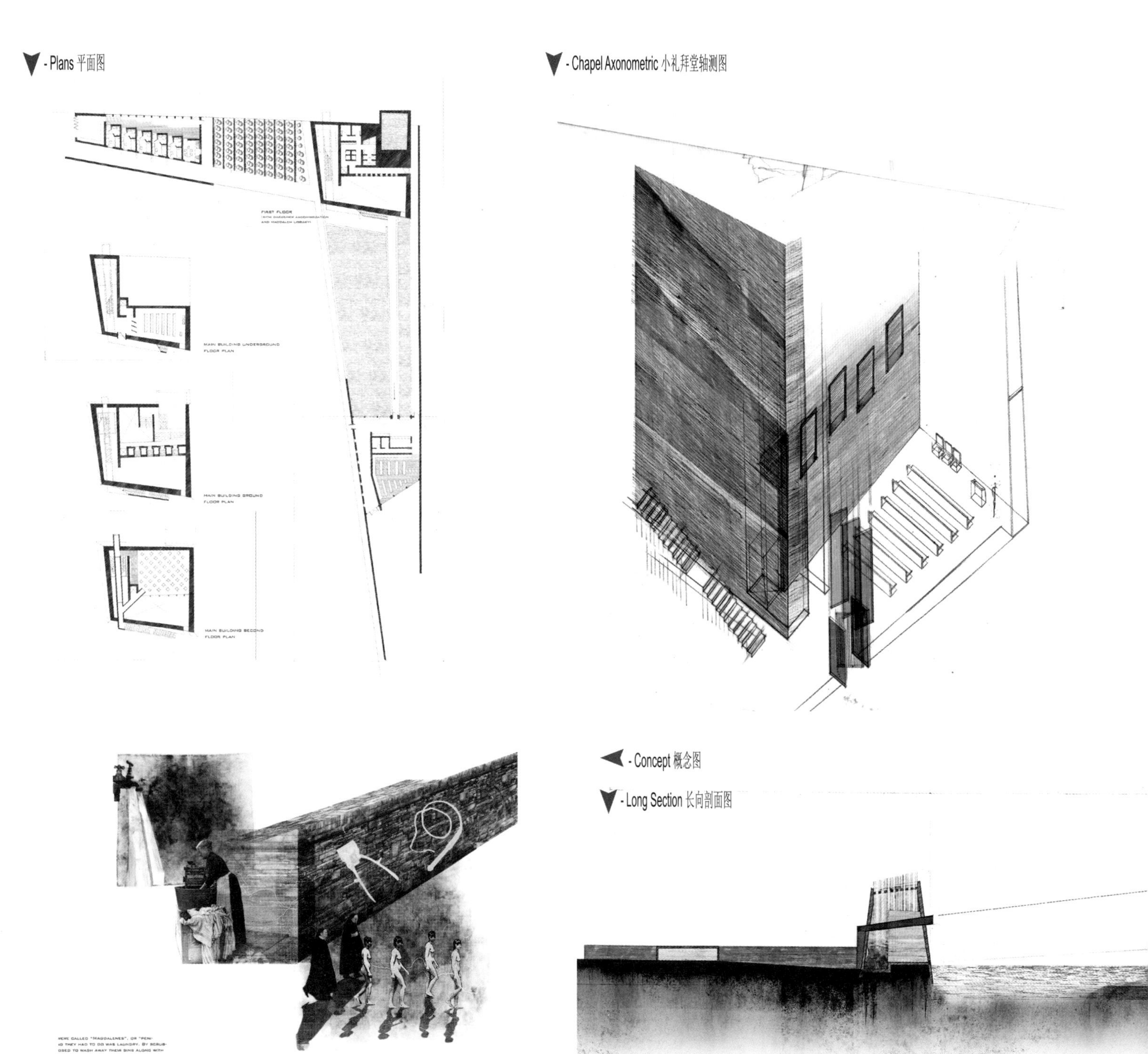

- Plans 平面图

- Chapel Axonometric 小礼拜堂轴测图

- Concept 概念图

- Long Section 长向剖面图

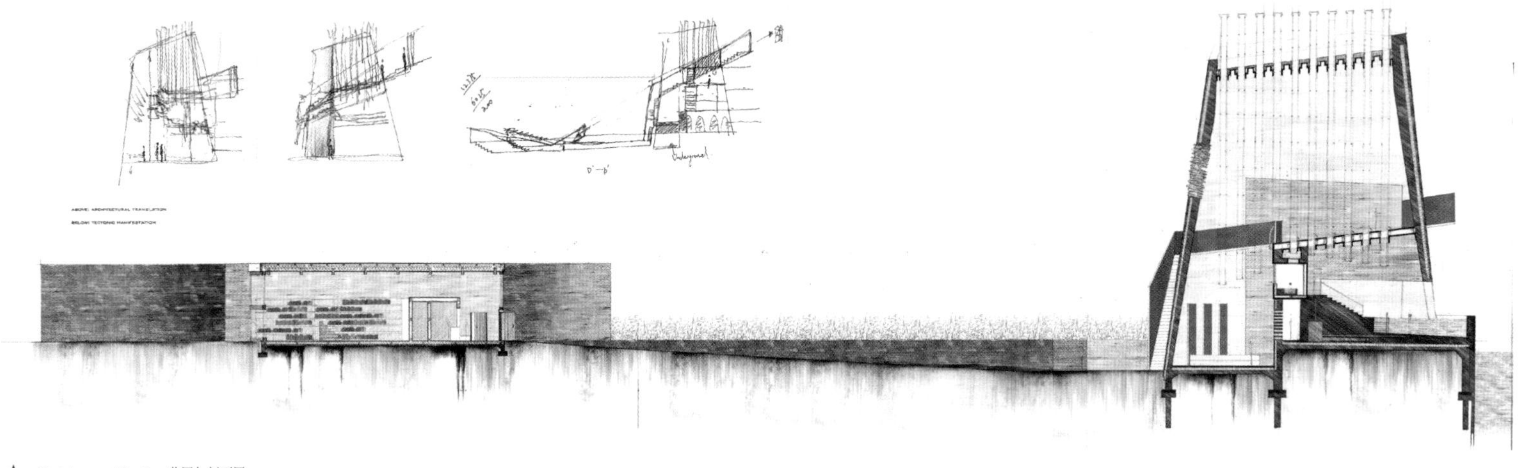

- Sketches and Section 草图与剖面图

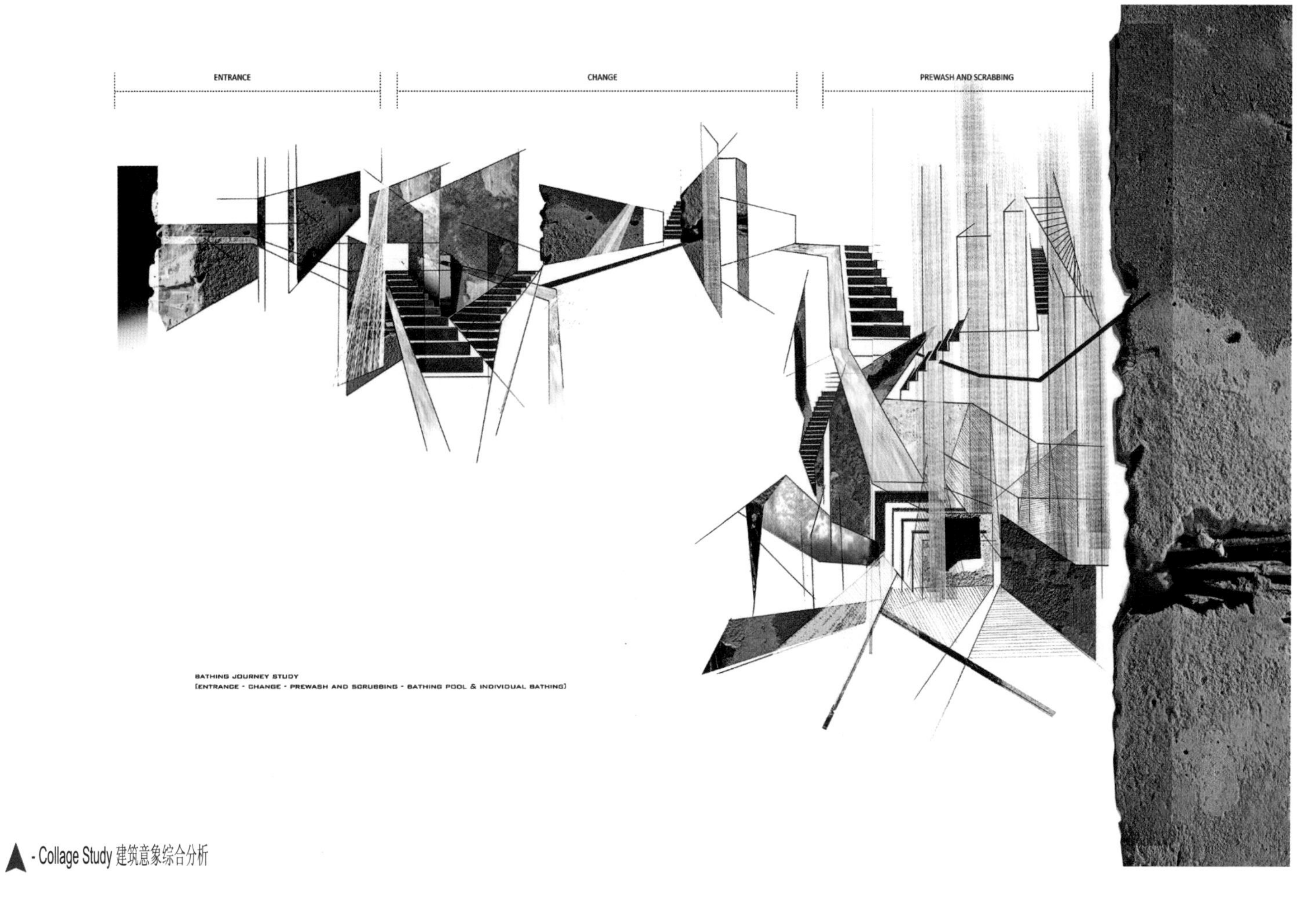

- Collage Study 建筑意象综合分析

012

DESIGNER-设计人: Greg Skinner

Evræ

In the future, an infinite source of energy may become a reality. Fusion is one option, but there may be others. Having limitless energy would be a potential wrecking-ball for the world economy as we understand it today. Indeed, presently all roads lead to oil (sooner or later). Within a short while, an endless energy source would challenge concept of ownership, which has stood at the centre of almost all world cultures since man became human. Concept of ownership defines our id, ego, and super-ego, and is inherent in our language expressed as the personal pronoun. Haizum, which is the phase through which humanity must pass in order to overcome the potential roughage such an energy plethora would bring, is not shown here. However, Haizum is - in one form or another - a certainty. If Haizum is overcome, Evræ is one possible outcome.

神龙

在未来，拥有无限的能源可能成为现实。核聚变是一种选择，但是还有其他方式。拥有无限的能量就像拆毁建筑时用的大铁球，可以击毁我们所理解的当代世界经济格局。当然，目前所有的方式都是依赖石油（迟早有一天）。在很短的时间内，取之不尽的能源将挑战现在所有制的概念。而自从人进化成为人类后，这种所有制就处于大多数世界文化的中心。所有制的概念决定了我们的身份，自我意识和超我意识，并以人称代词的表达方式存在于我们的语言中。神马Haizum暗示着一个阶段，人类身处其中就必须克服能源过剩所可能导致的粗放经济，不过这里并没有对它详细说明。然而，神马Haizum是从一个形式到另一个形式时的必要条件。如果神马Haizum被控制了，那么神龙Evræ就会出现。

Mutagenic Surgical Impliment
SC 1/2
PROJECT EVRÆ
URBAN DEVELOPMENT/Elevations 1:200

013

DESIGNER-设计人: Khyle Raja

The Question Concerning Architecture

In the age of increasing technological sophistication, humanities' position in relation to architecture and technology must be better understood. The direction of this thesis, 'The Question concerning Architecture', has led to a critical and highly creative approach through writing and drawing. The primary concern of the work is the articulation of a new relationship between architecture, humanity and technology. A series of essays explore the potentials of Protocell, a radical new language for architecture, and artifice, designing entire worlds and universes. The drawn project articulates a laboratory as a landscape, moving through the site of Sherwood Forest, England. This synthetic-biological environment acts as a technological and spatial artificer with various technologies interacting with site and the individual. The resulting architectural spaces alter with time and develop irregular and unexpected spatial outcomes, immersing the scientists in a living, reacting laboratory. The thesis details an alternative view on architecture, draws an alternative reality of space, and exposes a radical departure from the constraints of contemporary practice.

关于建筑的问题

在这个高新技术不断发展的年代，人类与建筑和技术之间的关系必须得到更好的理解。本方案的研究——“关于建筑的问题”——通过写作和绘画方式而引申出了一个富有批评性和高度创造性的方向。本作品主要关注建筑、人类和技术在新的关系中是如何相互联系的。先前所做的一系列短题不仅探索了“原细胞”的潜力，这是一种新式的基本建筑语汇，而且也探讨了“技巧”，可用以设计整个世界和宇宙。项目清晰地描绘出了一个类似景观的实验室，穿过英格兰的舍伍德森林。这个合成生物学的环境扮演了一个技术与空间技工的角色，它通过各种技术与基地和使用者产生互动。最后衍生出的建筑空间会随着时间变化，并产生不规则和意想不到的空间模式，使科学家可以沉浸在一个充满活力和互动的实验室里。方案从一个不同的视角讨论了建筑，描绘了一种不同的现实空间，这从根本上摆脱了当代实践的约束。

- The Crucibles – The entry of Protocell into existing technological systems
熔炉 - 原细胞入口，可由此进入现有技术体系

- The Laboratory – A Protocellular laboratory
实验室 - 原细胞实验室

- The Laboratory Door – A metabolizing portal through the landscape
实验室大门 - 一处通往周边景观，可自我新陈代谢的门户空间

- The Loom – An electromagnetic cradle, weaving the synthetic laboratories into the landscape
纺织机 - 电磁发生器，将实验室群编制入周边环境之中

- The Desert - A theoretical exploration of our current relationship with technology
沙漠 - 对当前人类与技术关系的理论探讨

TUTORS - 指导教师: *Phil Waston, Jonathan Morris, Projects 12-14 are from Unit X, more details can be found from http://slamshed.co.uk/*

DESIGNER-设计人: Chandni Modha

Decadentphoria Temple for Cosmetics and Decoration

Decadence is back in the age of excess commodities. This moral decline has lost ornament. It is in fact a material Labyrinth where 'she' is caught up. Her only solution is to conquer this new landscape with a set of jewelled accoutrements. Welcome to the new range of decadent Lizard tools, which are devices for setting out and planning an architectural site. Each piece of jewellery can be worn on the body or the shoe heel. The main function is to take measurement and archive the landscape to prepare for a bespoke building project. Each piece works in synchronisation with the ergonomics of a modern woman's body. The object can be worn as 'everyday commodity' and forms a new relationship to the landscape. It creates a geometrical labyrinth structure of co-ordinate, line and surface, which establishes a time and space for split-site architecture. Embodiment of the jewellery within the gardens locates permanent spaces for setting out the growth of a bamboo column in the form of caryatids.

梳妆打扮的颓废寺庙

在这个物质过剩的年代，颓废又回来了。这种道德的滑落已经失去了伪饰的装扮，事实上，我们只有在这个物质的迷宫中才能抓住"她"。对于她而言，唯一途径就是通过各种珠光宝气的装扮来征服这个新景观。于是，欢迎使用这一系列新型的颓废蜥蜴工具！它可以用来设计和布置建筑基地，而每一件珠宝都可以被穿戴在身体或是鞋跟上。这些珠宝的主要功能是用来测量和记录景观，以此来为特定的建筑设计量身准备。每个挂件都是参照现代女性身体，根据人体工程学设计出来，它们可以同步运作和计算。你可以像对待日常用品那样，穿戴这些挂件，从而与景观形成新的联系。这是一种几何式的迷宫结构，有坐标、线和面，共同组成了一个关于时间和空间的分散式建筑。而在花园中，这些珠宝装饰主要体现在一些固定的空间中，在那里，呈女神柱的形式的竹柱茁壮生长。

- Labyrinth 迷宫

- Experience photo 设计者自我体验

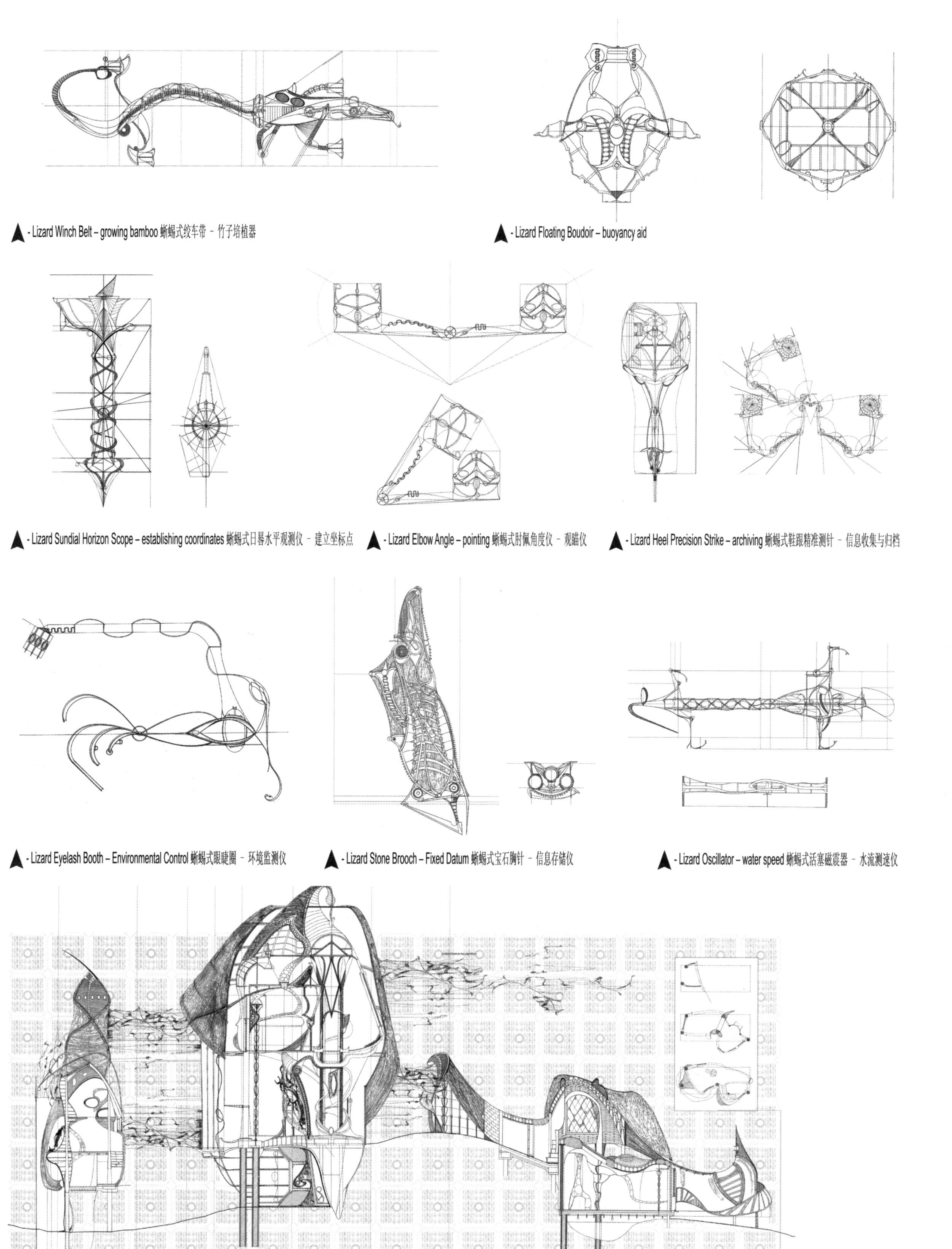

- Lizard Winch Belt – growing bamboo 蜥蜴式绞车带 – 竹子培植器

- Lizard Floating Boudoir – buoyancy aid

- Lizard Sundial Horizon Scope – establishing coordinates 蜥蜴式日晷水平观测仪 – 建立坐标点

- Lizard Elbow Angle – pointing 蜥蜴式肘佩角度仪 – 观瞄仪

- Lizard Heel Precision Strike – archiving 蜥蜴式鞋跟精准测针 – 信息收集与归档

- Lizard Eyelash Booth – Environmental Control 蜥蜴式眼睫圈 – 环境监测仪

- Lizard Stone Brooch – Fixed Datum 蜥蜴式宝石胸针 – 信息存储仪

- Lizard Oscillator – water speed 蜥蜴式活塞磁震器 – 水流测速仪

- Section of Caryatids 主体建筑“女神柱”剖面图

015

DESIGNERS-设计人: Ellie Atherton, Dale Muscroft

QMC Water Towers: A Grievance Process

Our project was situated around the water storage towers and pumping station on the edge of the Queen's Medical Centre site; a resource which has supplied the hospital with water since its opening in 1970. By studying the spatial and textural qualities of the towers - alongside the actual process of extracting, treating and supplying the water - 5 key stages were identified within the process. Referencing our initial architectural studies, we separated and explored the spatial qualities of each stage (i.e. volume, materiality and flux) and through the development of drawing and the crafting of maquettes, we were able to define an architectural language for our intervention. As the project progressed we layered a secondary process onto the building that was relevant to the site. This was the process of grieving; as defined by the Kübler-Ross model ('The Five Stages of Grief'). We proposed a new tower that linked to the existing storage towers and contained a variety of spaces such as contemplative seating areas, counselling rooms and personal exhibitory spaces; each relating to aspects of our architectural and process-related research.

皇后医疗中心水塔：体验悲伤

我们的项目坐落在位于皇后医疗中心场边缘的水塔与水泵站周围。自从医疗中心1970年开业起，这里就源源不断地为医院提供用水。通过研究这些水塔的空间与材质特性——这其中包括实际的取水，净化与供水工艺过程——五个重要阶段得以确认。基于最初的建筑研究，我们对每一个阶段的空间质量进行了分离与剖析（例如：容量、材料与流量等）；而通过绘图与制作模型，我们终能为这一建筑介入提炼出适当的建筑语言。随着设计的深入，我们在场地上现有的相关建筑之上叠加了第二种“过程”——一种悲伤的过程，正如库伯勒-罗丝模型所描述的那样（哀伤的五个阶段 - 注：由伊丽莎白·库伯勒-罗丝在其于1969年出版的《论死亡与临终》一书中提出的描述人对待哀伤与灾难过程中的五个独立阶段的心理活动模型。）。我们毗连现有的水塔来设计了一座新塔，其中包含一系列不同的空间，例如冥想静坐区、咨询室与个人展览空间，它们每个都与我们在建筑与过程研究中所揭示的不同方面相关联。

- Texture and Shadow Study 质感与光影分析

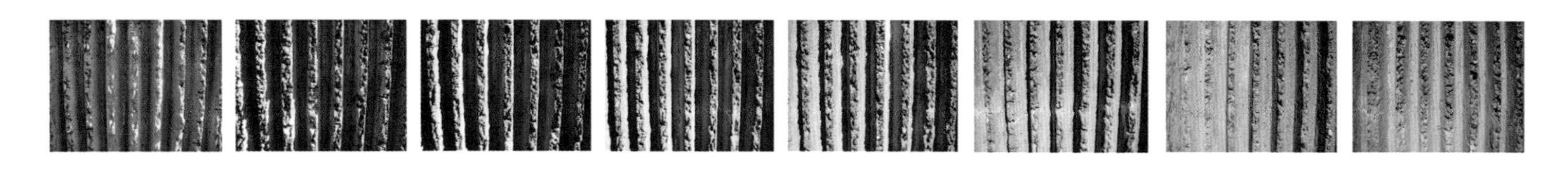

- Interior Lighting Study 塔内光环境分析

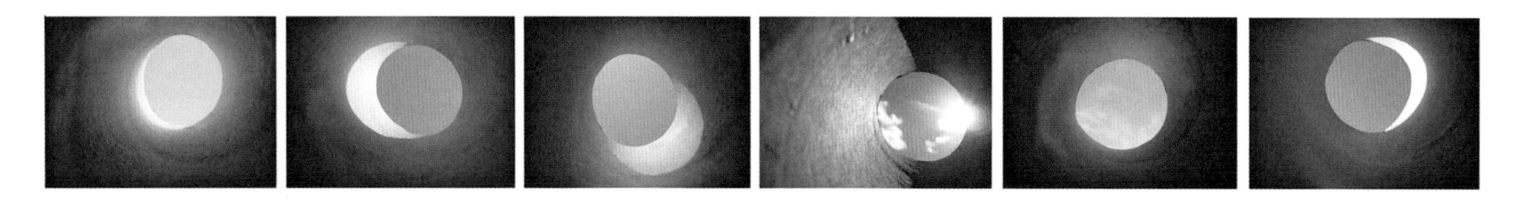

- Spatial Framing Study 空间构架分析

Materiality Study 材料特性分析

▲ - Hybrid Space Study 综合空间分析

▼ - Spatial Process Study

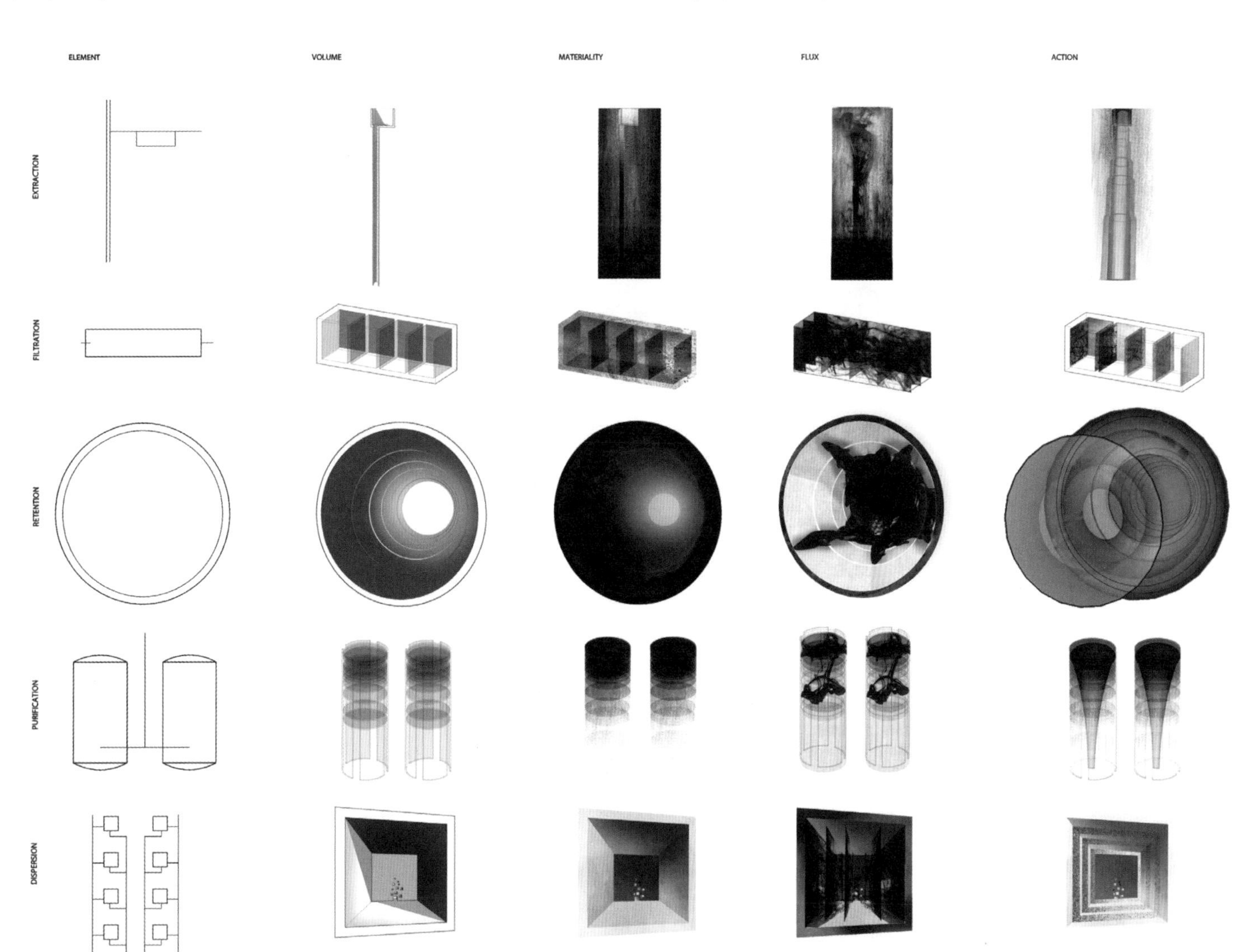

016

DESIGNERS-设计人: Ana Moldavsky, Justin Chan

Eastcroft Incinerator - Exhibiting Contrasting Narratives

This project sets out to explore how an experience overlaid within an existing architecture can tell its story without words, and can reveal or conceal notions of its past, present and future. We are interested in how spaces can speak and how we can interpret them in an individual reading. Our inquiry into the poetic nature of space, matter and architecture and their relationships, led to a series of phenomenological explorations related to the incinerator. We questioned the changing quality of our spatial readings when light, memory, scars and identification trigger in us a sensory or metaphysical response. We explored how instances when dualities and multiplicities occur can provoke a complex atmospheric response: lightness, mass, delicacy, sensory abundance, echoes and distant memory were some. We strived for an architecture that allows personal identification throughout a process, while at the same time questions its very existence. The complex set of conflicts arising from this endeavour made an architecture of contrasting narratives. At the end, it is up to us to choose what to keep and what to forget.

垃圾焚烧炉——展览的对比性叙述方式

这个项目试图探索凝结在一座现存建筑之中的体验如何才能够在不依靠文字的情况下来讲述该建筑的故事，并进而展示且同时掩盖它的过去、现在与将来。对于空间如何来诉说自己的故事以及我们如何在一对一的阅读过程中解读这种诉说，我们均很感兴趣。我们对空间的自然诗性、事件、建筑以及两者间关系的寻求进而体现为一系列关于焚烧炉的现象学探索。当光线、记忆、伤痕与认同感在感觉和形而上学层面上触发我们的回应时，我们会关注空间阅读的质变。而我们也力图去探索，当二元性和多元性相遇时，建筑实例如何才能迸发出复杂的氛围效应：明快、大量、精巧、丰富的感知、回声以及对遥远历史的记忆，等等。我们力图在建筑的整个过程中促进个人认同感，却同时又让人质疑它的存在。这一努力会产生一系列复杂的冲突，而这些冲突则会进而创造出一栋充满语义对比的建筑。当然，在整个参观结束后，该保留什么，该忘记什么，还是取决于我们自己。

▼- Exploded View 建筑分解表现 - 剖透视图

YEAR 5 PROJECT 2010-2011 —— 五年级学生作品 2010-2011

TUTOR - 指导教师: *James Alexander*

- Light & Material Study 光效果与材料分析

- Intervention 嵌入体

- Entrance Model 入口模型

017
DESIGNERS-设计人: Andrew King, Matt McKenna

River Room
Nottingham's Queens Medical Centre, a place of birth, death, healing, curing and transformation, but soulless and endless, with mile upon mile of corridor. Posited atop one of the many stairwells, and between the plant room and roof edge, the River Room overlooks the River Leen and allows for some significant human events to be interpreted. Housed within is a Thank You Card Archive, a curating and display area, a series of contemplation spaces, a rooftop garden and a votive candle repository hanging within the stairwell. One might sit, watch the river below or the mundanity of everyday working lives, imagine, and ponder someone else's events for a moment. Some light a candle in the repository, then ascend in order to look down, seeking the reflections that commemorate their sentiments. One might sit and read the archive on display, assessing life. You are not alone there were legions before you and yet more to follow, as the thousands of archived cards bear witness. Not everyone lights a candle or comes to think or brings a card, but for those who do this is their place.

观河室
诺丁汉的皇后医疗中心是一个汇集了出生、死亡、治疗和整形的地方，但同时透过那数英里长的走廊，亦可以体会到其缺少魂魄，漫漫无休。于是在众多楼梯间中一个的顶部，在设备房与屋顶边缘之间，设计一个可以俯瞰利恩河的观河室，便可以激发许多意义深刻的人文活动。室内布置了一间“答谢卡片”档案库、一个展览区、一系列冥想空间、一个屋顶花园，以及悬挂在楼梯井中的一个许愿蜡烛储藏室。进入到这个观河室，有的人可以坐下来，静静地看河水流淌，或是品味每天生活的平凡，想象并思考一下其他人的事。有的人会在冥想空间里点亮一根蜡烛，登高俯瞰，寻求对那逝去情感的回应。而有的人也许更愿意坐下来慢慢阅读档案室的展览，思考人生。你并不孤寂，因为那数以千计的，归档了的纪念卡代表了那永不停息的生死轮回，无数人先你一步，却又有更多步你后尘。也许并不是每个人都会或点一根蜡烛，或来此冥想，或带来一张卡片，但对于那些愿意这么做的人们而言，这里恰恰极为理想。

Model - Out Appearance 实体模型 - 外观

- Model - Light Study 实体模型 - 日照分析

- Model - River Room Plan 实体模型 - 滨河屋平面布置

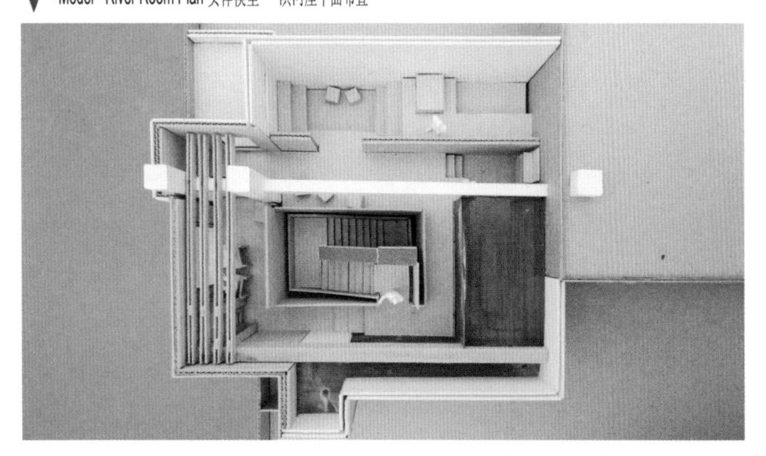

- Hospital Maquette 医院现有建筑空间环境工作模型

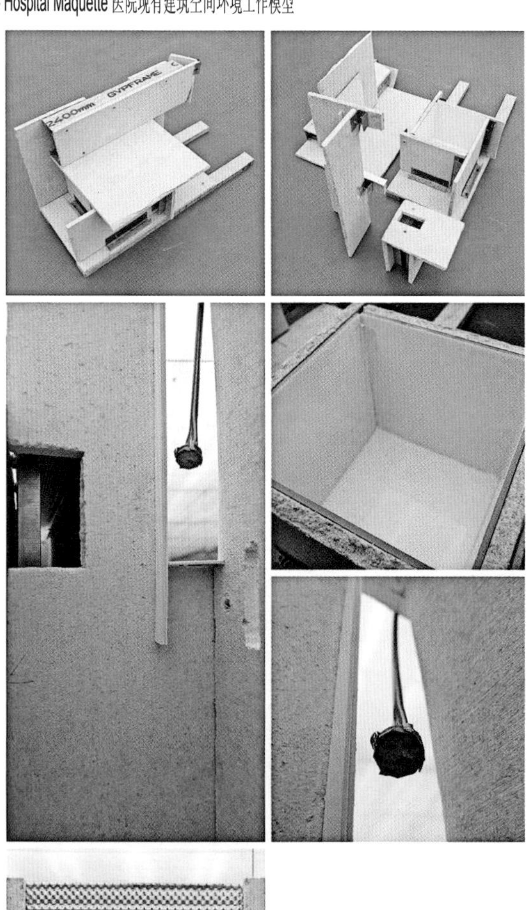

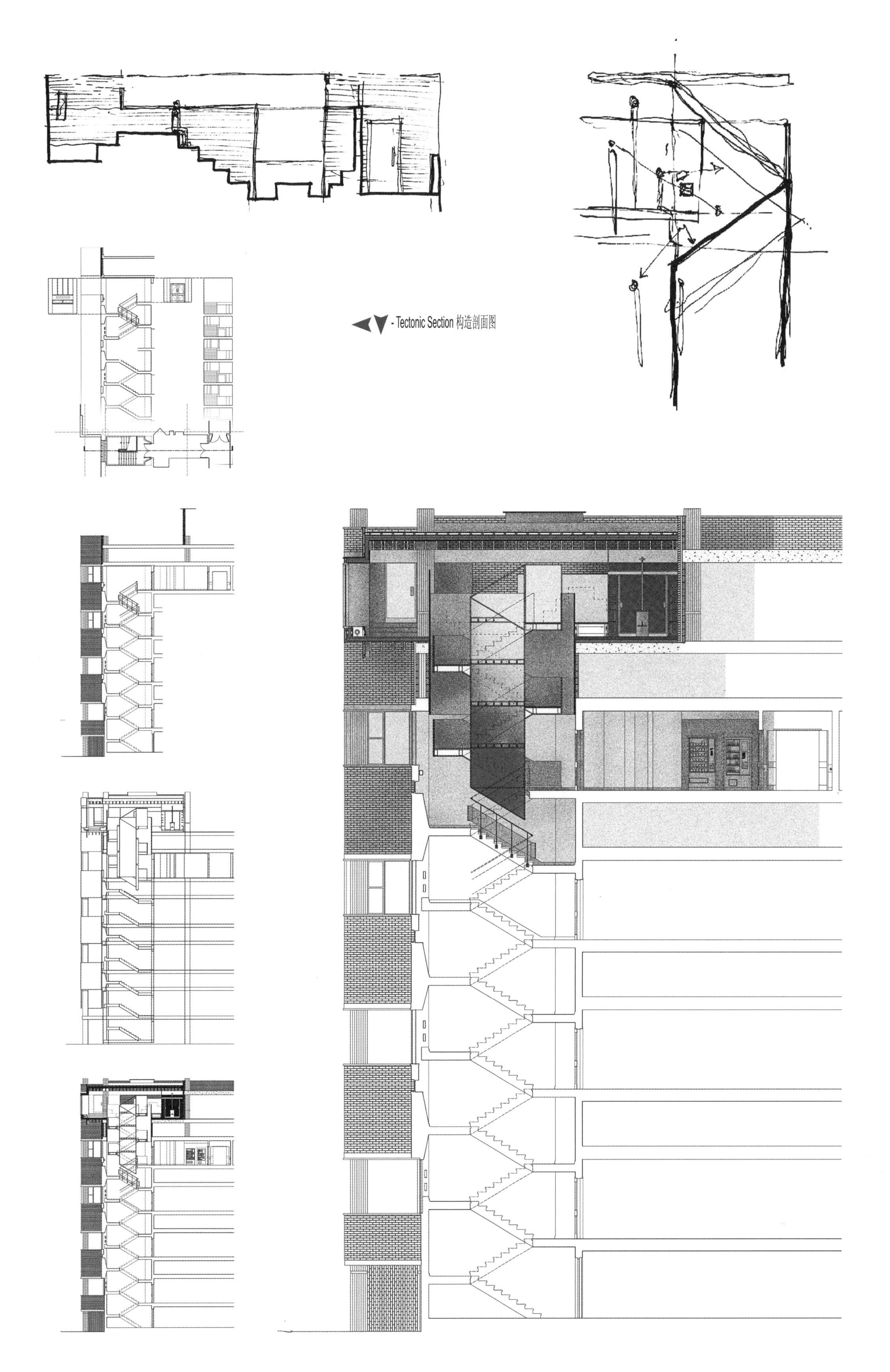
- Tectonic Section 构造剖面图

DESIGNERS-设计人: Zhang Licheng, Tong Jun, Xu Wen

Hurts Yard

Hurts Yard in Nottingham comprises a long and narrow alleyway bounded with an eclectic mix of low rise buildings from different eras. The unique fine grain of the alleyway sits in contrast to a monolithic post-war multistory car parking structure which tore though a number of similar alleyways in the area during the 1970's. Whilst a recent downturn in economy has left a number of vacant properties along the alleyway there still remain a strong sense of community. The ambition for the project is to preserve the existing vitality of the street, and to provide an architectural background in order to push the sense of community to a higher level. Through a detailed study on its history, materials, social interactions and daily use a flexible program was defined for the vacant rooms and spaces along the alley. Programmatically the new architectural insertions are centred on a series of small local gallery spaces. The symbiosis of different functions including working, resting and socializing spaces creates new and intriguing situations which also deal with the limited space within the alley.

赫茨庭院

诺丁汉的赫茨庭院是由一条狭长的小巷和不同年代的低矮建筑相连组成。这个小巷目前的独特肌理主要形成于20世纪70年代，一个单体多层停车场占据了当地原有的多条小巷，使得现有的赫茨庭院和众多低矮建筑都紧挨着这个巨大的停车场。尽管近来不景气的经济导致了沿街许多店面空置，但整个小巷依然保留着很强的社区氛围。这个设计就是力图在延续街道活力的前提下，用建筑特有的方式将社区氛围提高到新的层次。通过对基地的历史、材料、社会活动和日常生活的研究，我们决定赋予那些空置的房屋和空间更灵活多变的功能。这些新的功能主要由许多社区小展廊构成，并结合了工作间，休息室和社交活动场所。虽然同在一个屋檐下，但每个功能的使用时间不尽相同，它们相互协调，更有效地利用了空置的房屋。这个设计不仅策划出新奇的使用方式，也巧妙地解决了小巷空间有限的问题，为小区增添了生命力。

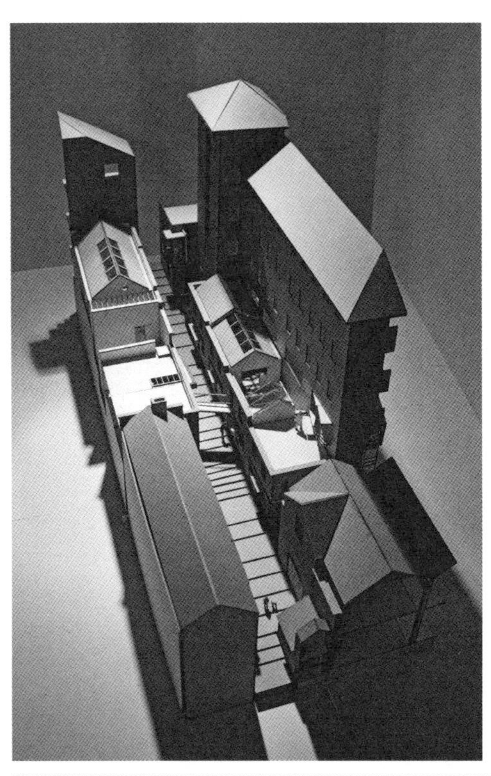

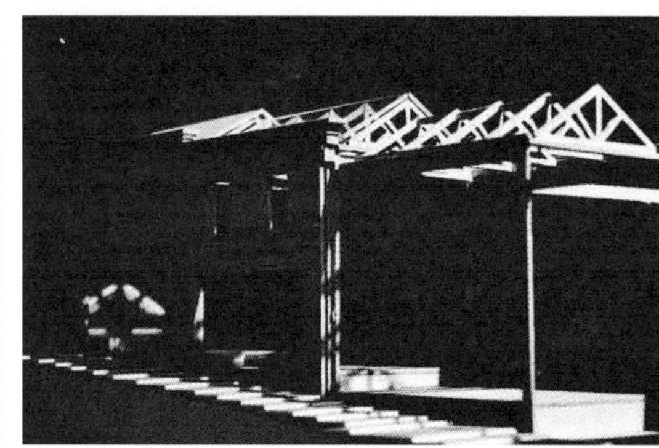

- Hurts Yard Model 实体模型

- Detail Intervention Model 嵌入体细致模型

- Street Elevation Model 街道立面模型

- Street Activity Analysis 街道公共活动分析

■ Record of Street Activity

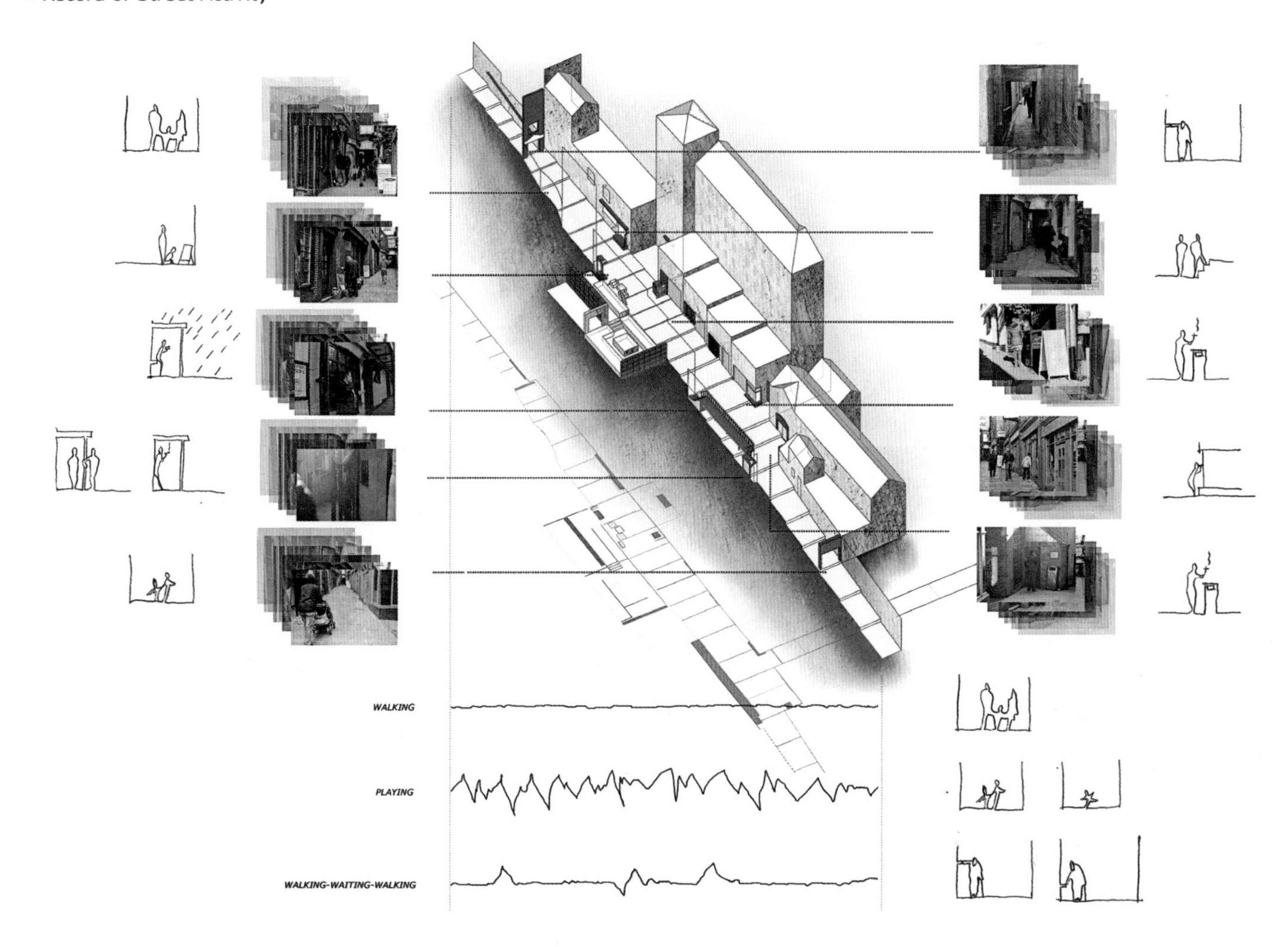

DESIGNERS-设计人: Nichola Finch, Toby Martin

Revealing Craft through Exhibition

The site occupies a neglected corner of the city, caught between the collision of Broadmarsh and Maid Marian Way. A concrete car park bridges the road; its exposed structure forming the soffit of a vast, abandoned space. A disused lift shaft attracts regular policing therefore no public inhabitation prevails on site. A more delicate, accidental space caused by this unplanned collision creates an intimate cave space, revealing an underlying craftsmanship in the car park structure. Exploring this pressurised condition through crafting a sectional cast brought forward the opportunities of sculpture in its inherent processes and display. Juxtaposing the intricacies of sculpture against the concrete allowed an awareness of previous craft embedded in the concrete car park. We sought to establish a relationship between residing sculptors and Nottingham's public. Exhibiting within the constraints of these spaces allowed independent routes, dictated by individual experiences, established by the visitor; enabling them to discover craft and re-orientate themselves within their city.

手工艺的展示

基地位于城市中一个被遗忘的角落，在这里布罗德马什商业中心与玛丽安女士大道相互碰撞。一座混凝土多层停车场横跨在道路上空，其暴露在外的结构体在下部形成了一个空旷的废弃空间。由于一个废弃电梯井需要定期的治安管理，所以基地里并没有什么公共活动发生。然而这处意外的碰撞也产生了一种更微妙，更巧合的，类似洞穴的私密空间，并展示出凝聚在停车场结构体之中的暗藏工艺。我们通过浇筑一个剖面模型来研究这一备受压迫的状况，并借此在其内在的过程与展示中逐渐找到了摆放和展示雕塑的可能性。将雕塑与混凝土相遇时展现出的复杂性加以并置，会赋予这座混凝土停车场一种对老工艺的眷恋之情。我们试图在当地雕塑家和诺丁汉民众之间建立一种联系，在这些充满限制的空间里布展可以鼓励独立游览线路的产生，它们以参观者的自我体验为主，由参观者自行建立，从而使他们探索发现那些被遗忘的手艺，重新在这个城市中找寻自己的位置。

- Sculpture Workshops Axonometric

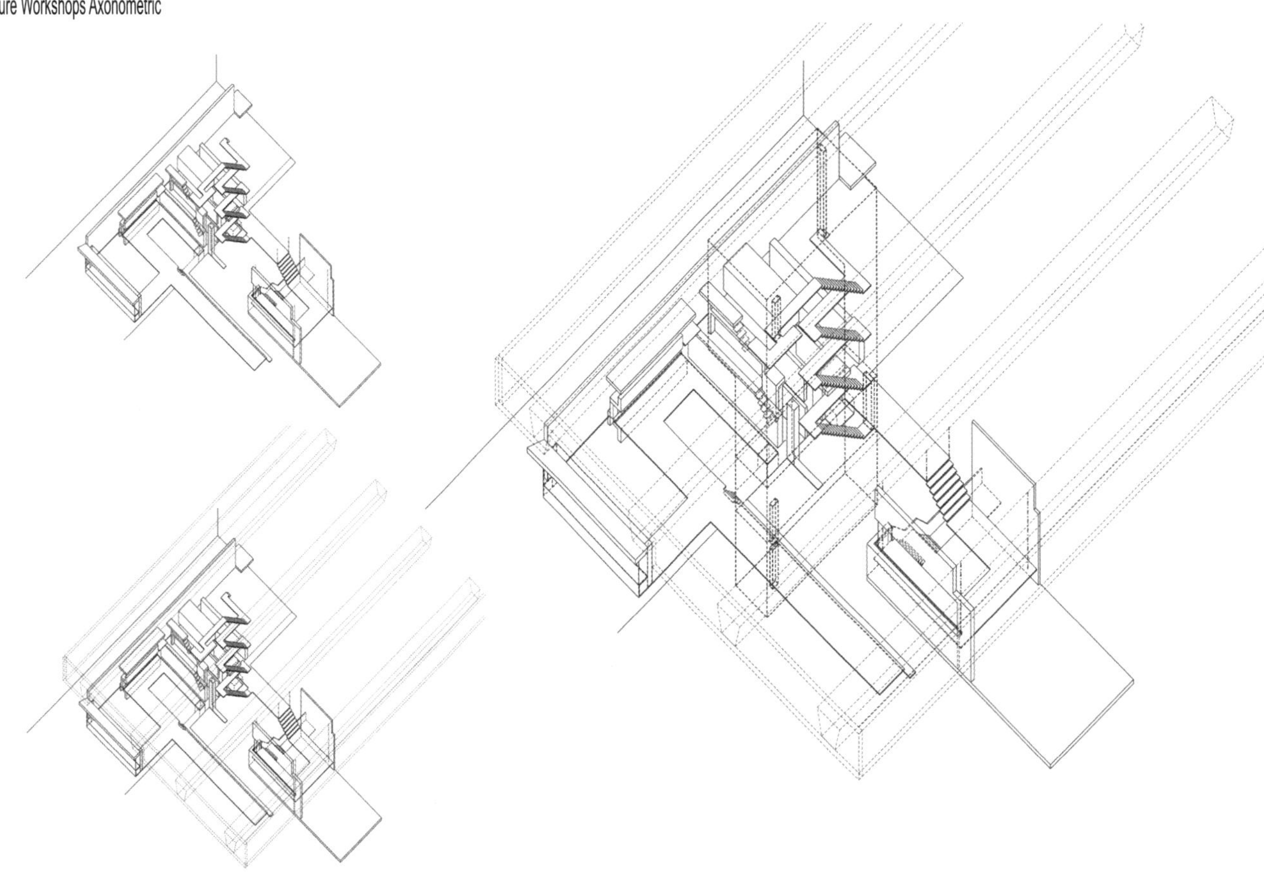

- Casting Negative 倒铸模型负模

- Casting Positive 倒铸模型正模

- Cast Material Detail

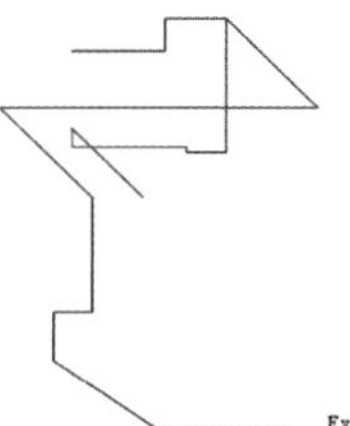

Exploring moment and experience on, through and around New Art Gallery Walsall

- Case Study: Walsall Art Galley Composite Drawing
案例分析 - 沃尔索尔艺术馆元素分析

020

DESIGNERS-设计人: Nichola Finch

Book Depository West End Arcade, Nottingham

The Bench in the West End Arcade began a dialogue between the shop units and the street. Establishing a more accessible library block became the overall project ambition to further amplify these conditions. The Arcade offered found pockets to become an informal book depository. Projecting from the Arcade's facade, a new reading room addresses the street and provides a glimpse of the activity above. This renewed, activated street continues within the arcade where the language of the bench is employed. Steel stringers reach down to receive rising timber steps. Whilst vertical structure offers up rooftop reading rooms, the brick of the back wall serves to locate the arcade within the wider urban block. The topography of the Arcade's rooftop allows a relationship to be established with the beer garden of the adjacent pub, engendering a social familiarity within these informal spaces and facilitating a place to pause within the city.

韦斯特恩德商业长廊存书处，诺丁汉

通过一条位于韦斯特恩德商业长廊的长凳，我们将为您开启沿街店铺单元与这条街道的对话。为了进一步强调这些条件，营造一处拥有更高使用率的图书馆区是整个项目所追求的目标。在长廊中设有很多装书的小袋子，从而形成一种非正式的图书储存处。从长廊外立面上探出来的是一处新的阅览室，既装点着街道又为上述的阅读活动提供了内外视觉交流的机会。这一重新更新的，充满活力的街道设计也同样，通过引入长凳的模式语言，而被延伸入商业长廊之中。钢桁向下延伸至木台阶，同时垂直结构又支撑着屋顶阅览室，而背面的砖墙则将长廊融嵌入更宽泛的城市街区脉络之中。长廊屋顶的起伏形式与隔壁酒馆的啤酒花园形成了一种微妙关系，进而在这些很随性的空间中营造出了一种社会熟悉感，并为城市提供了一个设施完备的场所。

- Book Depository Entrance 图书存储处入口

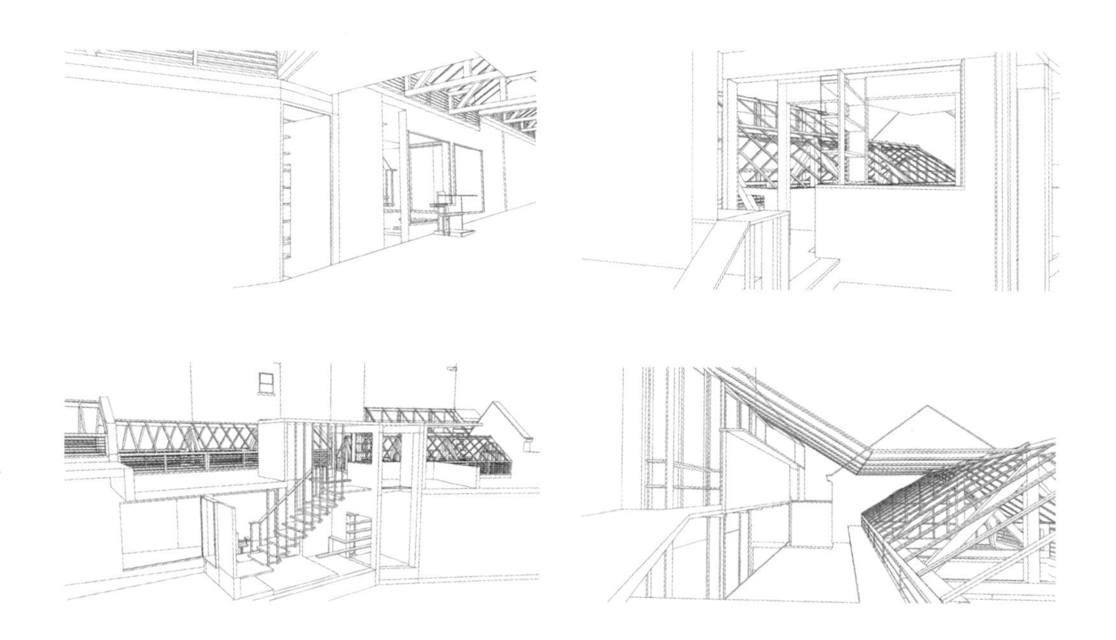

- Room Perspective View

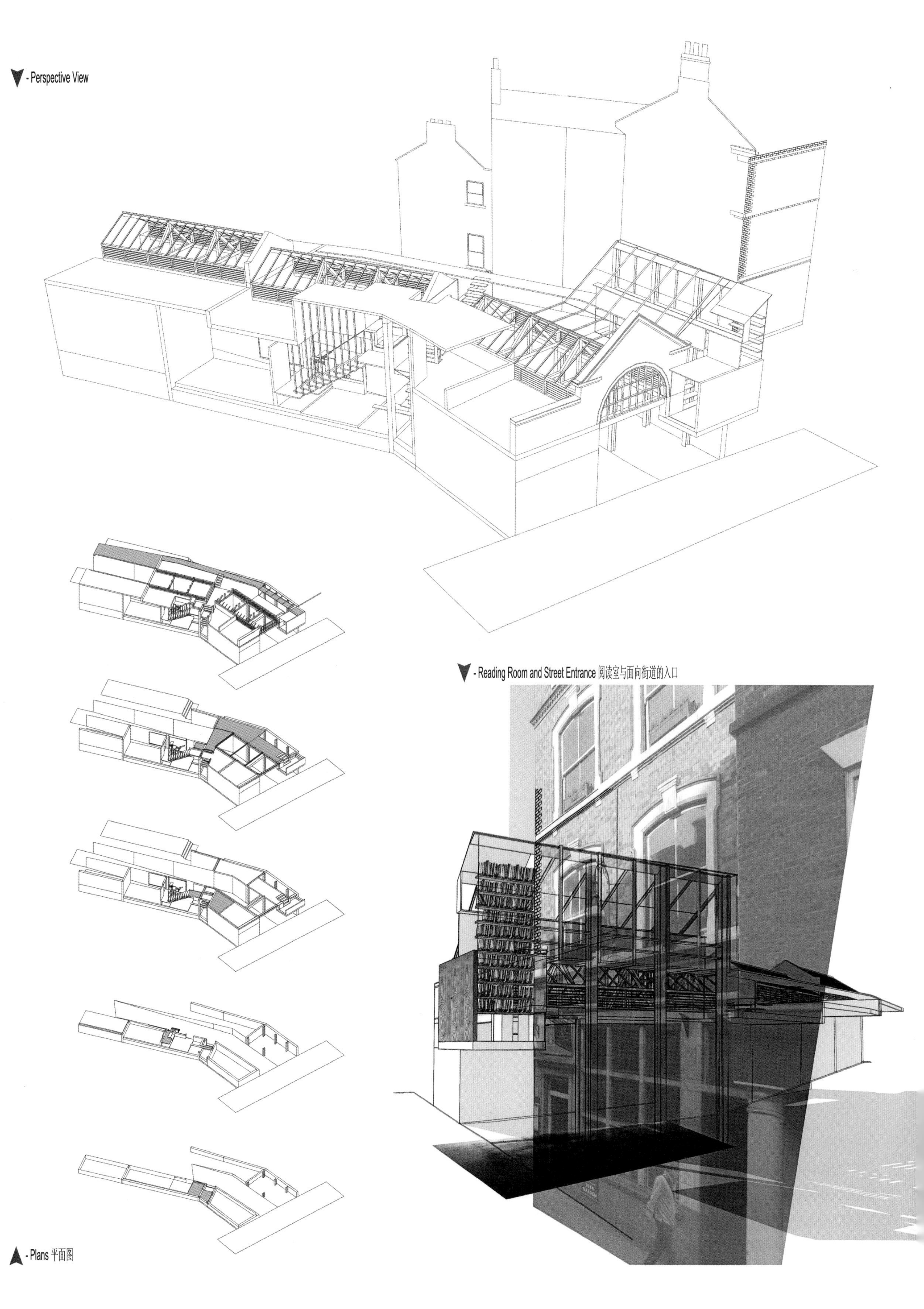
- Perspective View
- Reading Room and Street Entrance 阅读室与面向街道的入口
- Plans 平面图

021

DESIGNER-设计人：Ji Soo Han

Scrap Metal Refinery

The Geometric Abstraction Drawing demonstrates the characteristics of the site context, expressing architecturally through the devices' notation and tectonics of both residential and industrial conditions. Fragments of geometries from both sides combine to construct an abstract composition of architectural space. Situated on the boundary and extending into both sides, the intervention creates a point of connection through the insertion of a Scrap Metal Refinery, an art studio run by two artists who collaborate with both residential and industrial community. The artists work closely with Dunkirk's residence to produce art work. Cars are dismantled and elements are extracted for resale or recycle. Then artists use recycled car parts to create art work, displayed on the residential side. Unused metal at the workshop is then sent to the local scrap yard. The proposal aims to mediate the two disconnected sides of Dunkirk by creating new opportunities for the residential community to be involved in the creative process of metal works.

废旧金属精炼厂

这幅抽象的几何绘画表示了这个基地文脉的特点，装置用建筑的方式表达了这个区域居住区和工业区的符号和构造特点。来自这两个区间的几何碎片结合为建筑空间中的抽象组成结构。位于边界之上却又并向两个区域内延伸。新的介入体通过引入一所废旧金属精炼厂，一所由两名与当地居民与工业社区合作创作的艺术家所运营的工作室，创造了一个连接点。艺术家与敦刻尔克的居民在此紧密合作，共同进行艺术创作。汽车被拆解为基本零件，进行转售和回收利用。艺术家也可以使用回收的汽车进行艺术创作，并在居住区举办展览。最后，工作室中没用的金属将会被运到当地废料场。通过为居住区提供新的机会来融入极具创造性的金属艺术加工中来，这个计划希望最终能够将敦刻尔克区内互不联系的两个社区调和为一体。

- Geometric Abstract 建筑几何形体分析图

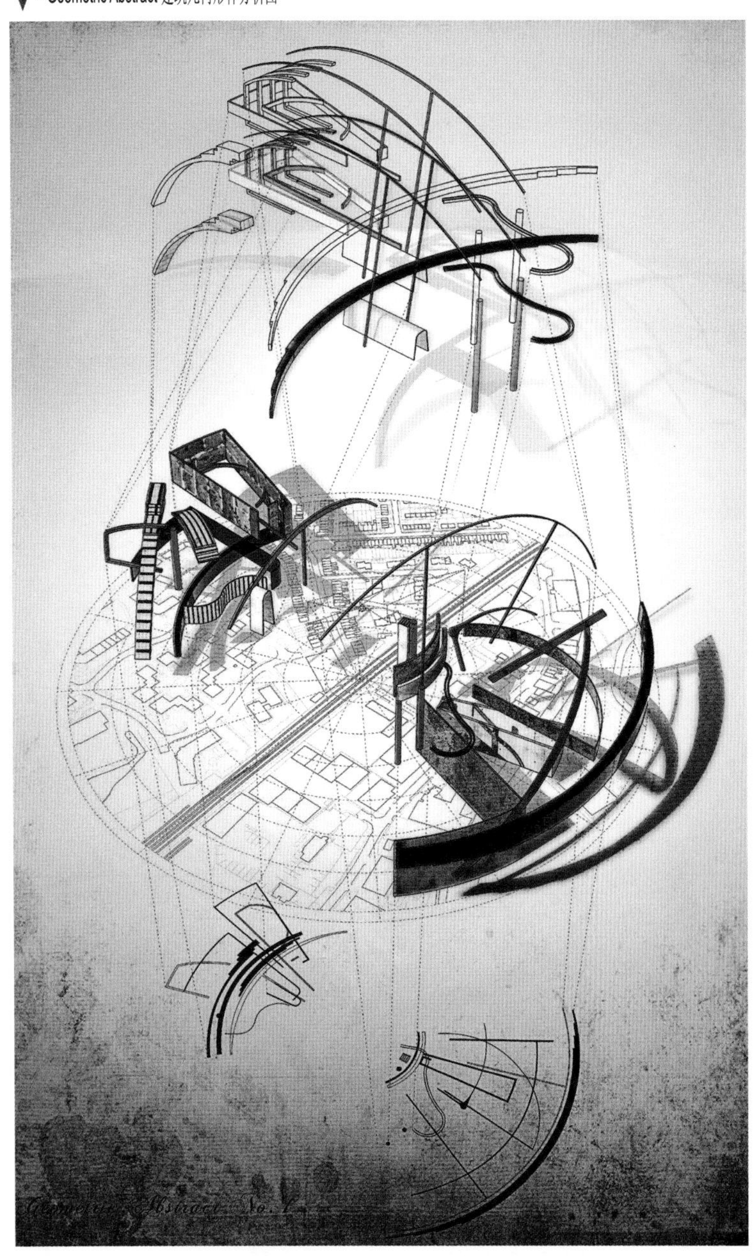

- Section 剖面图

- Perspective View 透视图

- Exterior View 外观透视图

- Section 剖面图

DESIGNERS-设计人: Oscar Bond, Grant Giblett

Evolution: A New Tectonic in the Lace Market

This project is a study on conservation and the effects on the development of the city. The proposals attempt to demonstrate the possibilities of an alternative approach to architectural conservation in both existing buildings and new developments. From a detailed analysis of the existing context a design process was developed that takes typical fragments from an area of central Nottingham, and through a method of abstraction, uses them as generators for a new tectonic. These abstracted fragments ensure a strong contextual relationship due to the structural system and construction method adopted, going beyond the aesthetic mimicry encouraged by the local conservation guidelines. Thus, an alternative architectural language is developed that allows urban change and development, whilst still retaining an inherent connection with the immediate surroundings.

进化，位于蕾丝花边市场中的新构筑

该项目旨在强调建筑保护以及其对城市发展的影响。设计本身试图展示一种新的，涵盖已有旧建筑与新建建筑的建筑保护手段。基于对现有文脉的细致分析，在设计过程中，源自诺丁汉市中心的一些典型片段被提取出来，并通过抽象化使其衍生出新的构筑体。借助采用结构系统与构造方法，这些抽象化的片段保证了坚实的文脉联系，并超越了地方建筑保护规范中所鼓励的在美学方面的简单模仿。从而创造了一种新式的建筑语言——它既允许城市改变与发展，又同时与周边的环境时刻紧密相连。

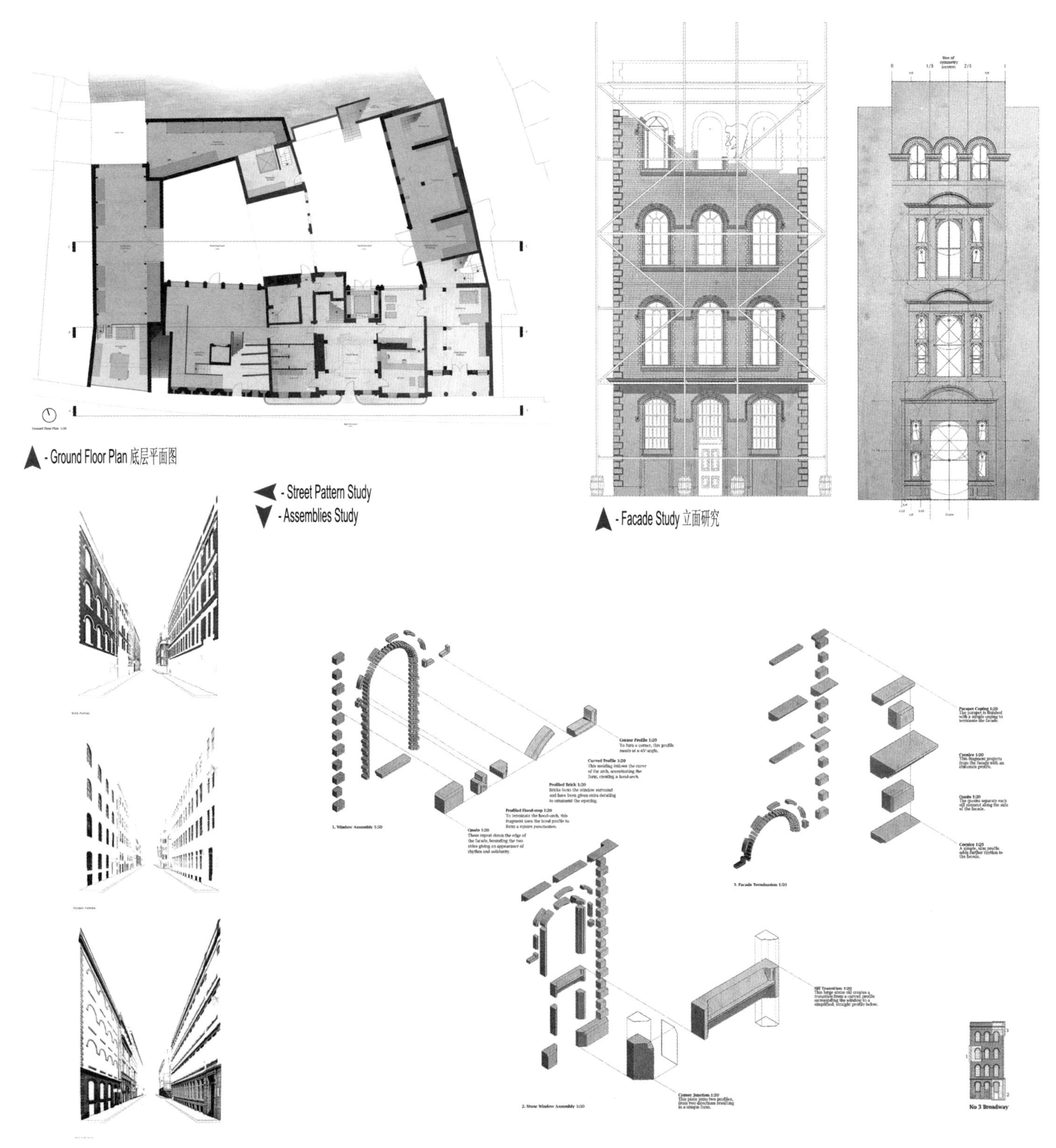

▲ - Ground Floor Plan 底层平面图

▲ - Facade Study 立面研究

◄ - Street Pattern Study

▼ - Assemblies Study

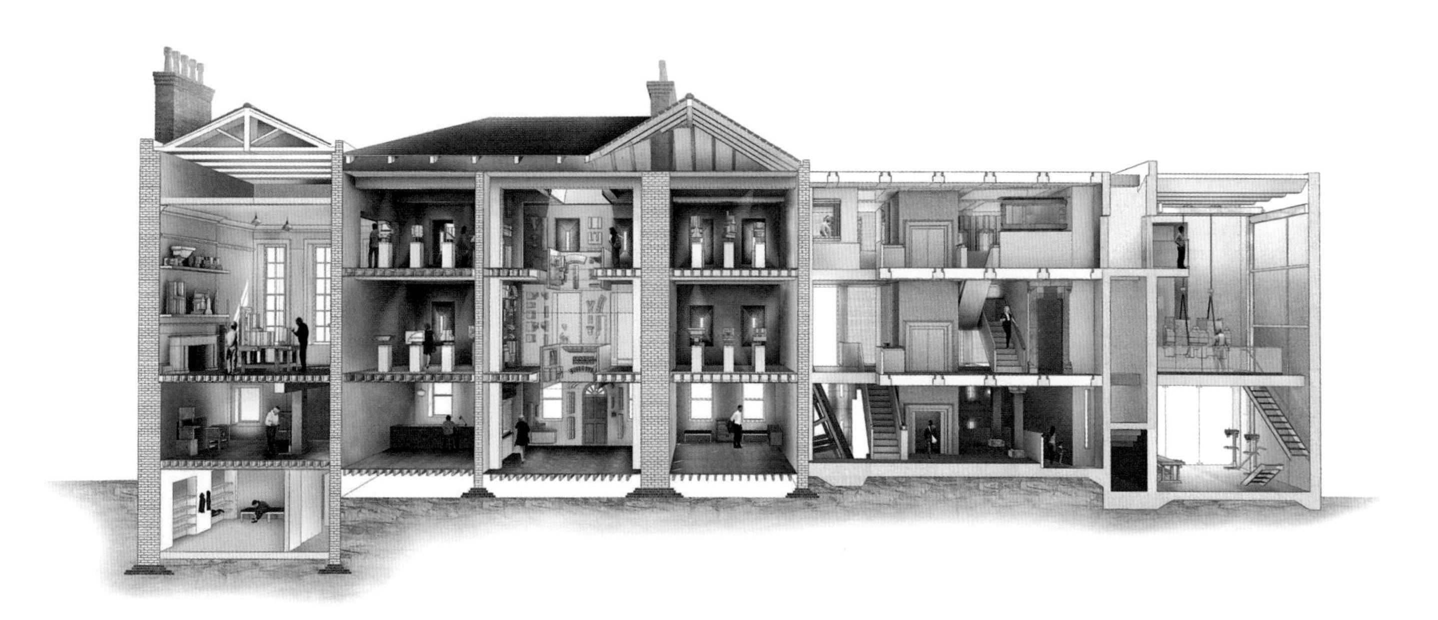

▲ - Section 剖面图

▲▼ - Internal Views 室内效果图

▼ - South Elevation 南立面图

DESIGNER-设计人: Imogen Lesser

Ex Libris ‡ The Sleeping Cemetery

The Cemetery of books allows books to be safely stored along with their memories, preserving the written word in its physical format. These books can never be removed, once deposited, but are able to be read by those who visit. These books are not all dead some are merely sleeping deep within the rock. The depth of the Cemetery protects the books from the harsh sunlight, yet without readers the books die. In order to live they must be read, worn and damaged. Individuals can escape the darkness in the tallest towers, taking their precious burden with them up the long spiral staircase to observe the city. It is the physical act that enhances the memory, the contrast between light and dark, cool and heat. The reader undertakes a new journey with every visit, with every book and the reassurance of the foundations of knowledge allows the city to continue in the digital age.

废弃的图书馆 ‡ 沉睡的墓地

这座书的墓地使承载着记忆的旧书得以安全地保存下来，将书写文字以其物理形式延续下去。一旦被存储在这里，这些书将永远不能被带离墓地，不过它们随时都可以被那些慕名而来的读者阅读。事实上，这些书并没有死亡，它们仅仅是长眠在这一石窟深处。墓地的深度足以保护这些书不受强烈阳光的照射，但是，如果没有人来阅读，那将意味着书本的死亡。所以，为了重获新生，这些书籍必须被阅读，被使用，甚至被损坏。单独的参观者可以站在最高的塔楼里来远离黑暗，沿着螺旋楼梯扶摇而上，带着他们的"宝物"来观察这个城市。借助明亮与黑暗，阴冷与温热之间的反差，这种实实在在的行为巩固了他们的记忆。每一次参观，每一本典籍，对读者来说，都是一场新的旅途。知识的基础被牢牢地夯实，城市将迎来数字时代的新纪元。

▼- Plans 平面图

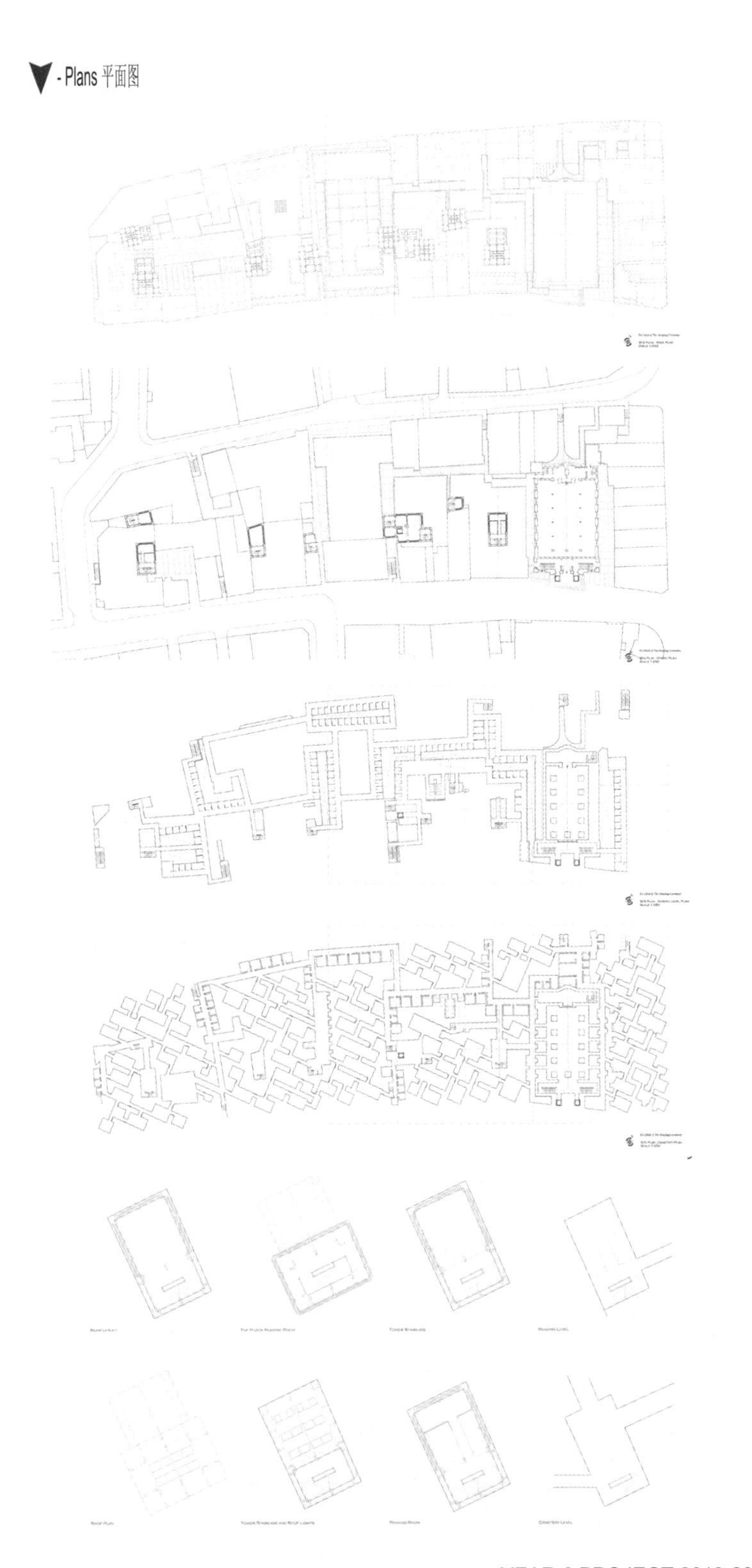

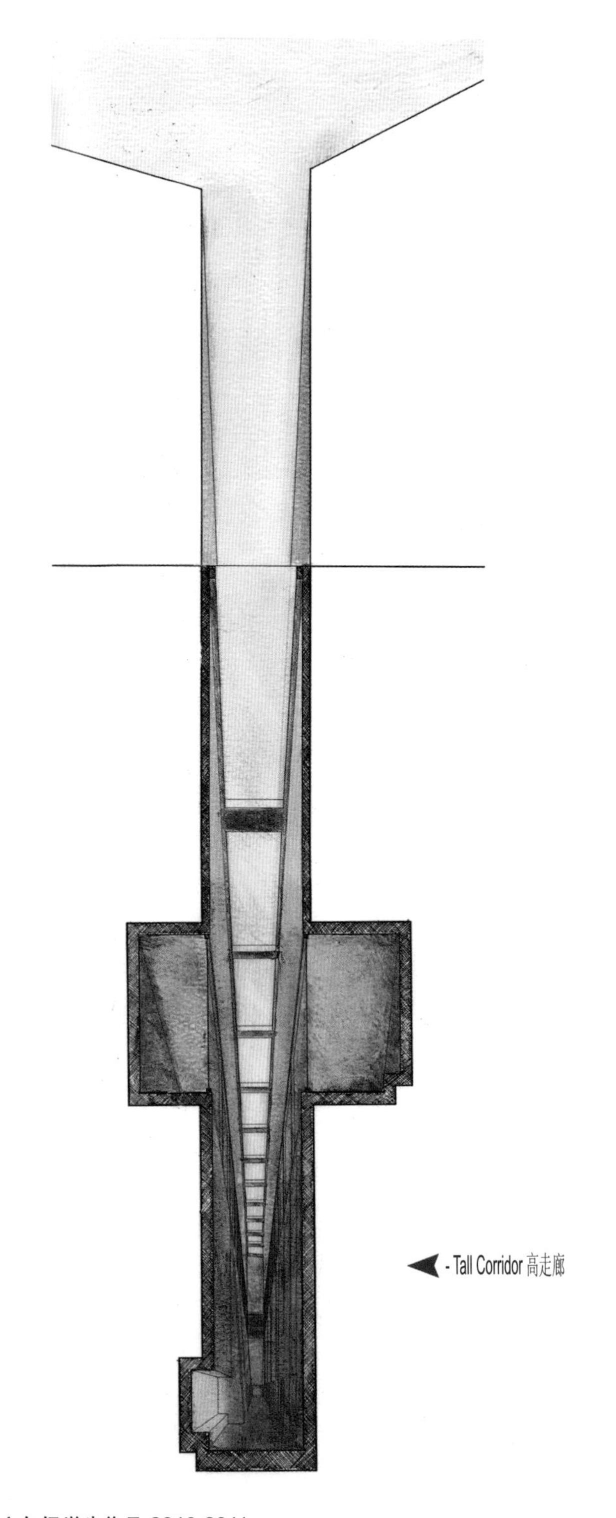

◄- Tall Corridor 高走廊

▲ - Concrete Book Concept Model
混凝土书概念模型

▼ - Main Hall Perspective
主厅室内透视图

▲ - Book Pile Concept “书堆”概念图

◀ - Concrete Model 混凝土实体模型

▼ - Section 剖面图

DESIGNER-设计人: Lok Ting Vickie Cheung

Beyond the Wall: The Final Rite

The thesis project stemmed from an initial study on the embodied experiences within the city and the role architecture and the city's ambience play in scripting our urban experiences, evoking our emotions and memory. In the fast-paced, developing city we inhabit in today, we often overlook the details and voids which compose the city's memory. They are often old, disused buildings which are abandoned for years or demolished and replaced by new commercial developments. Our experience within the urban realm gradually becomes over-sanitized and inhuman. Taking the abandoned Smithfield Annex Market building as its site the project proposes an Urban Promatorium, a new funerary typology similar to a crematorium where the deceased is frozen and transformed into powder-form 'premains'. The promatorium act as the final threshold of the journey, the shift from life to death, from body to 'premains', from presence to memory, thus reinterpreting the site as a place for public congregation and celebration of life.

超越墙垣：最后的仪式

城市内的具体体验，建筑所扮演的角色以及城市周边环境在编写着我们的城市经历，激发我们的情感与记忆，而该项目则正是源于对这种情况的初步研究。在我们现在所居住的快节奏发展的城市中，那些形成城市记忆的细节与空白点往往被忽视。它们经常是一些老旧无用的建筑，或废弃多年，或被拆毁并由新的商业建筑所取代。在城市环境中，我们的体验正在逐渐变得过于纯净以致毫无人情味。因此，本方案希望将废弃的史密斯菲尔德附属市场建筑改造为一座城市转魂堂。这是一种新的丧葬方式，与火葬类似，在这里遗体通过深度冷冻而转化成粉末状态的“体灰”。转魂堂就像人生旅途的最后一道门槛，从生到死，从身体到“体灰”，从存在到记忆的转化点，从而使这里变成了公众的圣堂，对生命的礼赞。

- Promatorium Layout 转魂堂平面图

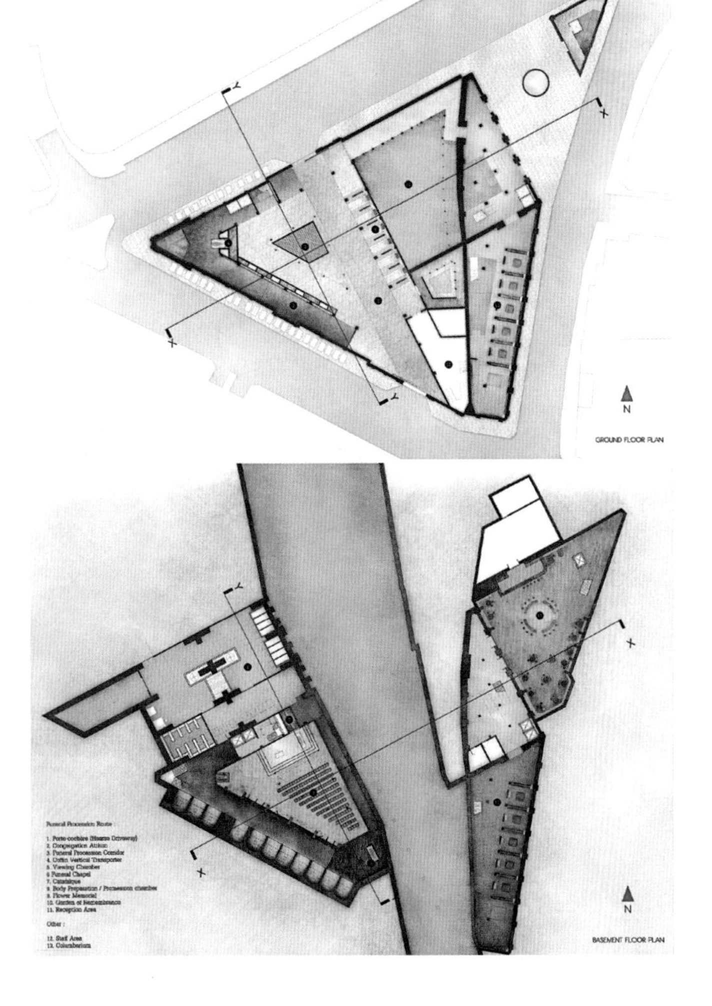

- Detailed Section 详细剖面图

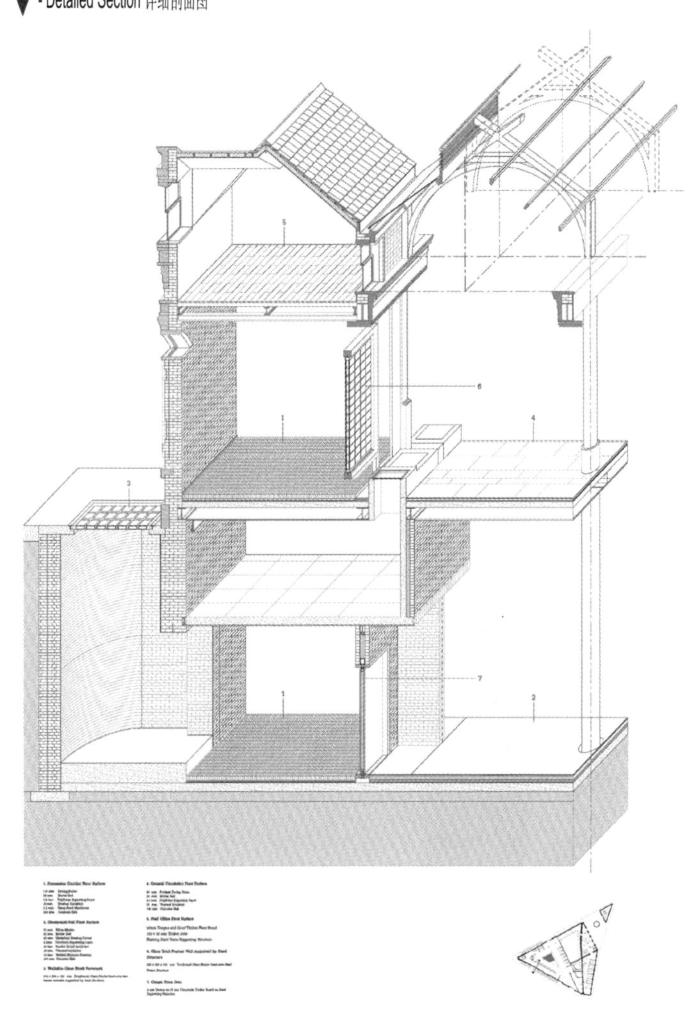

- Smithfield - History and Materiality 史密斯菲尔德 - 历史与材料分析

- Funeral Procession 葬礼

- Viewing Room 遗容瞻仰室

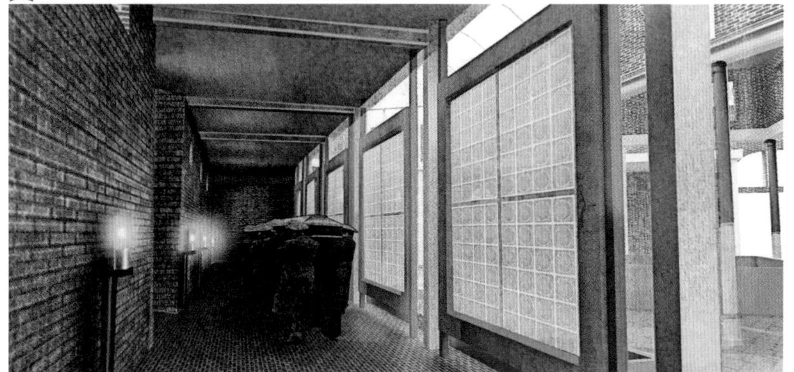

- Procession Corridor 走廊

- Columbarium“体”灰堂

- Promatorium - Building as Procession 转魂堂—建筑过程

- Section Perspective 剖透视图

DESIGNER-设计人: Rand Al-Shakarchi

Urban Stitching: Ruins Reconstructed, London

Remnants of the past are usually left to the tourists and their "ruinous gaze". This proposal is to reuse and accommodate new architectural interventions within a currently neglected site of historical importance - fragments of the existing Roman Wall in London. The project proposes to use the ruins in an untypical way and give them a new function; not merely as an "exhibit" for viewing history as in a traditional gallery or museum environment. The architectural design reconstructs the missing fragments of the wall through a walkway. The building acts as a "prosthetic" attachment. The user is encouraged to strengthen the body for rehabilitation by using the wall as a means of support. The fabrication of bespoke prosthetics in the underground spaces draws upon the relationship between the building, the body and the Roman Wall. The site's history has strong correlations with the body and healing which is intertwined within the program reconnecting the past with the present.

城市缝合：遗址重建，伦敦

历史遗存通常都仅供旅游者们来参观并“凝视它们的凋落与毁灭”。而该项目则试图对一处被忽视的重要历史建筑——伦敦现存罗马城墙残段——进行重新利用并在其中融入新的建筑干预。方案计划用一种非典型的方式利用废墟，并赋予其新的功能，而不仅仅像在传统的画廊或博物馆环境里那样，将其作为一个“展品”来审视历史。通过设计建立一条走道，损毁消失的墙段得以恢复，而新建筑则如“假肢”般附着在原有结构之上。我们鼓励使用者将墙体作为支持手段来加以使用，从而加强其建筑体（身体）的复苏与康复。而位于地下空间中的那些“定制的假肢”所构成的机理则明确了建筑、身体，与罗马城墙之间的关系。该项目用地的历史与身体及康复有着很强的相关性，而正是后者交织于设计之中，重新建立起过去与现代的联系。

▲ - Ground & Walkway Plan 底层与走道平面图

▼ - Basement Plan 地下室层平面图

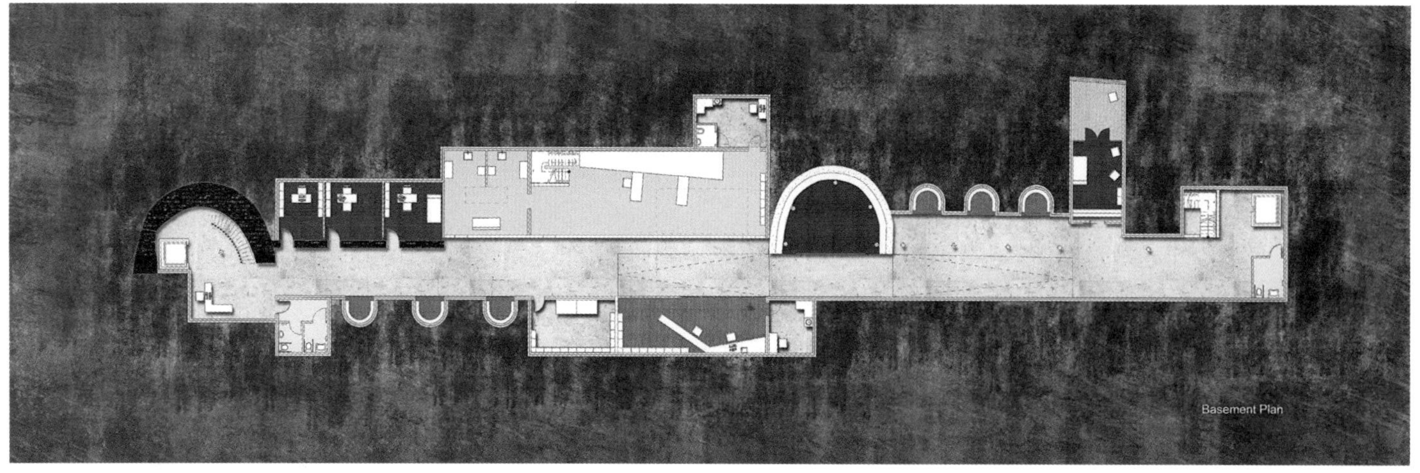

▼ - Roman Wall Studies 罗马城墙基地分析

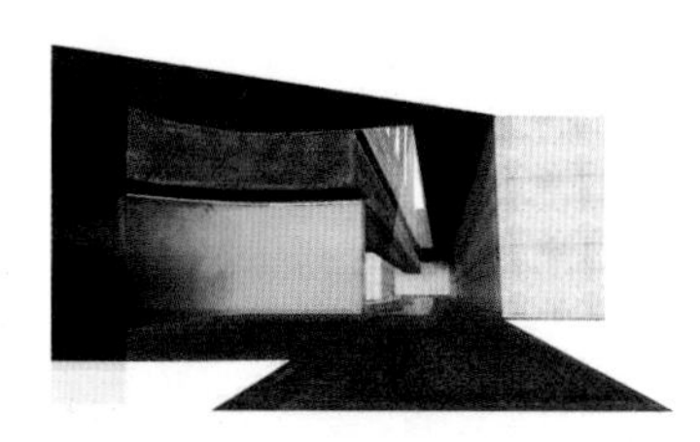
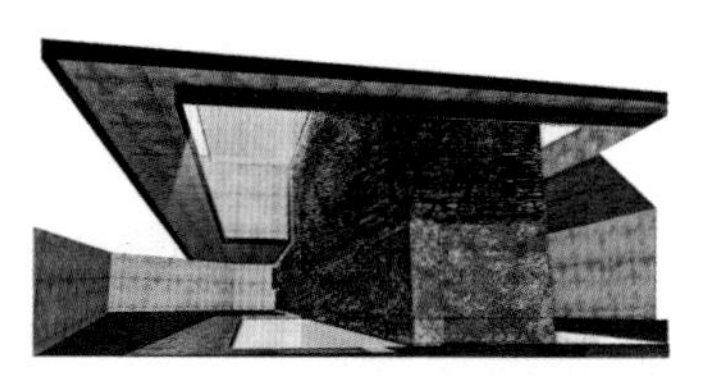

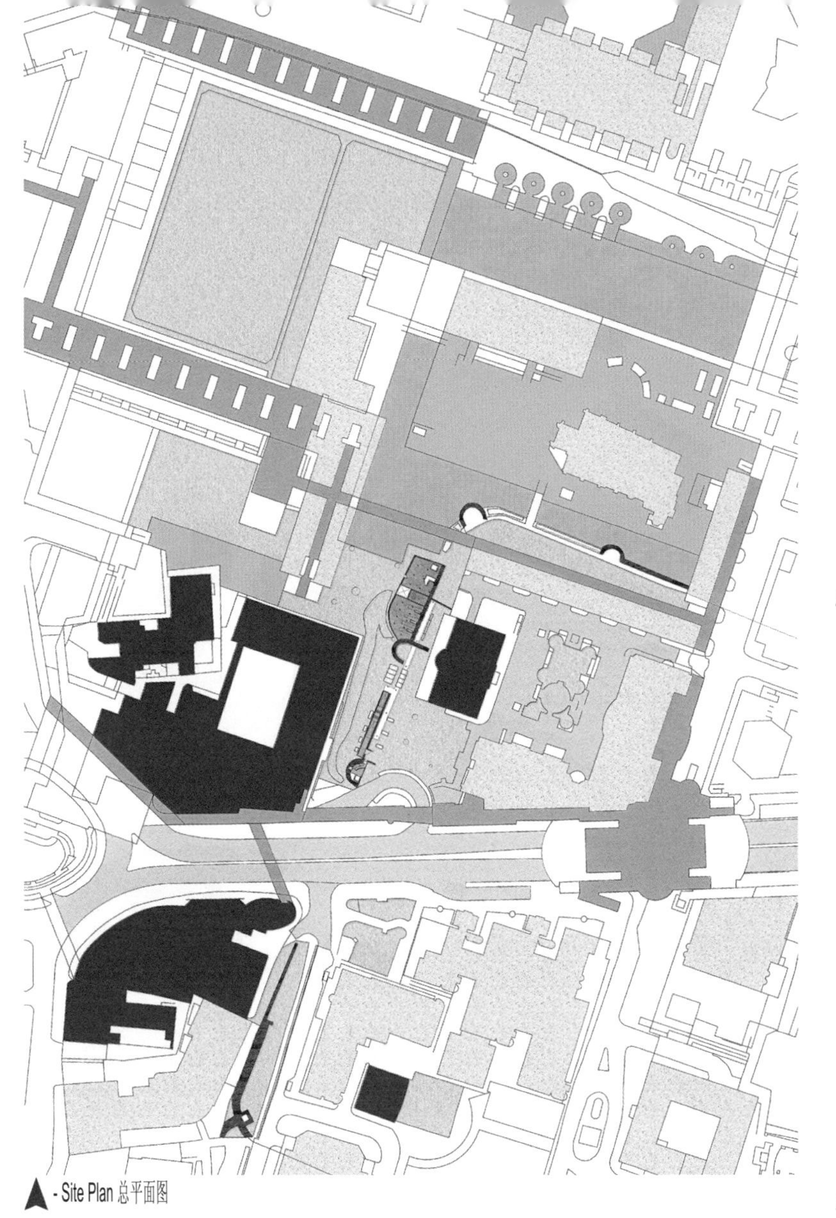

▲ - Site Plan 总平面图

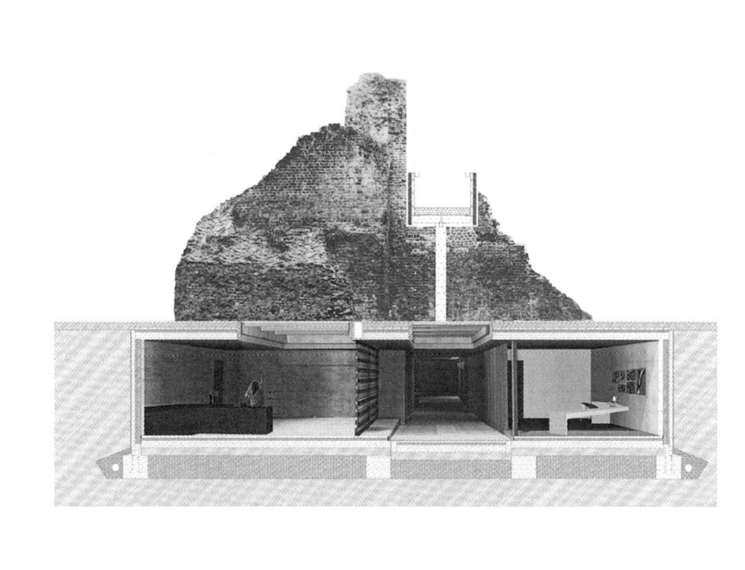

▲ - Workshop Detail 工作室详图

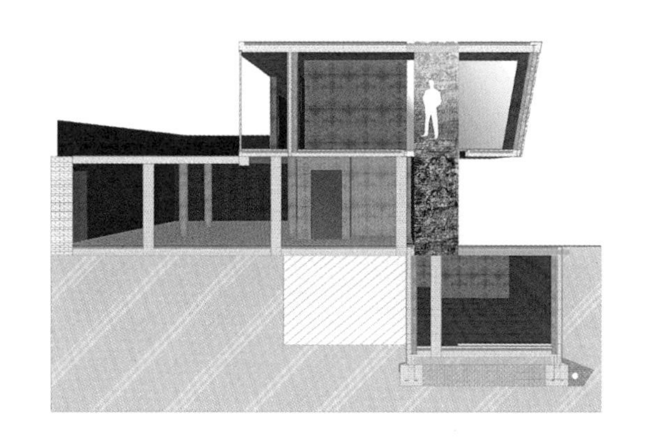

▲ - Rest Room Detail 休息室详图

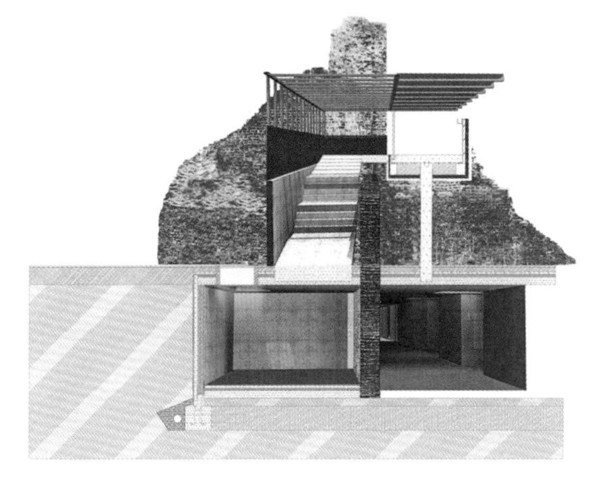

▲ - Walkway Detail 走道详图

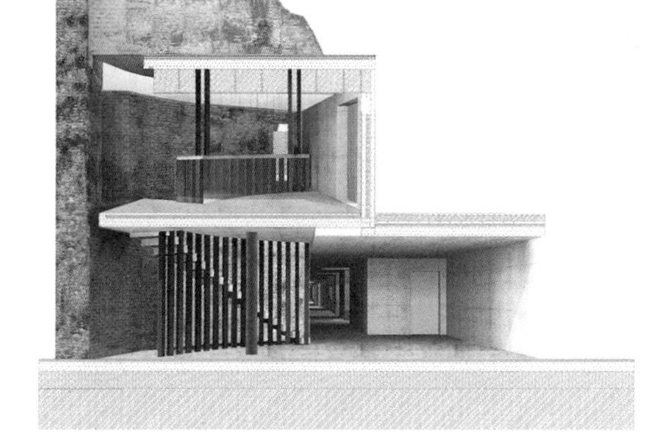

▲ - Bastion Entrance Detail 墙堡入口详图

▼ - Section 剖面图

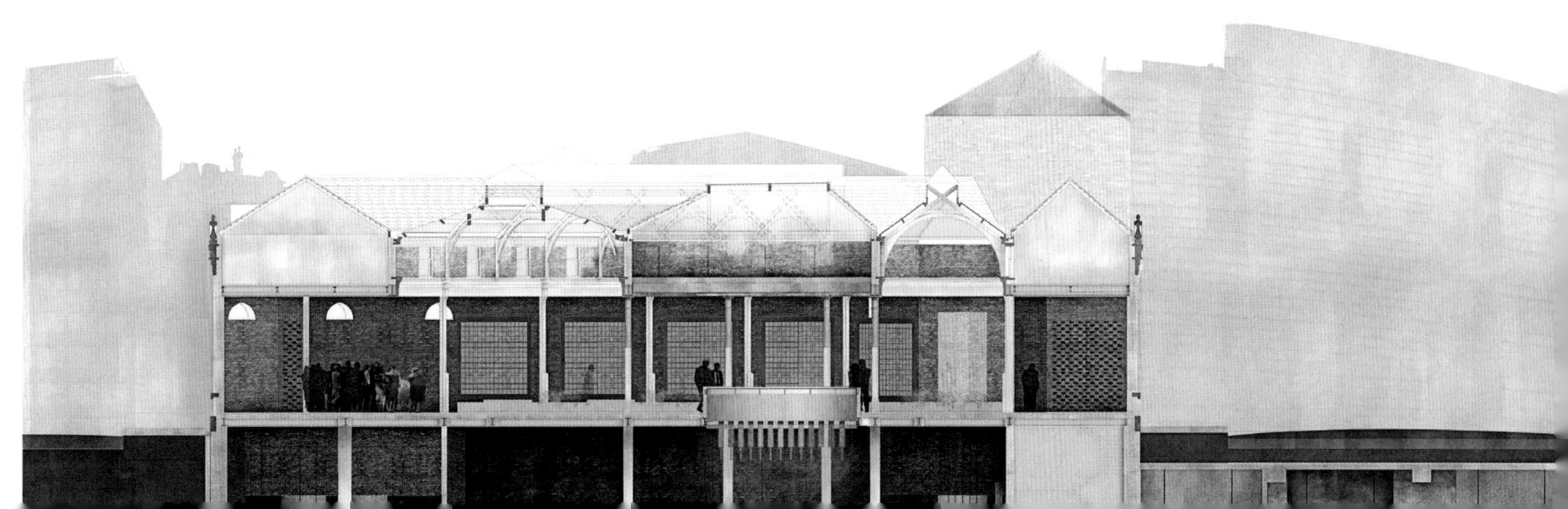

DESIGNER-设计人: Richard Woods

Neglected Domains - A Retro [Per] spective

The project began with an exploration of the city and the discovery of its lost and hidden space. I began to document them using photography in an attempt to capture the essence of what made them unique. This developed into an interactive mobile device and web-based catalogue of the changing neglected domains of Nottingham, utilising overlays from similar, non-neglected spaces in Venice as a mediator. The investigation then progressed as an exploration of the potential of photography, which act as an architectural tool and focus on one particular neglected domain. The project is an experiment in perspective and photographic principles applied to architectural design. Speculative interventions play with the principles of perspective to produce illusions or distortions to the designed space. Each was developed through a unique process of drawing, the construction of dioramas and photography, using vintage and home-made cameras, in an attempt to test different approaches to their use as a design tool.

被忽略的领域——反[透]视

该项目始于对城市的探索以及发现那些失去和暗藏的空间。为了试图抓住使这些空间变得独特的本质，我开始借助摄影来记录它们，并且在后来被开发成为一种互动式移动设备和基于互联网的编目，这里面记载了诺丁汉的那些遭到改变却被忽视了的地方，并同时将威尼斯的那些类似的，但没有被忽视的空间作为对比层叠加于上。随后的研究则如同钻研摄影术的潜力，既作为一种建筑手段，又聚焦于一处特别的，被忽视的场所。该项目就是这样一种将透视学与摄影学原理应用于建筑设计之上的实验。随机的干预行为与透视学原理一道制造出对所设计空间的幻觉或畸变。其中，通过使用葡萄酒和自制照相机，并基于独特的绘制过程，对全景视窗的建立以及对摄影术的掌握，每一个方面都得到开发，而所有这些发展都试图，从不同的角度上，去测试如何将它们作为设计工具来加以应用。

▲ - Plans 平面图

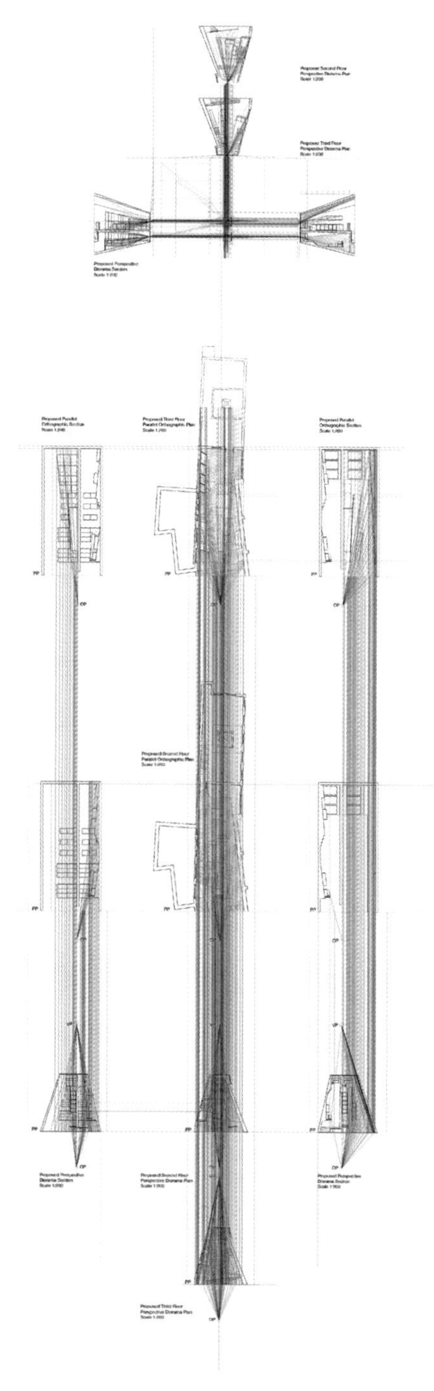

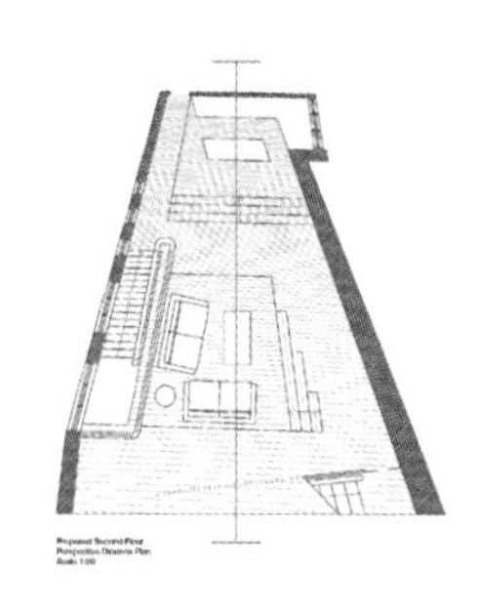

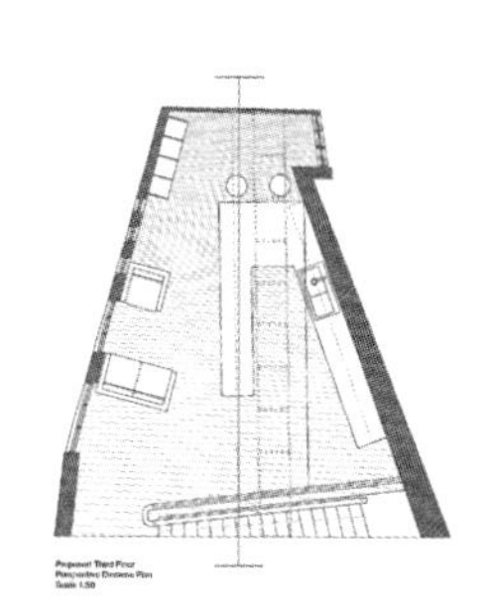

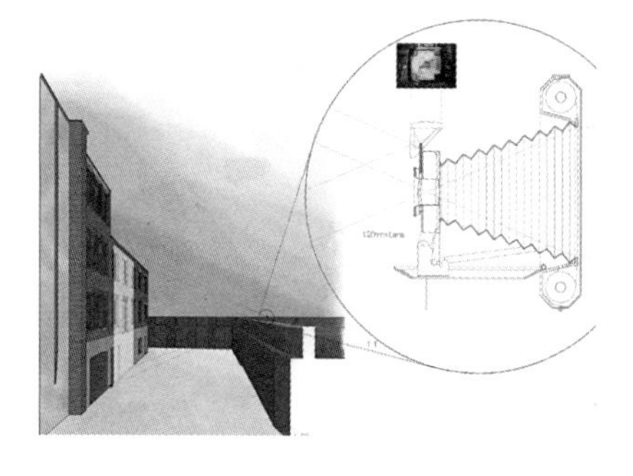

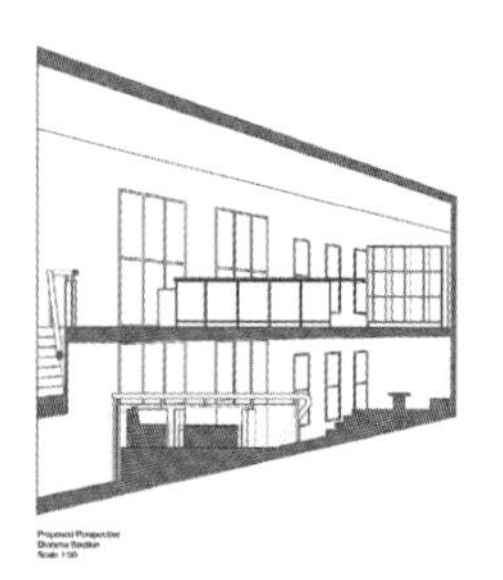

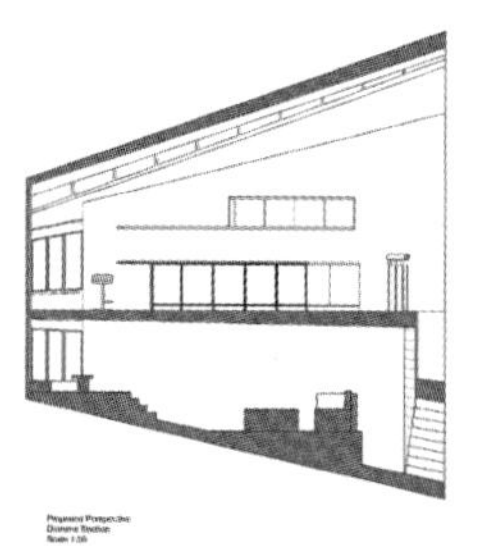

▲ - Kodak six - 16 Folding Camera Study
柯达 6-16型折叠照相机原理研究

◀ - Diorama Room Plans and Sections
全景视窗室平面图与剖面图

▶ - Compound study through the physical 1:20 Diorama room model
1:20全景视窗室实体模型综合分析

- Bridge and Pinhole 桥与针孔视窗效果

- Context Photos 基地文脉照片

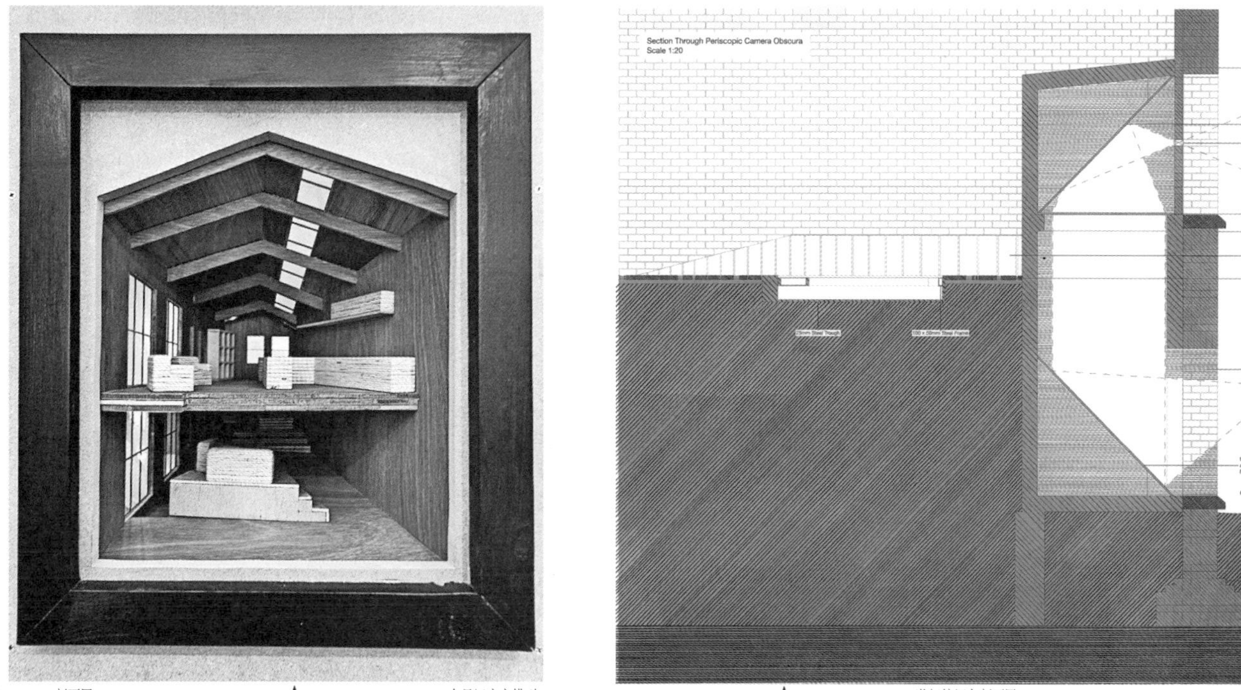

- Diorama Room Model 全景视窗室模型

- Periscope Section 潜望镜视角剖面图

- Section 剖面图

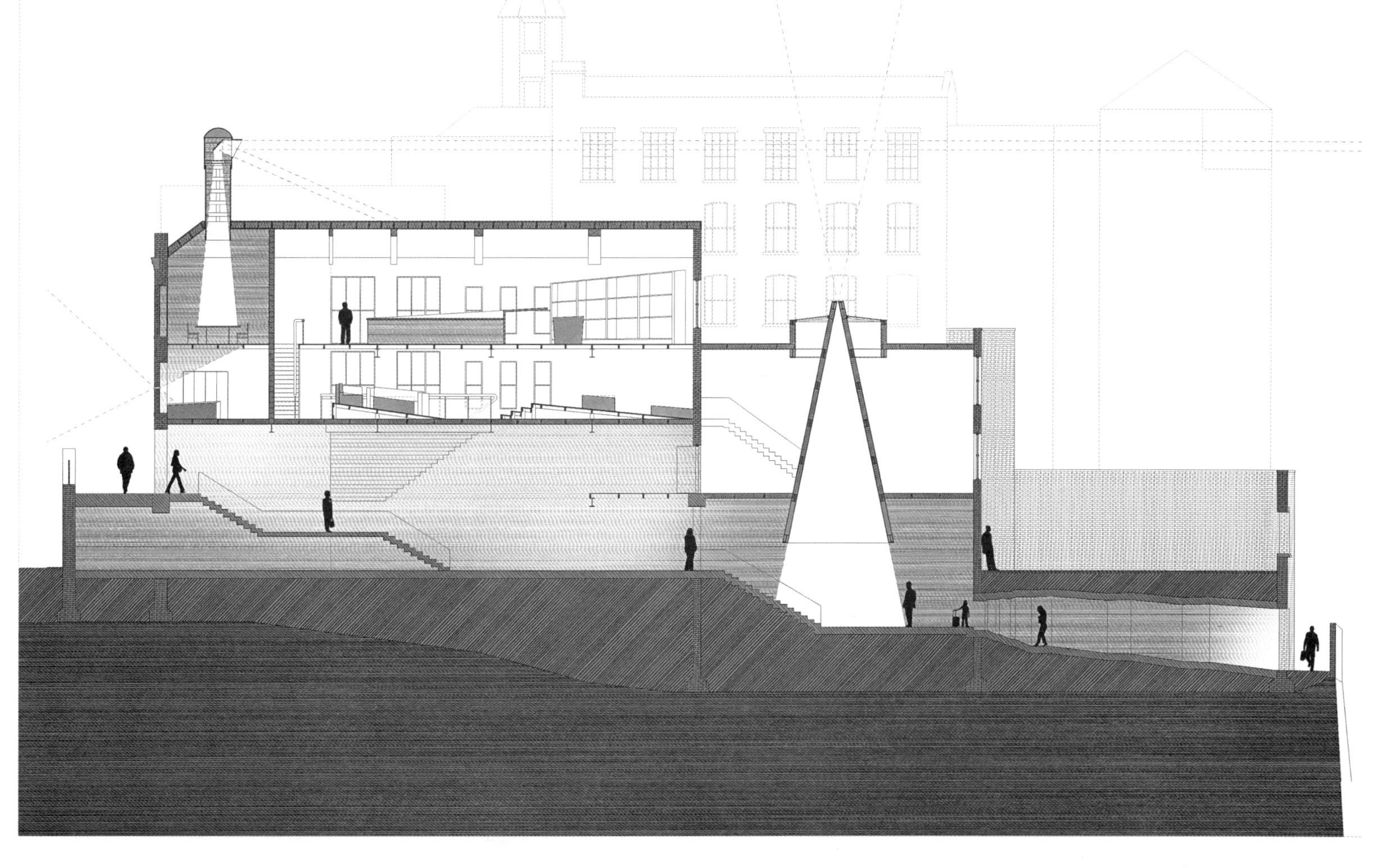

027
DESIGNER-设计人: Anthony Lee

Mixed Realities

As technology continues to be integrated into our everyday lives, the way in which we perceive and react to the world around us is changing. Our digital inputs are changing our reaction to spaces, creating a digital overlay of both external information and emotional response which has a great impact on our perception of the space around us. The Parasite explores the impact that mobile technologies are having on the sensual reaction to space and was designed to be worn during a journey through Nottingham city centre. During this journey through the city the Parasite fed digital information and directions to the wearer and created a mixed reality for the user. Mixed reality as a conceptual idea is focused upon the direct reaction of the digital technology to the human body. The Urban Parasite acted to create an architectural system capable of changing and reacting to human influence within an existing space but also able to overlay digital information onto its surface. This then takes the idea of mixed reality and the idea of Milgram's continuum and places it into an architectural form.

混合的现实

随着科技不断融入日常生活，我们对周围世界的感知与反应也在随之改变。对数码化信息的接收改变了我们对空间的反应，进而创造了一种对外来信息和情感反应的数码叠层，并对我们对周围空间的感知力产生了巨大的影响。设计这一名曰“寄生体”的装置旨在探索移动科技对空间感官反应所带来的影响。人们可以在游览诺丁汉市中心的过程中佩戴它。在此游览过程中，寄生体可以向佩戴者传输数码化信息与指令，并创造出一种混合的现实。作为一种概念，这种混合的现实集中体现出数码科技对人体的直接反应。而“城市寄生体”的概念则可进一步创造一种建筑系统，从而既可在现有空间中对人类的干预活动进行改变和回应，亦可在其建筑表面叠加数码信息层。因此即可进一步将混合现实的概念与米尔格拉姆的连续介质理论延伸至建筑形式之中。

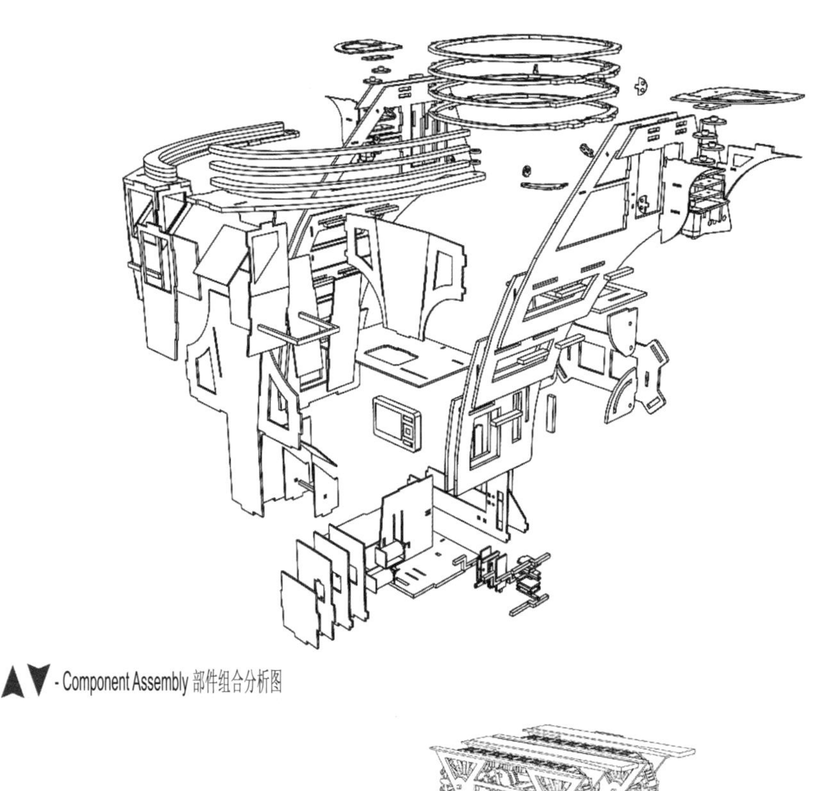

▲▼ - Component Assembly 部件组合分析图

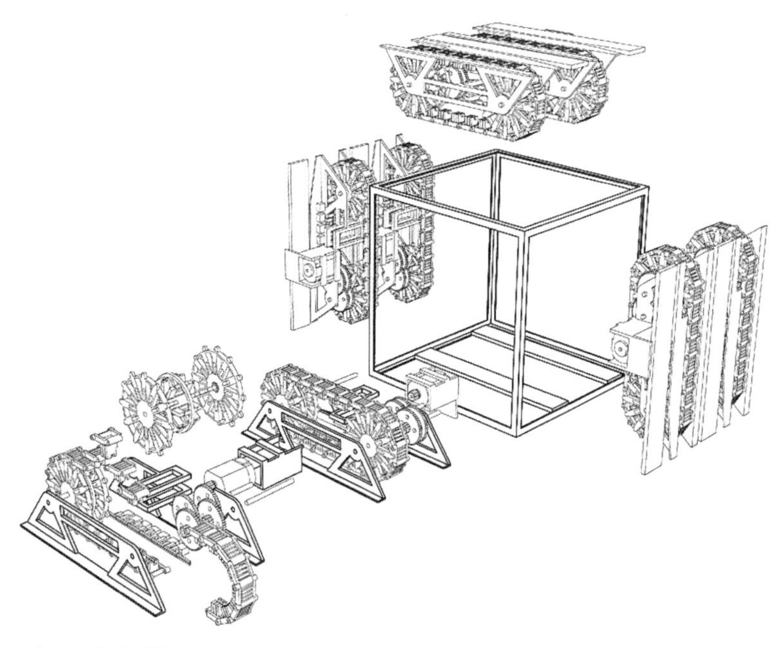

- Layers“层”概念

- Perceptual Reader“寄生体”感知器

DESIGNER-设计人: Anna Pichugina

Psychometric Hostel & Archaeological Promenade

The project combines two programmes. The first related to spatial perception & cognition, and the second derived from the actual site condition. The resulting synthesis is a hostel that seeks to optimize the cognitive benefits of exposure to new stimuli and questioning of schemata associated with travel and displacement, and an exposition of the stable strata of archaeological heritage existing in situ. Satellite programmes underpin the attempt to create a playful social node that grows from within a neglected block in the historic downtown of Athens.

心理测试学青年旅馆与考古学漫步

该项目由两部分方案组成。第一部分与空间洞察力与认知力有关，而第二部分则源自对实际场地条件的分析。它在三个特殊方面：通过面对新时代所特有的外部刺激而获得的认知收益，质疑与旅行和位置转换相关的企划，以及场地上现存的，稳固的考古学文化层，进行最优化处理，其最终的综合结果生成了一间青年旅馆。其周边的一系列卫星方案则致力于在雅典老城区内的消极街区中去创造一些有趣的社会节点。

▲ - Swish 文脉概念

▼ - Section 剖面图

- Phaser 建筑阶段分析图

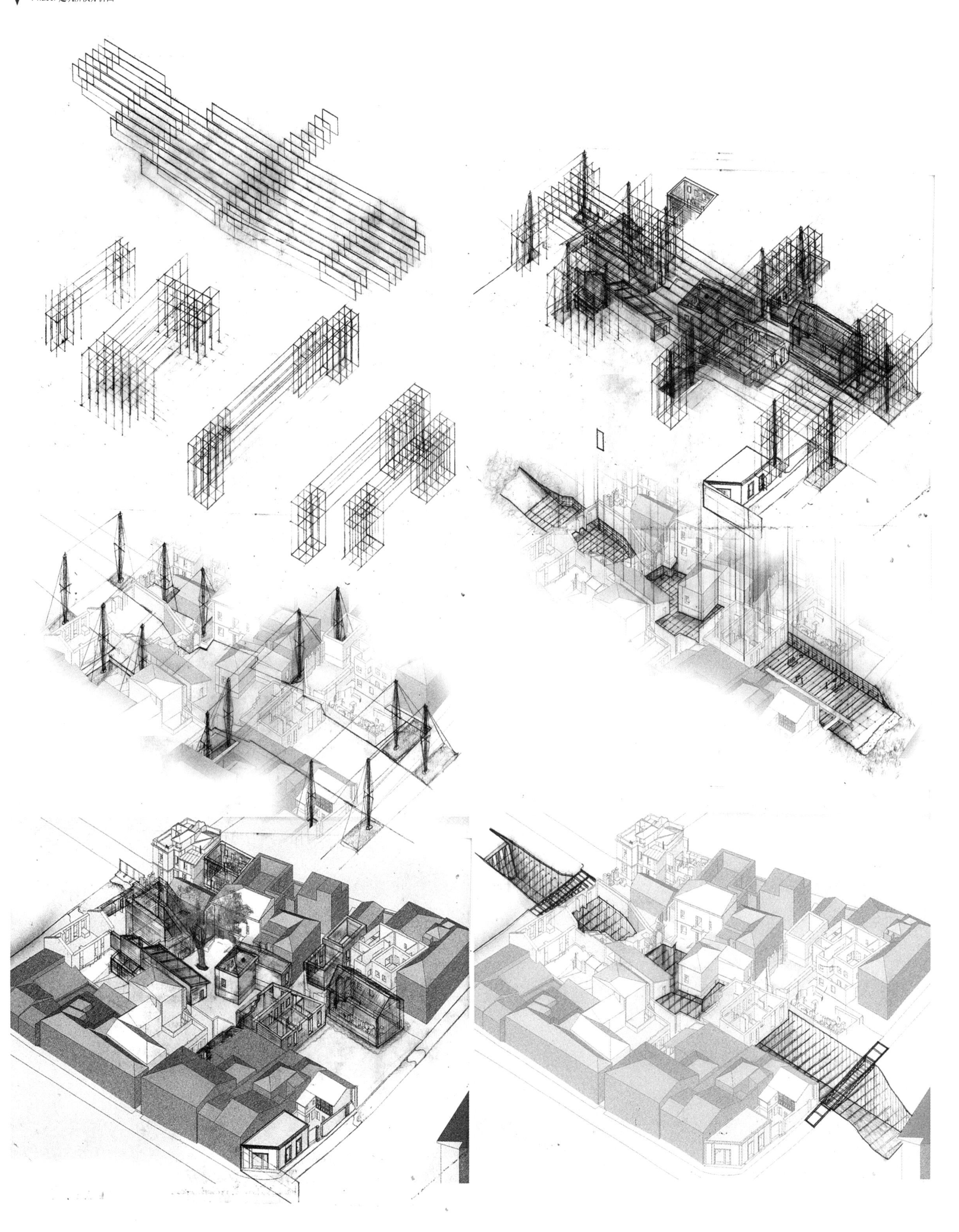

DESIGNER-设计人: Sally Emery

The Urban Equine

For many hundreds of years man and horse have worked together. We now realise the horse possesses a unique temperament and character, proving highly beneficial in personal and professional development. Financial restraints regularly force urban based equine organisations to share facilities, with one party loosing out to unsuitable architectural conditions. This situation is eliminated through combining the equine based therapy sector with a highly lucrative equine assisted team building scheme. The facility I provide will primarily function as an equine therapy centre, however the spaces provided will also be suitable for professional skills development courses as their spatial requirements are very similar. The ultimate vision is to create a bespoke environment with an urban setting, in which the public are exposed directly and indirectly to the unique therapeutic nature of the horse, providing benefits to the whole community.

城市之马

人类和马已经共同合作达数百年之久了。我们知道，马拥有独特的性情特点，与马匹接触，可以对个人与专业领域发展提供很多益处（西方称之为“马疗法”）。然而由于经济条件的限制，城市中的驯养马匹的机构不得不时常分享各种设施，从而使建筑条件不见得能适应各个机构的不同要求。因此我建议，通过将马疗中心与一个可以盈利的马匹辅助团队建设项目相组合，进而来消除这种情况。我所提供的设施主要是作为马疗中心，不过由于对空间的要求相似，这些空间也适合进行专业技能培训。其最终的蓝图是基于当地城市布局而为之量身定做的一处环境，在这里，公众直接或是间接地接受来自马匹那独特的天性，享受其特有的疗效，从而为整个社区提供福利。

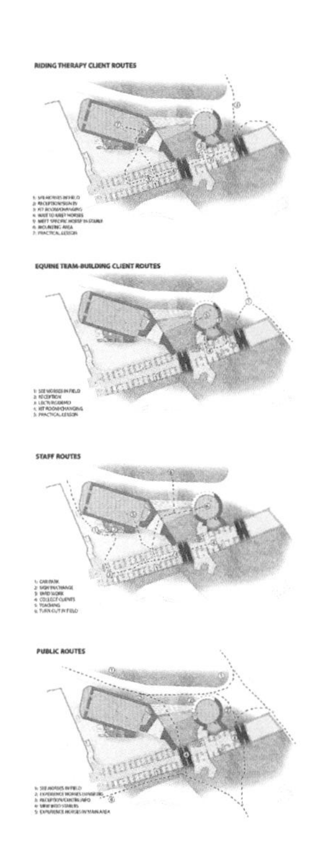

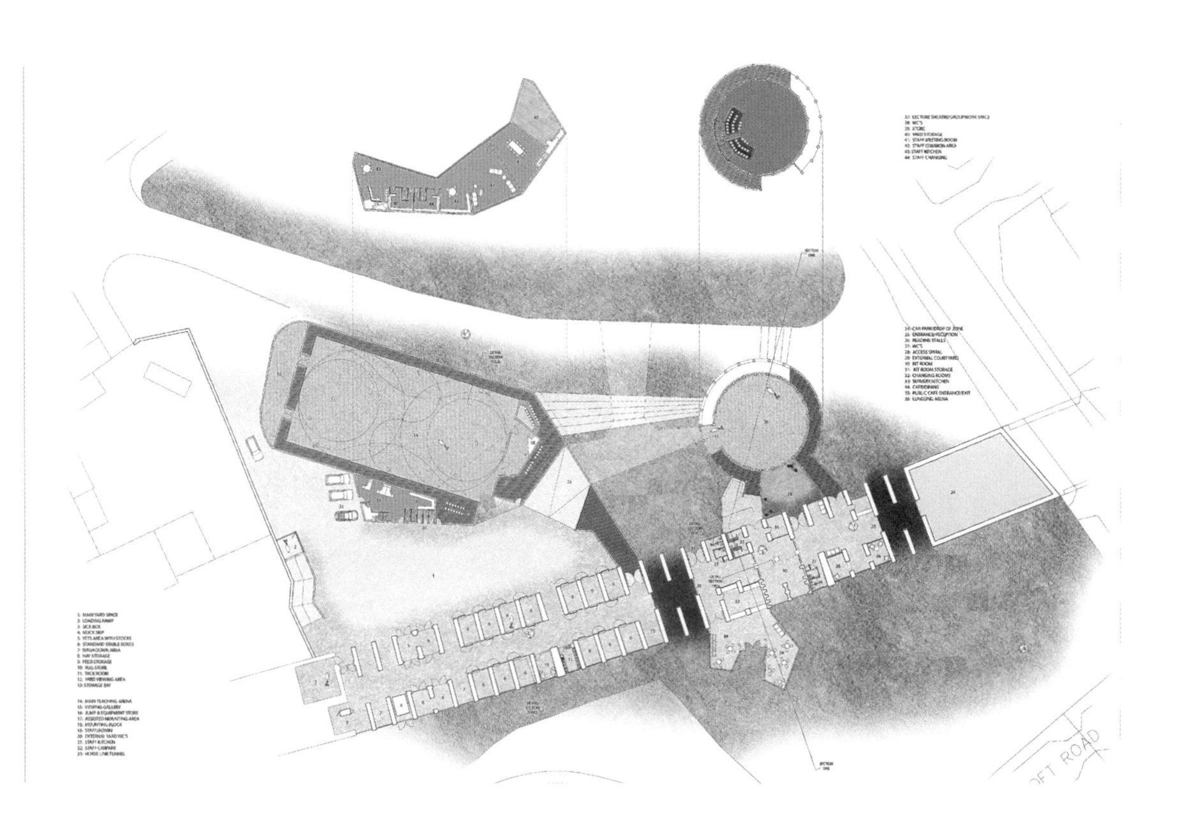

▲ - User Routes & Plans 游览路线与总平面图

▲ - Perspective View 透视图

- Facade Design 立面设计

1 THE THREE BASIC EQUINE PACES

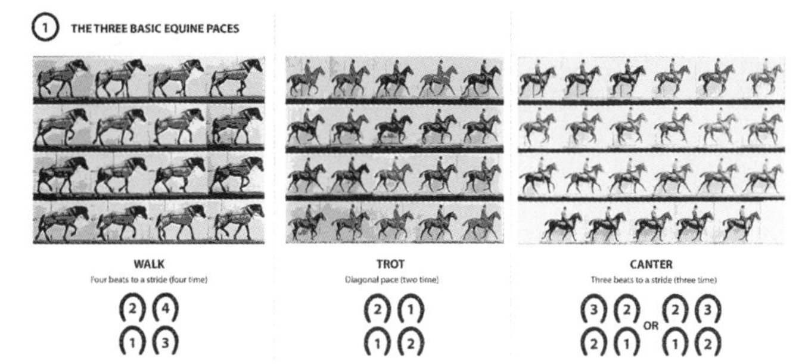

2 EXTRACTING THE KEY MOVEMENT

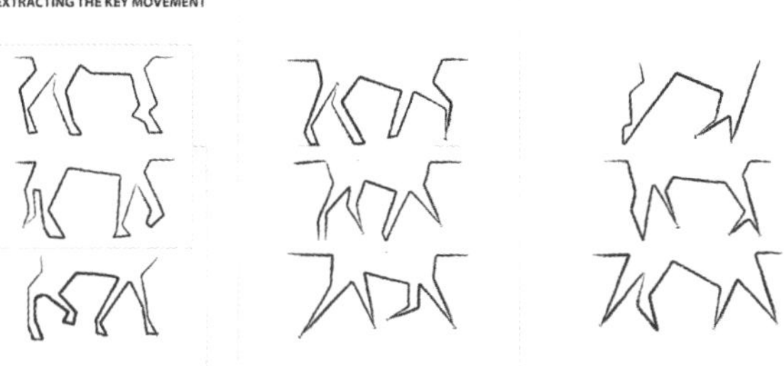

3 EXTRUDING THE SHAPE

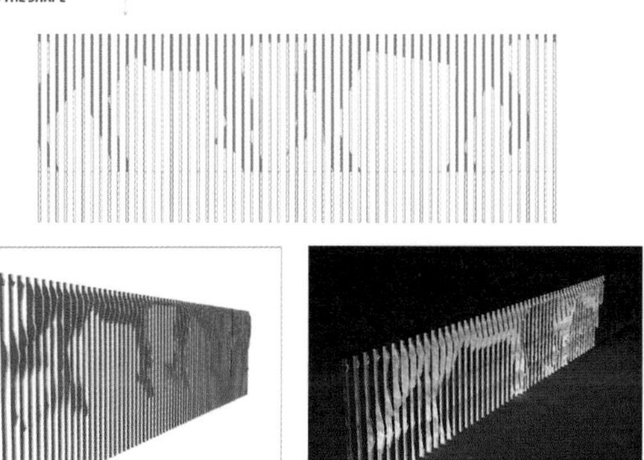

4 TESTING THROUGH MODELLING

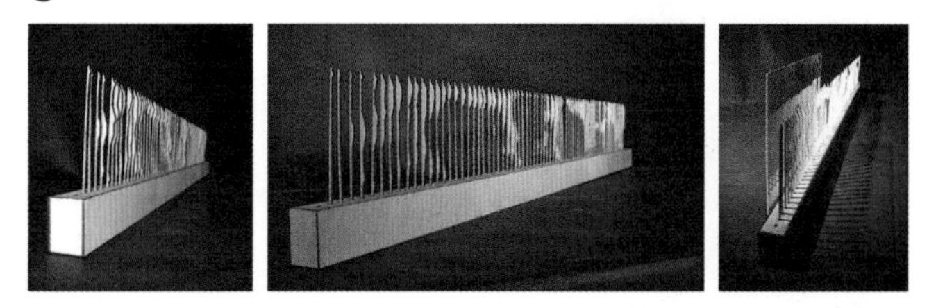

5 THE PROPOSAL IN CONTEXT

- Section 剖面图

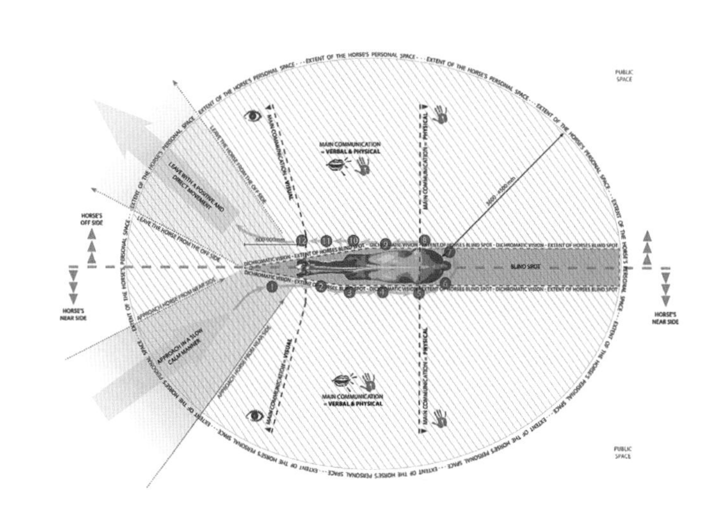

- The Horse Stages of Approach 人与马交流感知分析图

DESIGNER-设计人: Phillip Etherington

Agronomy: Community Vineyard

Portland's unique geological composition, location and genius-loci determine an agronomic system of vineyards as a programmatic catalyst to restore an extended field condition. As the centre of a new métier (skill base) the proposal aims to stimulate embodied encounters with the physical landscape thus restoring identity, value and a sense of place to post-industrial quarry settings. The programme revolves around a centralized cooperative where individual small-holdings are leased to local wine producers to cultivate vines. The idea is to recover common land stolen from the community by the military through a principal public building known as a communal growers' club, which serves to enrich the extended field and relationship of the territory to urban society. Overall the proposal can be actualised as a macro-threshold between urbanity and agriculture.

农业经济学：社区葡萄园

波特兰那独有的地理环境，位置和场所精神决定了可以将葡萄园种植业作为一种项目催化剂来复苏对大范围土地的利用。作为一处推广一种新职业的中心，该计划旨在鼓励与土地实实在在的接触，从而恢复地方特点，价值与一种后采石业时代的场所感。方案围绕着一个中心合作社系统来运行。在这里，家家户户的小块土地都可租给当地的葡萄酒生产商来种植葡萄。此外，通过成立一处名叫“公社种植者俱乐部”的主要公共机构，人们可以加强对大块土地的管理，提高地力，以及增强周边土地与城市社区的联系，进而也可使社区逐步将那些曾被军事工程所“偷走”的公共土地讨要回来，物归原主。总的来说，该项目可被作为一道联系都市风貌与农业风貌的门户来加以实施。

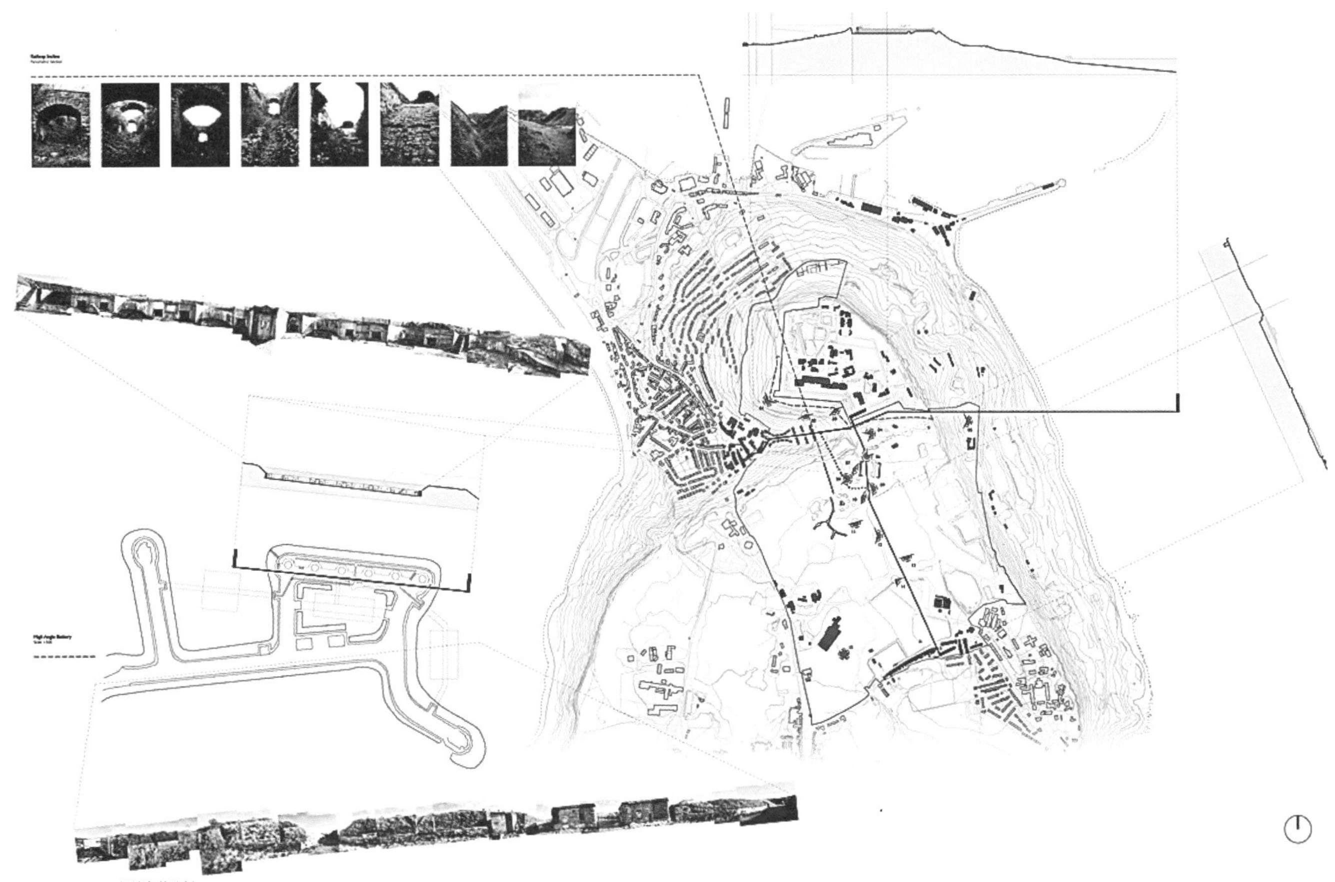

▲ - Analysis of Field Condition 场地条件分析

► - Club Dining Room 俱乐部餐厅

- Entrance Ramp 入口坡道

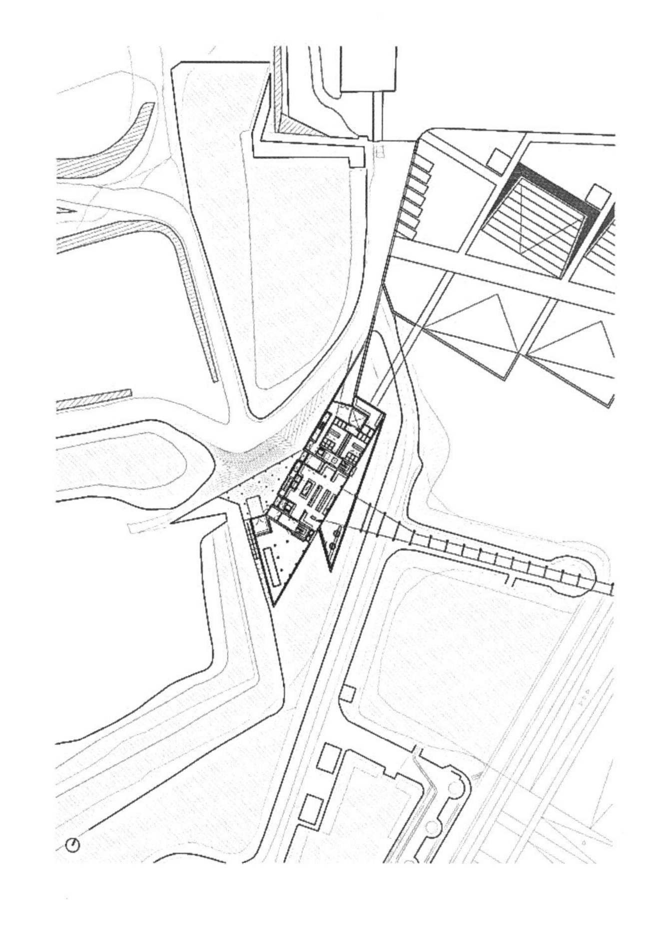

- Clubhouse in Context 俱乐部总平面图

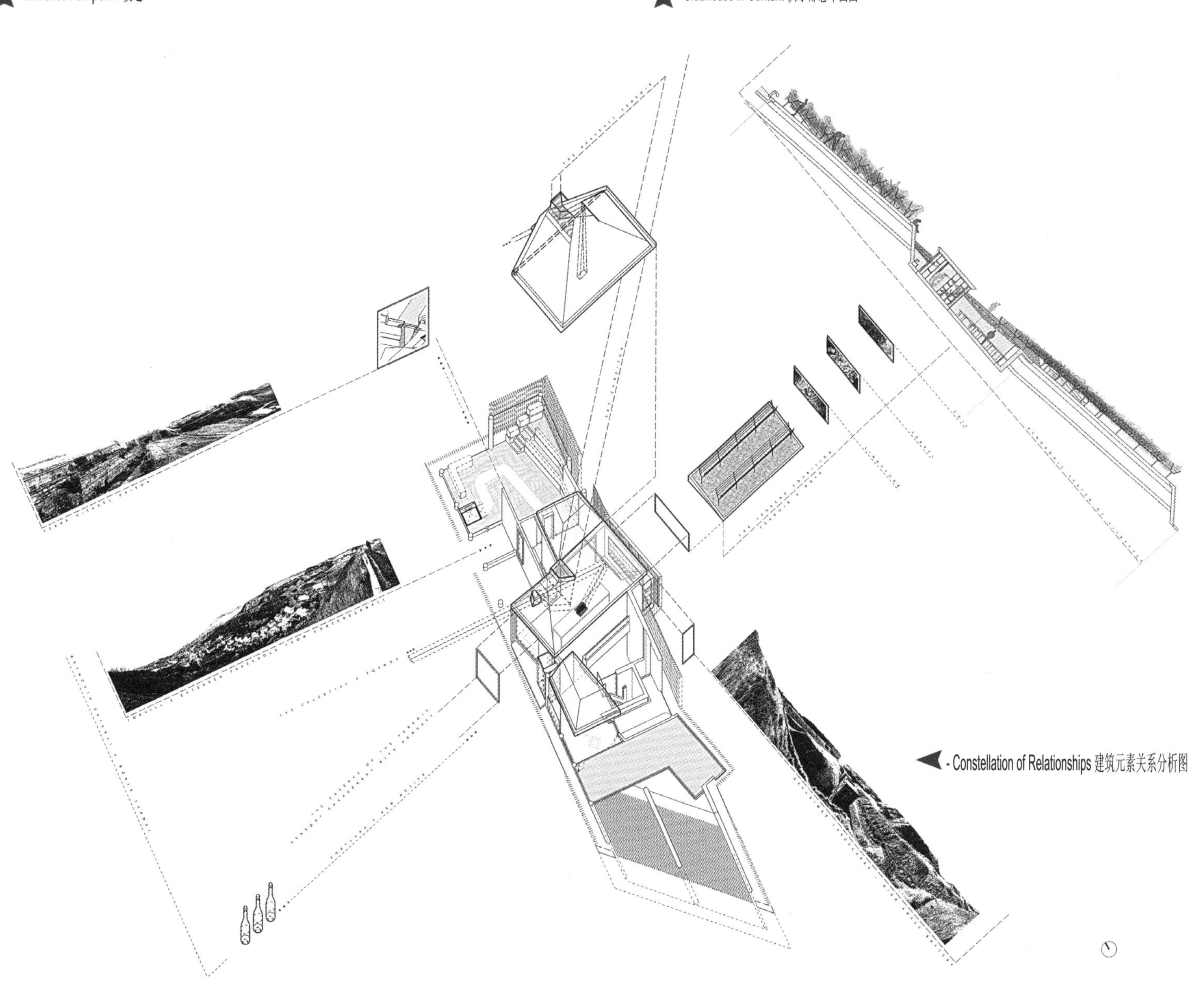

- Constellation of Relationships 建筑元素关系分析图

DESIGNER-设计人: Madeline Pope

Middle Ground: Between Incarceration and Freedom

Portland presents a unique social situation; 9% of all residents are incarcerated. Confined behind impenetrable walls, the prison sits physically and metaphorically on the periphery. By addressing this awkward edge condition as a critical juncture to be both manipulated and inhabited, the proposal stimulates interaction across the threshold. Originally convicts worked in the quarries which stimulated a relative cohesion between freedom and incarceration. With the decline of quarrying on Portland these once interrelated elements today sit in relative isolation. As a critique of the deficient mediation witnessed today between a young offenders prison and the domestic scale of the adjacent community, the project manipulates the prison wall, incorporating it within the heart of the building. Strategically the project reunifies the desolate space between the walled enclosure of St Peter's Church, and mediates back into the surrounding context by offering facilities to both sides of the wall.

中间地带：服刑与自由之间

波特兰存在着一种非常独特的社会状态，这里9%的居民是犯人。在那防守严密的高墙后，是一圈颇具隐喻性的牢房。通过将这一圈不怎么讨人喜欢的边界看作看守与犯人之间的一道重要接缝，本方案试图激励一种跨越门槛的互动。在原来，犯人们在采石场劳动，这其实意味着某种自由与服刑之间的融合。而随着波特兰地区石料的逐渐贫乏，这些曾经相互关联的元素如今变得相对孤立起来。今天，我们对少年犯监狱与周边普通社区之间的融合颇有诟病，而这一方案对狱墙进行巧妙的处理，使之与监狱的核心部分相互联系。从策略的角度看，该方案将圣彼得教堂那高墙环绕下的荒芜场地整合起来，并通过为墙两边提供共用设施而使其重新融入周边文脉之中。

▼ - Re-developed Prison Complex Plan 监狱改造计划平面图

▲ - Cafe 咖啡厅局部透视图

▼ - Interior 内景局部透视图

DESIGNER-设计人: Jayesh Mistry PRIZE-获奖: Shortlisted RIBA Silver Medal 2011

Local Government Chamber, Portland

A new political institution here becomes an island meeting place in which dialogue and discussion can be encouraged. The proposal aims to provide a representative voice for Portland, UK, through appropriate architectural settings that range from the informal (cafe) to the formal assembly chamber, exhibition spaces, library, and finally the official space of the council chronicler. Strategically, the intervention is interested in the hill-topography between the ruins of Rufus Castle and St Andrew's Church. In response to this in-between condition, the building addresses both relics and takes on a sense of institutional responsibility. Anchored to a path at the top, its geometry turns to respect the old church ruins and then responds to the exposed cliff edge; utilising the topography to reveal itself as series of embedded plinths. From this base, the 'top-down' political order of the assembly chamber is placed, echoing the character of the castle above. The external space around the chamber and council office is imagined as a place for the expression of the latent casual play of the site, and on occasion, illicit activity.

地方政府议会，波特兰

这是一个新的政治机关，一个拖延许久的岛状聚会地点。在这里，多方对话与公开讨论希望得到鼓励。本方案旨在，通过对诸多功能——如非正式的咖啡厅到正式的议会大厅、展览空间、图书馆，以及政府编年史书记员的官方办公室等——进行适当的建筑安排，从而提供一处为英国波特兰地区所呐喊的场所。从整体规划角度上看，这一插入原有文脉的项目尽量利用了位于鲁弗斯城堡遗址与圣安德鲁教堂之间的山地地形优势。作为对这种居中的条件所作出的回应，该建筑既照顾了两处建筑遗址的特色又满足了政府机关的职责。与山顶的小路紧密锚连，建筑的几何形体在尊重了老教堂遗址的同时又与位于前方的悬崖边缘相呼应；利用特殊的地形将自己暗喻为一系列深埋入山脊中的粗壮柱础。基于此概念，并与位于上方的城堡特征相互呼应，那种在议会大厅中所宣扬的自上而下的政治结构在某种意义上得到暗示。但与之相反的是，围绕着议会大厅与政府办公室的外部空间则明喻着可能在此发生的各种随意活动，甚至在有的时候可能是非法的活动。

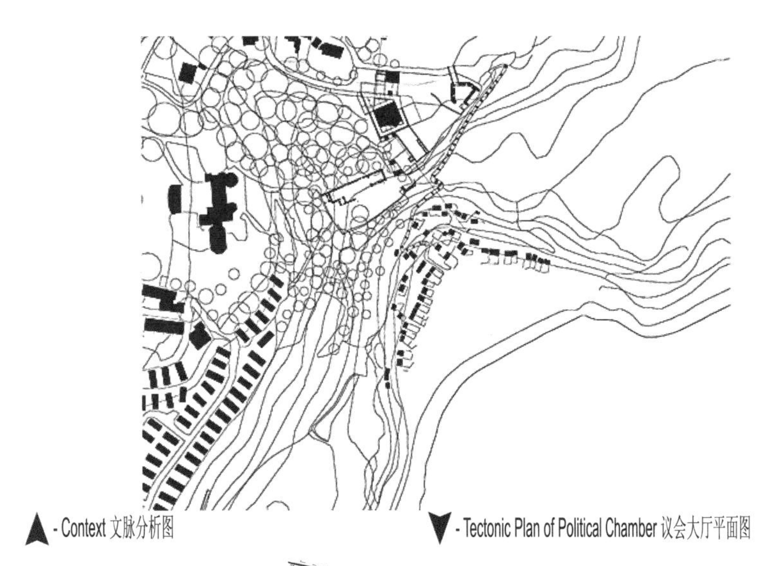

▲ - Context 文脉分析图

▲ - Massing Strategy 体块分析

▼ - Tectonic Plan of Political Chamber 议会大厅平面图

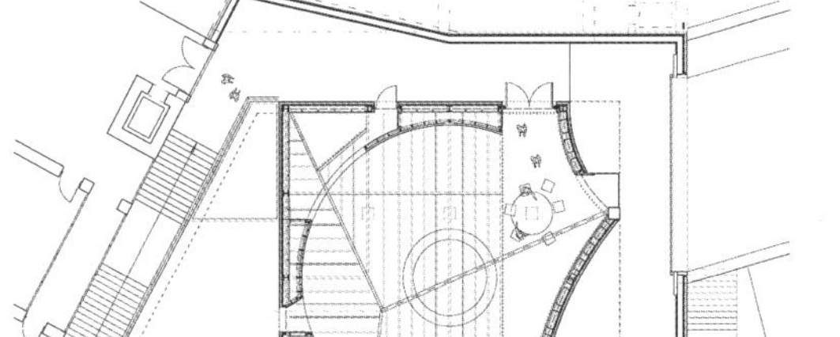

▼ - Site Composition Study 基地综合分析

▼ - Interior study 内景透视图

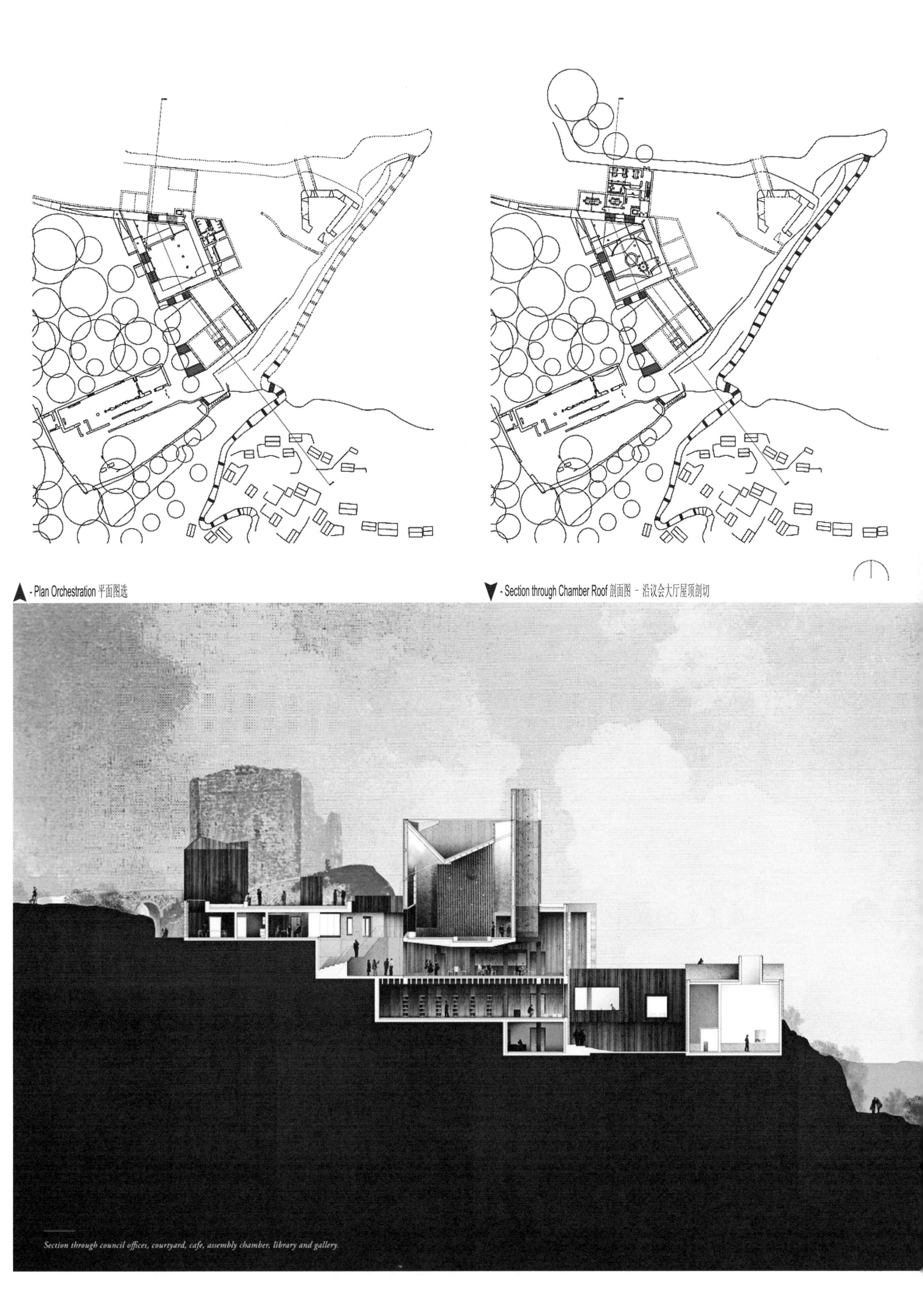

- Plan Orchestration 平面图选
- Section through Chamber Roof 剖面图 - 沿议会大厅屋顶剖切

Section through council offices, courtyard, cafe, assembly chamber, library and gallery.

DESIGNER-设计人: Joesph Bamber

Olympic Legacy: Therapeutic Crab Reconstruction

The 2012 Olympic legacy plan for Portland (UK) proposes a new zero carbon, high quality-housing scheme to house the Athletes of the 2012 Olympics sailing. After investigation it was revealed that these homes would be unaffordable to the local community. The Olympic legacy plan also intends to prohibit crab fishing during the event, the oldest traditional industry of Portland, therefore putting doubt into the future affects of the legacy plan on local industry and the identity of the place. The proposal intended to merge new developments to existing infrastructure in order to establish an appropriate amalgamated urban grain. The final proposal begins to soften the affects of the Olympic legacy plan by preserving the crab industry, creating a new remedial landscape, blurring the line between affordable housing and the athlete's village, whilst connecting local Portlanders specifically to their main industry and effectively their identity.

奥林匹克遗产：治疗性海蟹捕捞业再生

英国准备在波特兰为参加2012奥运会帆船比赛项目的选手建造崭新的零碳、高质量住宅。然而，基于2012奥运会建设再利用计划，调研结果显示当地社区无力在会后购买这些住宅。此外，在奥运期间，波特兰将禁止捕捞螃蟹，而这是本地最为古老的一项传统渔业。这两方面影响使得人们开始怀疑奥运会在未来会对本地工业与地方特色造成冲击。本方案意图将新的发展与现有基础设施融为一体，从而建立一种适当融合的城市脉络。最终方案试图从软化奥运会所带来的影响方面入手，保护海蟹捕捞业，设计规划补救式的环境，模糊经济适用型住宅与奥运选手村之间的差别，并特别将波特兰本地人与他们的主要工业、他们的地方认同感联系起来。

▼ - Mapping the Lifeworld of Crabs 螃蟹生存环境及区域研究

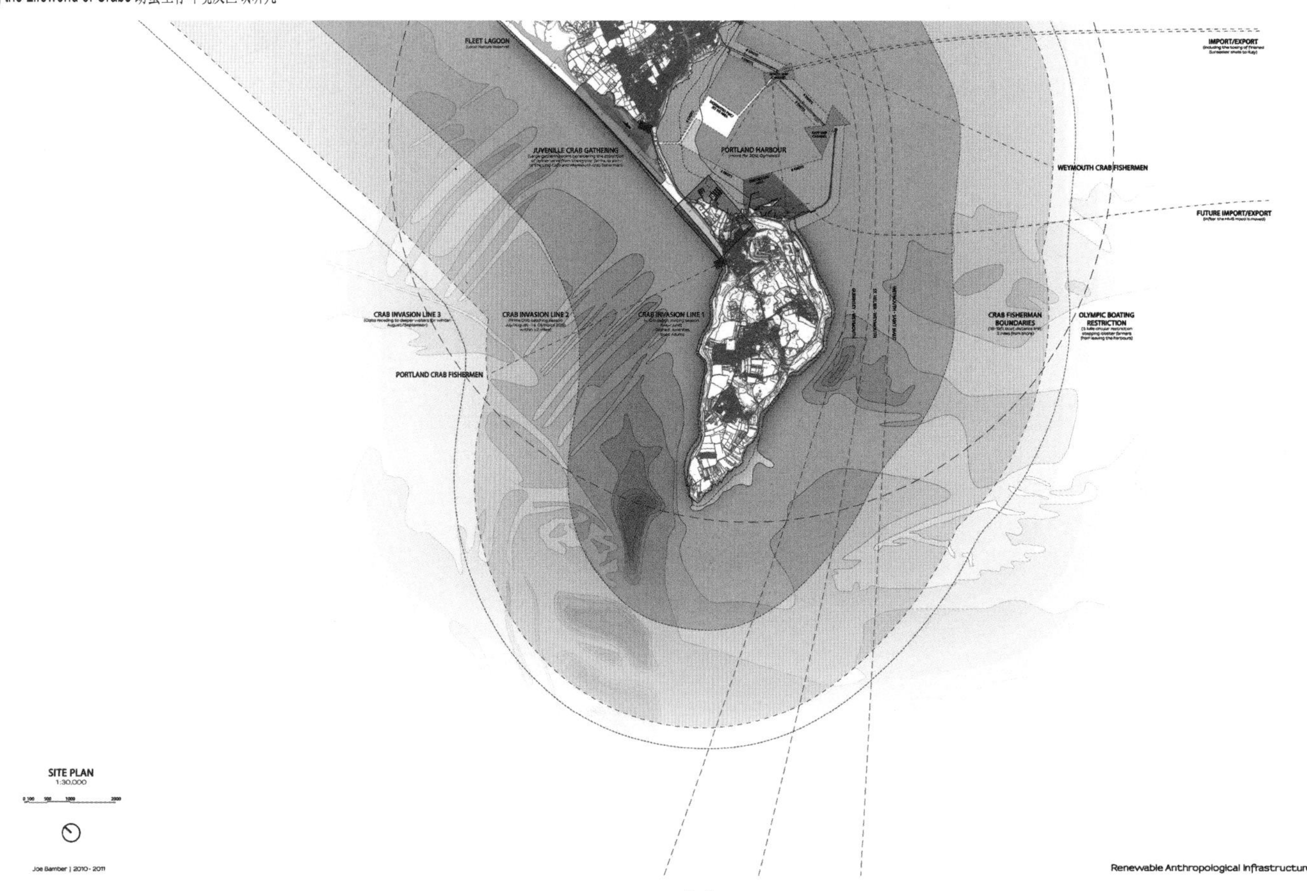

▼ - Entrance to Research Centre 研究中心入口处透视图

▼ - Site Evaluation 场地评价

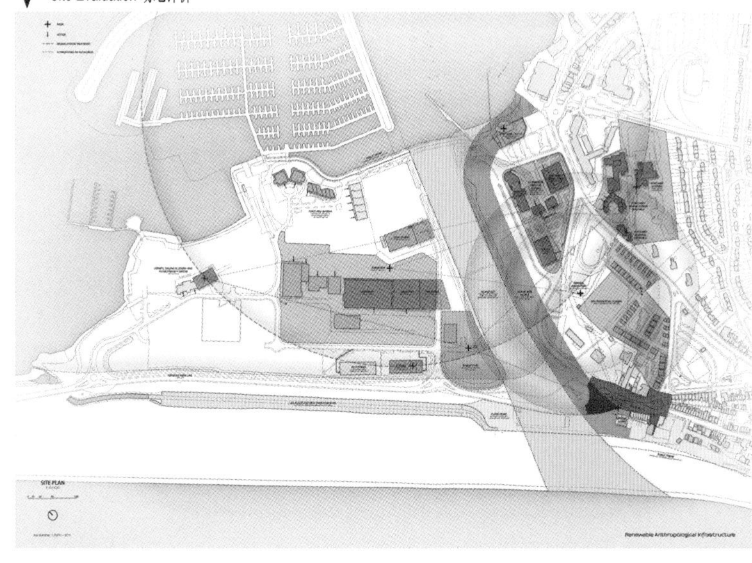

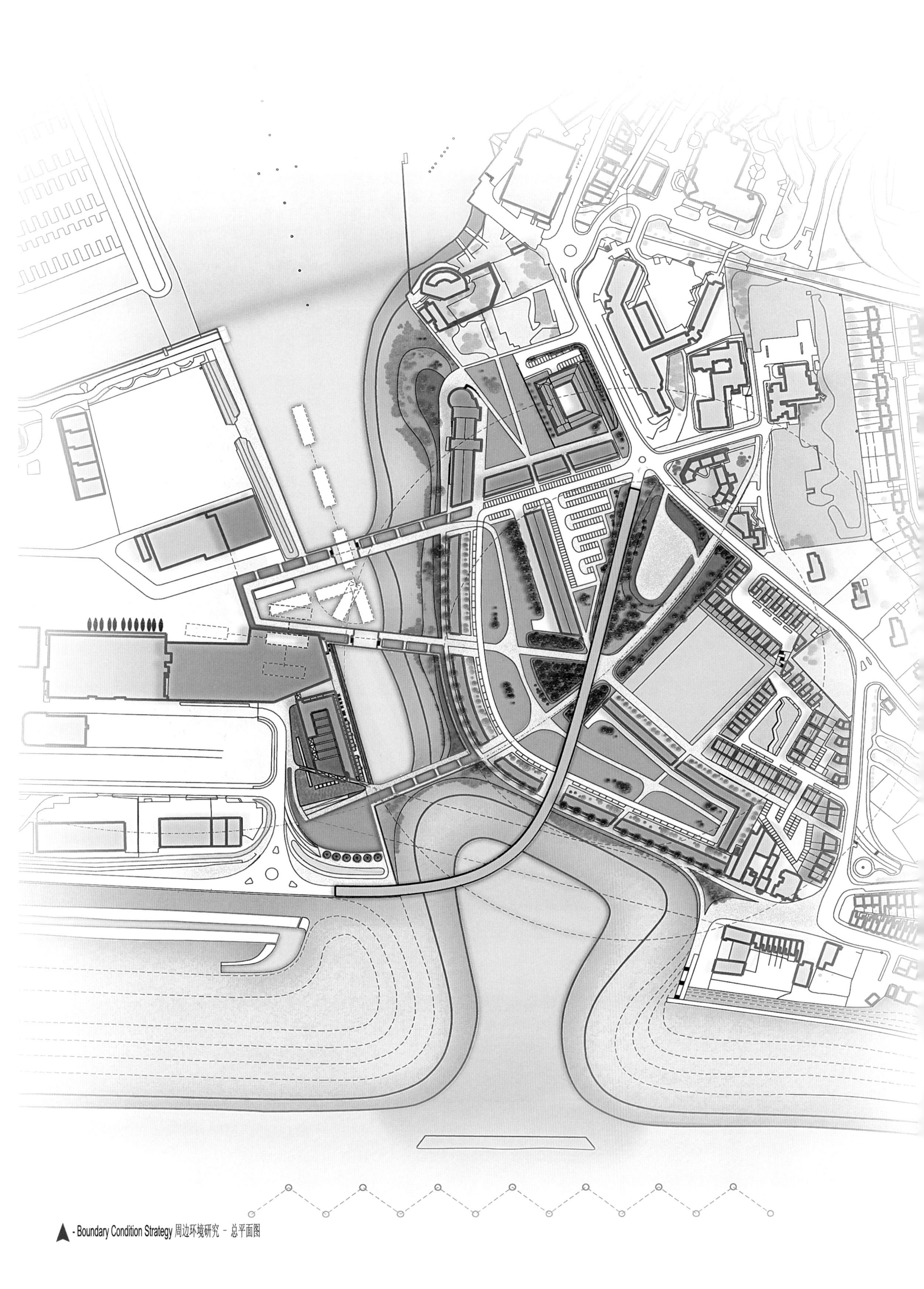

- Boundary Condition Strategy 周边环境研究 - 总平面图

DESIGNER-设计人: David Calder

Reactivated Landscape

This scheme begins by reinstating an industrial process in the derelict quarry which collects waste products and transforms them into usable aggregate, slowly revealing the natural landscape which lies below. Overlooking this coarse industrial reclamation lies a research centre in which a more refined process occurs – the attempted 'growth' of artificial stone which would enable sea defences such as those found in Weymouth harbour to be constructed without the mass quarrying traditionally associated with such terraforming. This inorganic chemistry focuses on the use of 'protocells' – simple cells which can be 'programmed' to shed a skin of calcium carbonate – essentially limestone – while responding to certain inputs such that control can be asserted over the growth of these artificial stones. Thus the interaction between two contrasting processes – destructive and regenerative – is mediated in part by an extension of laboratory into workshop.

再生的地貌

该方案希望在一处废弃采石场中恢复建立一种可以收集废石料并将它们转变为可用复合人工石的工业过程，并进而逐渐展现隐藏在下面的自然原始地貌。为了掌控这一粗放型的回收性工业，一处蕴藏着更加细腻加工过程的科研中心被放置进来——在这里，不断增加的人工石可以用来建造如威茅斯港那样的防波工程，而同时避免了那种传统的，可改变地形地貌的大规模石料开采。这种非有机的化学过程主要使用了“原细胞”——一种简单的、可以形成碳酸钙，特别是石灰石，表皮的“细胞”。这种“细胞”可以对特定的影响产生回应，从而使我们可以对人工石的生长过程加以严格控制。于是，存在于两种相互对立的过程——分解与再生——之间的互动在此被一处由实验室拓建而成的工作车间中部分地中和了。

- Quarry 采石场

- Photographic Metaphor of Industrial Site 工业用地之隐喻

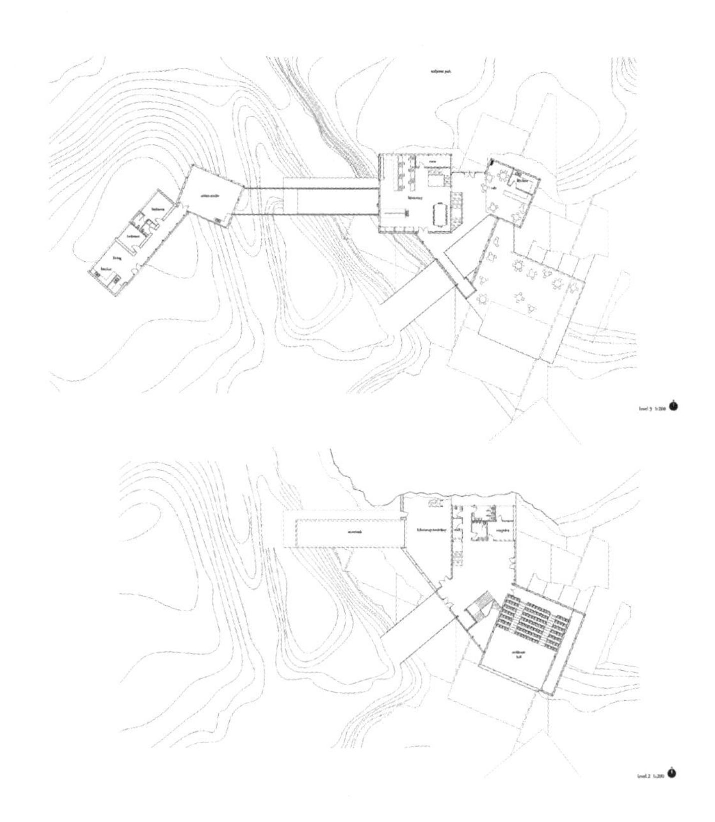

- Plans in Relation to Industrial Landscape 与工业性周边环境相联系的平面图

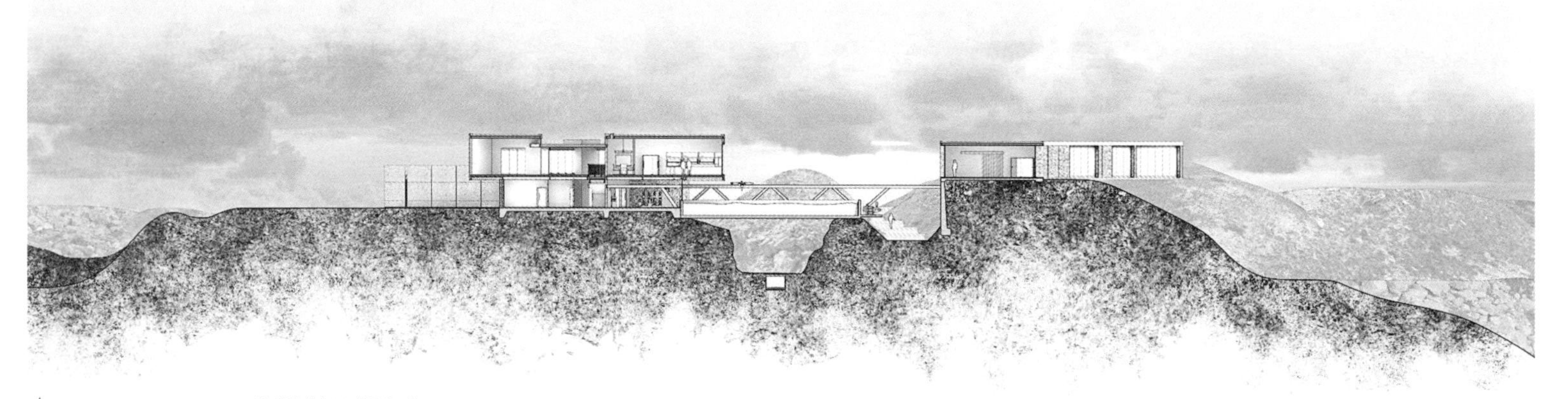

- Section and Scarred Landscape 剖面图与伤痕累累的周边环境

- Edge-Spoil 荒蛮的周边环境

- Perspective View 透视图

035

DESIGNER-设计人: Rebecca Harrison

Refectorium

The juxtaposition of spiritual/ceremonial refreshment with physical refreshment has historical origins in the 'refrigeria' of the ancient Romans. This historical link between physical and spiritual refreshments is explored and reinvigorated through the cross-programming of the Refectory (Refectorium). Using the site to combine the facilities of a ceremony hall, and a modern day refectory, brings together the two distinct user groups who form the two main communities of Portland (UK). For visitors to Portland, many of whom are walkers, the building acts as a restorative 'pit-stop' providing refreshment, shelter, comfort, changing facilities and tourist information. As a modern day refectory, these facilities draw from their historical counterparts, as monastic refectories also included nourishment for the soul through readings of scripture. Monks were also restored through the 'calefactory' or warming room. This has been re-interpreted as changing facilities, passively warmed through a trombe wall system.

复苏堂

在精神层面/礼仪层面上恢复，与在身体层面上恢复，这两种境况并置在一起，其历史渊源可追溯到古罗马时代，当时人们将其描述为"refrigeria"。通过对餐厅的交互运营程序，这一身体与精神层面恢复之间的历史联系可被了解且重塑。于是，通过将项目用地与一间礼仪大厅的设施，与一间现代餐厅相结合，来自英国波特兰两种主要社区的人们将会被联系在一起。来波特兰的旅行者多数都是步行，这一建筑即可作为恢复体力的"停靠点"，这里提供茶点、遮风避雨、十分舒适、可以换洗衣物，还可以领取各种旅游信息。作为一间现代餐厅，所有这些设施都能从历史找到影子，因为就算是一间古代清苦修道院的餐厅，在那里也至少可以通过诵读经书而为灵魂提供精神养料。修道士们也曾借助取暖室来重新振作。而这一点则通过向阳设置被动式双层墙体被重新加以诠释，新的功能则是换洗衣物的房间设施。

▼ - Urban Threshold Strategy 城市门户空间策略－平面图

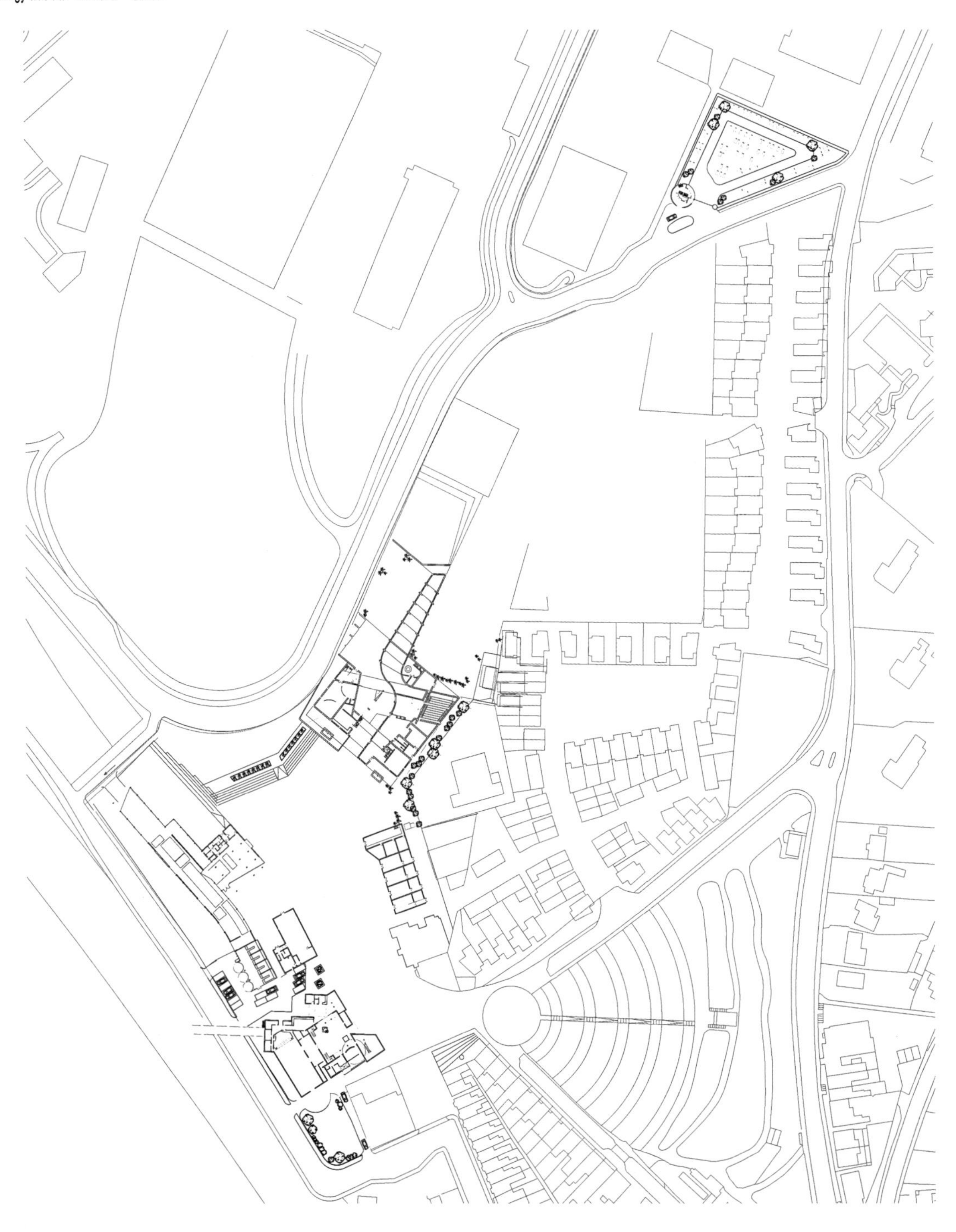

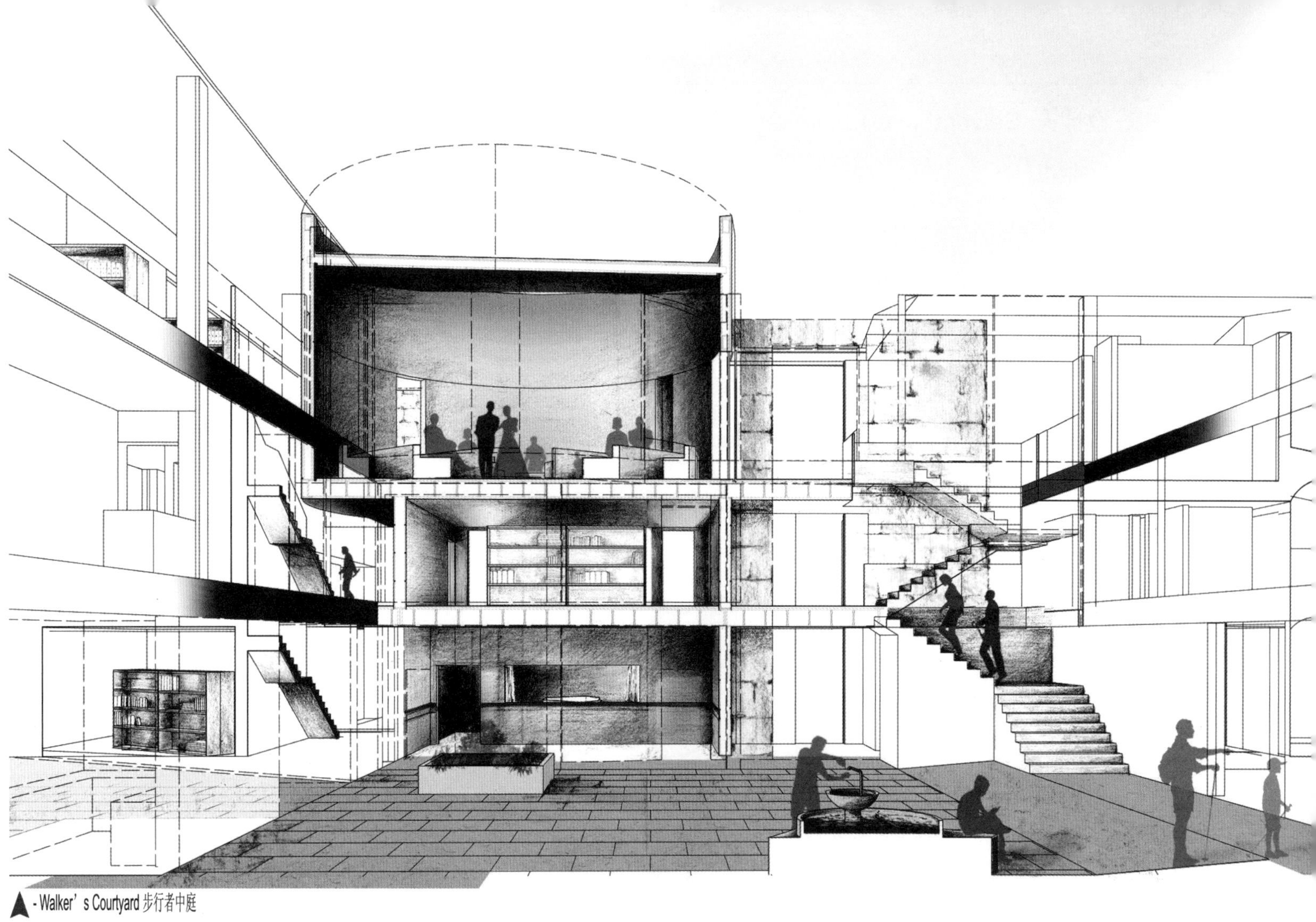

- Walker' s Courtyard 步行者中庭

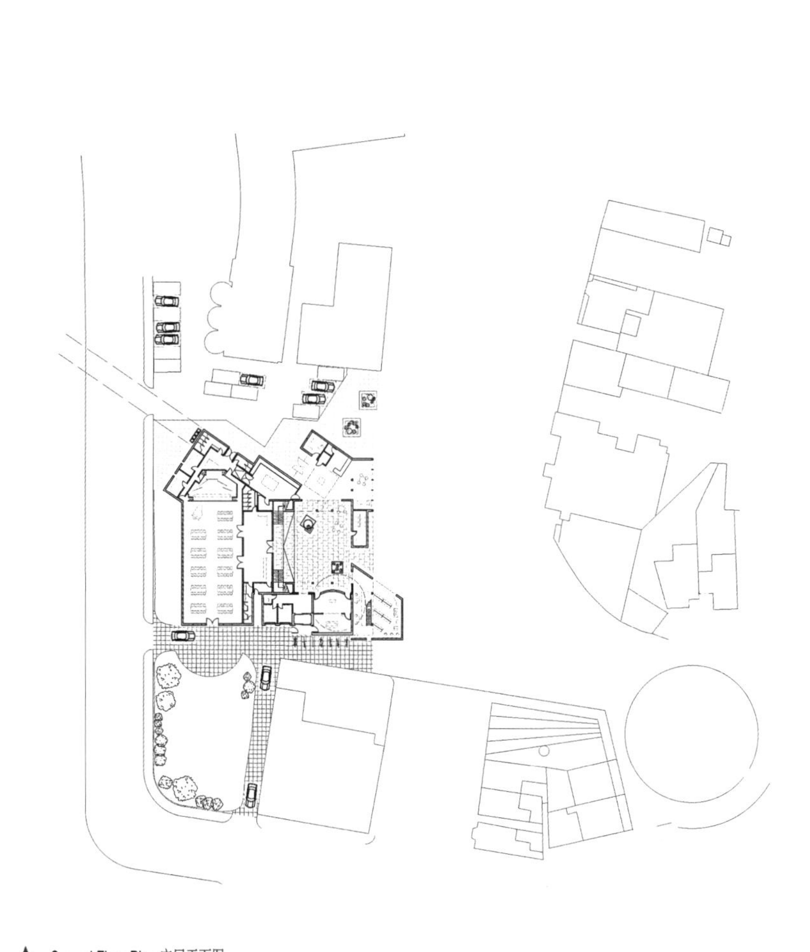

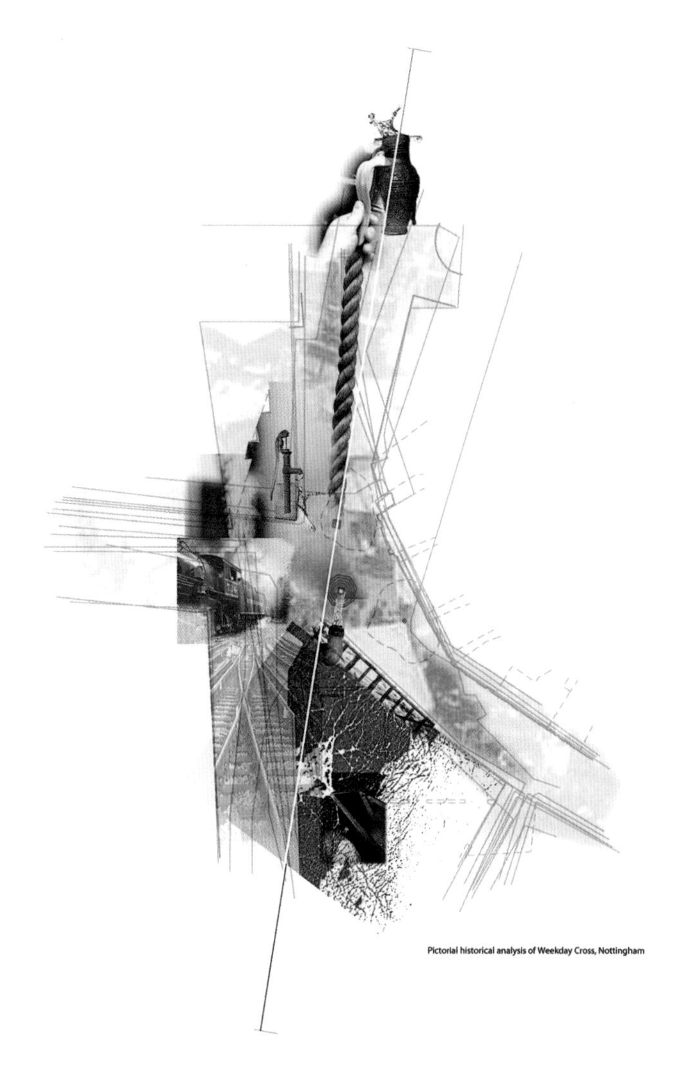

- Ground Floor Plan 底层平面图

- Poetic Study of Urban Fountain 城市元素“喷泉” - 写意研究

DESIGNER-设计人: Selvarajah Gauthaman

Sport as Reconciliation

In trying to reconcile the social rift between the penal institution and the community, this building has brought them together through the latent recreational history of the site. As they converge, the result is a blend of public entrance into the prison and a reactivated sports centre in a nearby limestone quarry. The Sports ground was renovated by the Young Offenders into an external football pitch and used to play matches against members of the community. The weekly "Grove Boys vs. Borstal Boys" matches took place every Saturday and were a focal point for the community as a whole. A linear masterplan encourages the prison to spatially dissolve into the community, thus synthesising civic and penal typologies. Allowing for the communal matches to resume once again, it also helps to give the local summer fetes a permanent home. The two distinct sides of the building are reconciled through an undulating roof bringing a sense of continuity to the building, which is treated as displaced ground and viewing platform for the sports programme below.

用体育实现调解

通过对建筑场地那潜在的娱乐历史的研究，该方案试图消解刑事部门与公共社区之间的隔阂，并将二者连在一起。两者的交集集中体现为一座综合性监狱入口，以及重新恢复对一座位于不远处石灰石采石场内的体育中心的使用。运动中心的场地过去被少年犯们改造为一处室内足球场，在这里，他们曾经可以与当地社区进行比赛。以前每周六，"果园队"与"少管所队"都会踢一场比赛，这曾经是包括监狱在内的整个大社区的共同焦点。如今，线性的总体规划鼓励监狱在空间上与当地社区融为一体，将公民性建筑类型与刑事性建筑类型综合起来。这一方案不但将给足球赛得以再次恢复的机会，也将给当地的夏季游乐节提供一处永久性的举办地。波浪形的屋顶为该建筑提供了连续感，从而将那截然不同的两边调和在一起，除此之外，它还代表着一处位移了的地面和一座看台，俯瞰这下面正进行着的运动项目。

▼ - Material Study 材料研究

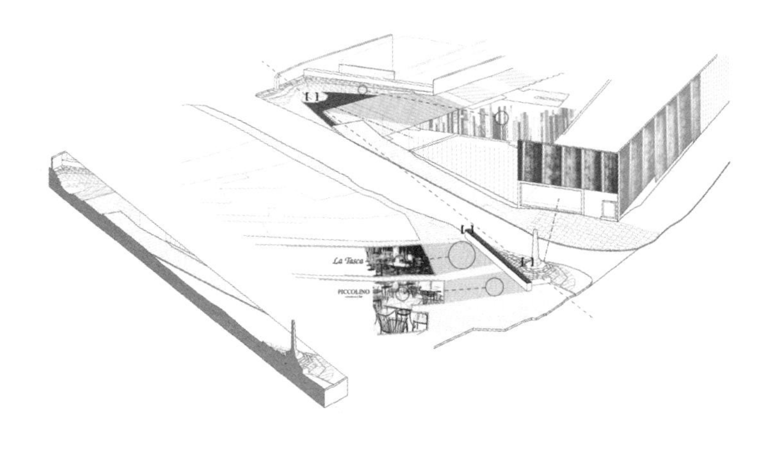

▼ - Urban Surface Drawing 城市表面元素分析图

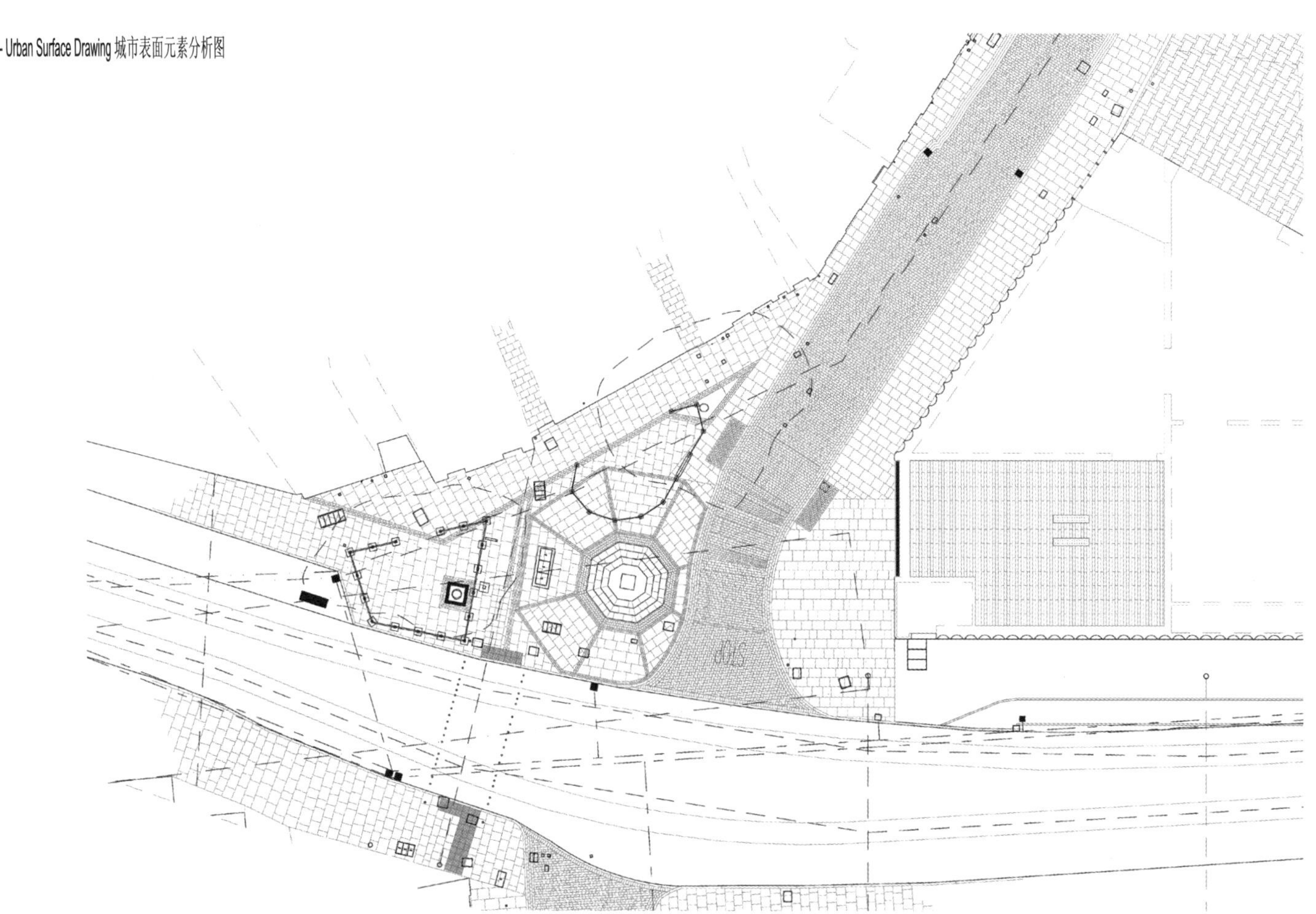

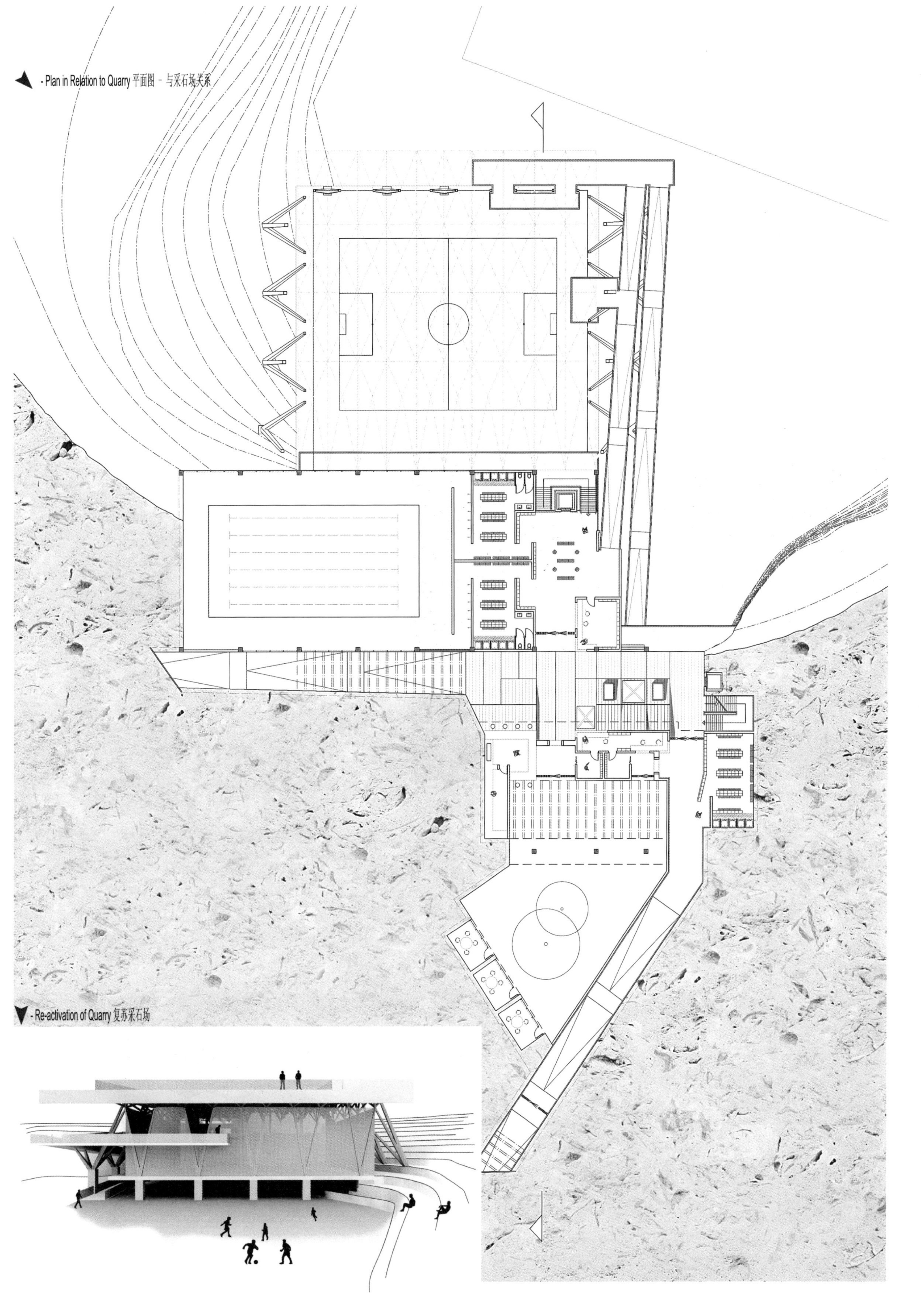

- Plan in Relation to Quarry 平面图 - 与采石场关系
- Re-activation of Quarry 复苏采石场

037

DESIGNER-设计人: Ana Moldavsky

Endless Street: Archaeology in Contemporary Salisbury

The project addresses an urban block forming the northern edge of Salisbury's Market Place, and is anchored to an adjacent Coroner's office, mediating between archaeological finds and the contemporary public realm. It aims to reactivate the continuity between the city's past and its contemporary condition both programmatically – by making public the archive of objects speaking of its past and publishing local knowledge – and physically – by restructuring the old alignment to the Old Sarum hill fort where the city originated. A set of urban 'moves' embody the project's programmatic and morphological intentions: a public, active entrance fronting onto Market Place housing a local bookshop and publishing a view deep into the block, leading to a new elevated courtyard, energised by its surrounding fabric, with the archaeological society offices and archive bookending this axial view. The processional experience of the city is hence engaged with, offering again the once firmly instated panoramic links between origins, route and destination.

无尽街：当代索尔斯伯里考古

本方案着眼于位于索尔斯伯里老市场北边缘的一片街区。这里毗邻一处法医办公室，在当地考古发现与当代公共领域之间形成一处中和区域。它旨在重新激起隐藏于城市历史与当代之间的连续性。从过程角度看，它为公众创造了一所文物档案馆，诉说它们的历史，传播当地的文脉；而从物理角度看，它重新规划了与老塞勒姆山罗马要塞之间的呼应关系，强调了城市的最早起源。一系列城市“运动”体现出本方案的过程性与形态性意向：这是一处公共的、积极的入口空间——直通老市场，且包含一所当地书店，展现了对该街区的深入理解，通向一处新的空中庭院，从周边机理中汲取能量，并由考古学会办公室与档案馆作为整个轴线的收尾。于是，对于城市的整个过程体验便因此而得到加强，从而在开端、发展与结果之间再次形成了明确的全景式联系。

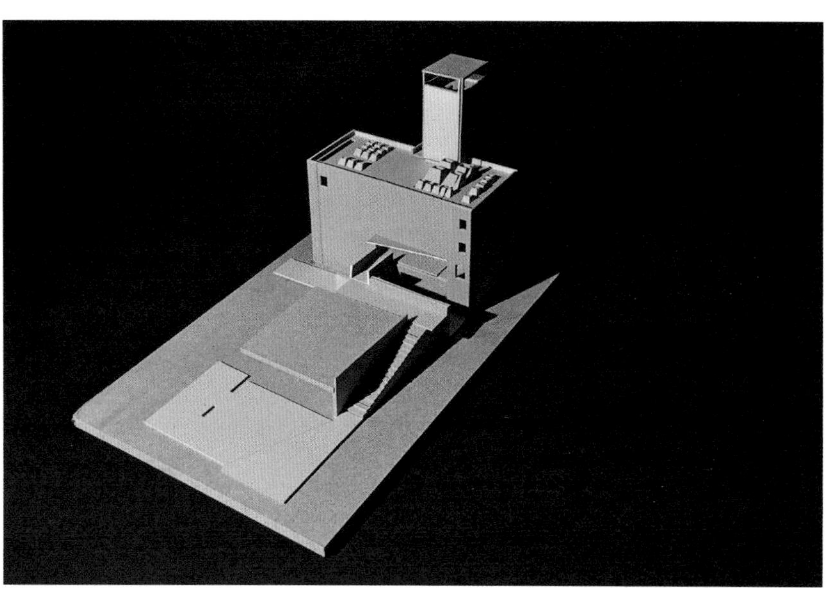

▲ - Massing Model 实体模型

▲ - Master Plan 总平面图

▼ - Re-incorporated Fragment 新旧建筑元素再组合效果

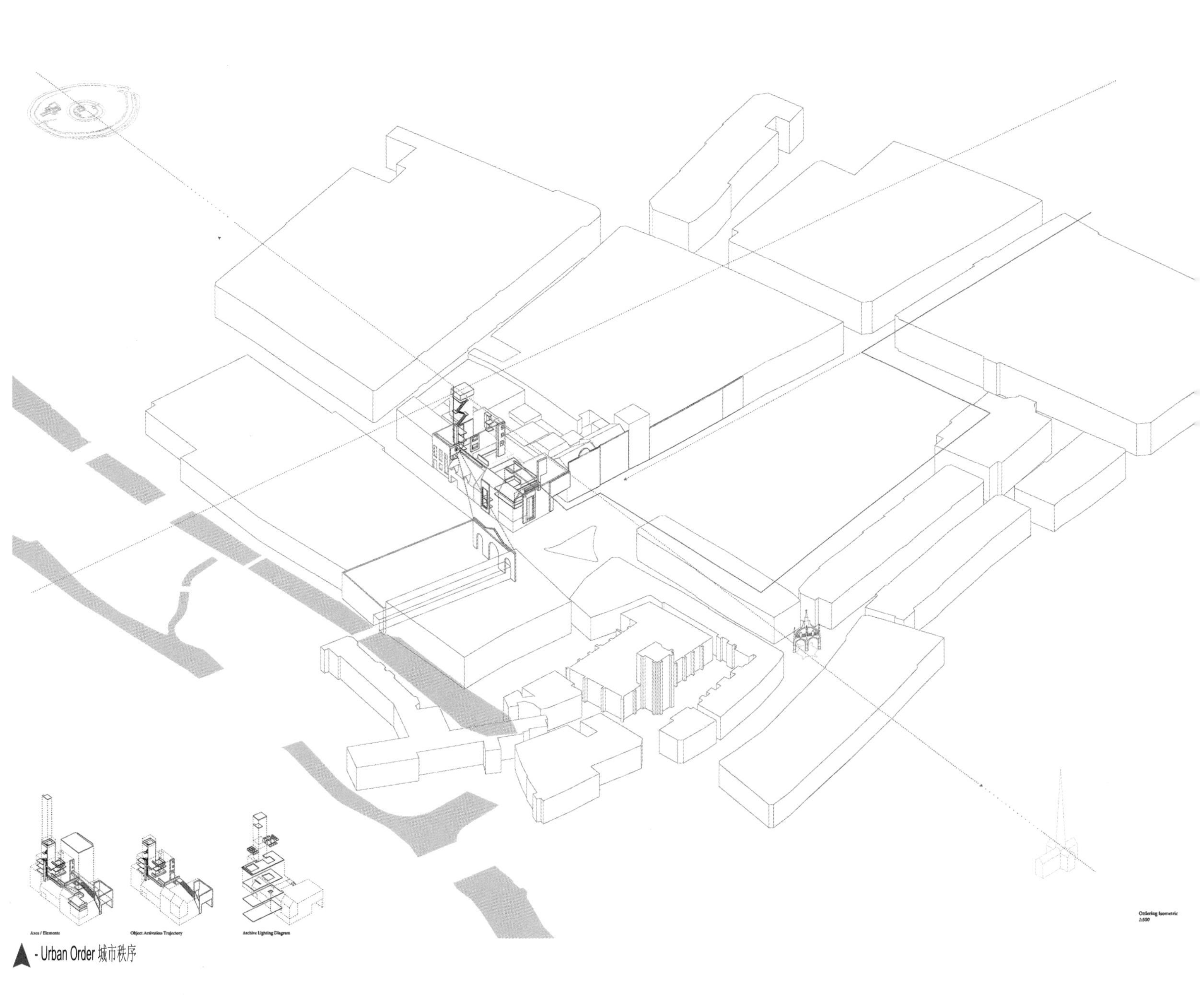

- Urban Order 城市秩序

- Reading Room
- Lobby

038

DESIGNER-设计人: Matt McKenna

Anti-Room

Situated adjacent to Salisbury cinema, the proposition posits a new political agora where cultural activity comes into contact with both formal and informal public debate. It explores the fractured space which has formed between the citizens of Salisbury and the various elected bodies with delegated powers. By enabling dialogue between political groups, limited vision is avoided and collective decision takes a lead role. The agora takes the form of a debating chamber consisting of a circular drum that is partly trapped and partly belonging to the city. These fragments overlap through occupation, occasionally becoming unified into a whole, eventually extending into peripheral urban conditions. The positioning lies to the south of the Guildhall (the City Council offices) on Market Square. This new context within the city is formed through the strategic removal and insertion of a new lane into the urban block. This reconfiguration reveals the sides of an encased mock Tudor cinema (Grade II listed) and its anteroom (Grade I listed). The Hall of John Halle, an important historical and political fragment is then re-appropriated by the "anti-room" proposal.

反房间

该方案位于索尔斯伯里电影院旁边，旨在建立一处新的政治集会中心，在文化活动与正式或非正式的公共辩论之间建立起联系。它试图去探讨存在于索尔斯伯里普通市民和他们选举出的不同权力代表之间的空间断层，坚信通过政治团体之间的对话，个体的目光狭隘性可以被避免，而通过广泛意见征集而得出决定的方式成为主流。集会中心采用了传统辩论大厅的形式，呈扁圆柱形，部分陷入周边建筑之中，部分则暴露出来，属于整个城市。这些建筑片段借助对其的使用而相互重叠，时而会变成一个统一的整体，却最终将延伸入城市外围空间中去。基地同时也位于金色大厅（市政厅）南边，在老市场中。而它所形成的这一新城市文脉主要取决于一次战略性挪移，以及在城市街区中嵌入了一条新的小巷。这一改造方案将一座原本被包围起来的仿都铎式影院（二级保护建筑）的侧墙以及其前厅（一级保护建筑）展示出来。而约翰•哈雷大厅——一处重要的历史性与政治性片段——则通过这一“反房间”项目而得以重生。

▼ - Interior Sequence 内景分析

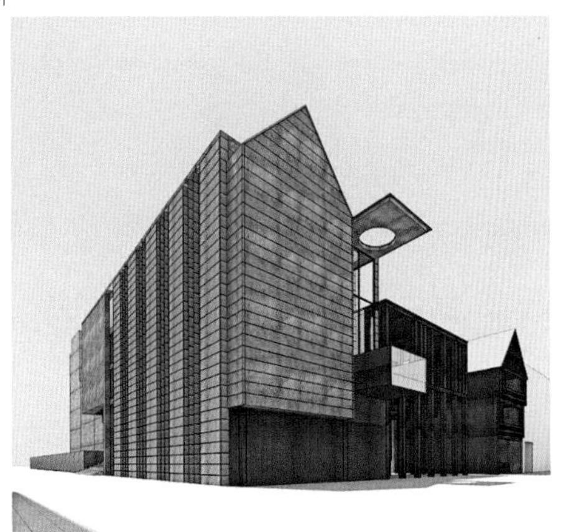

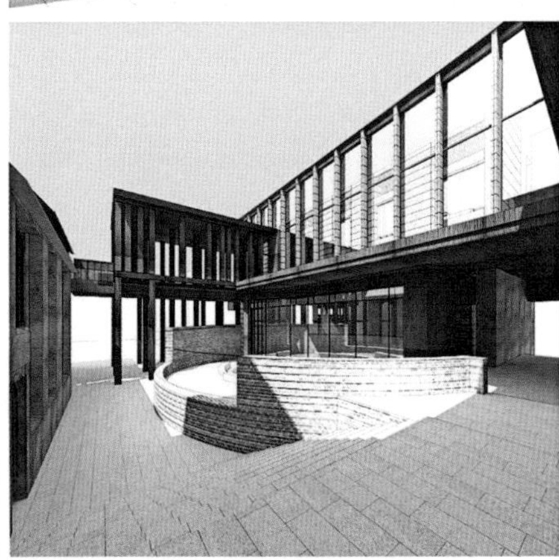

▼ - Urban Ground Plan 城市机理中的底层平面图

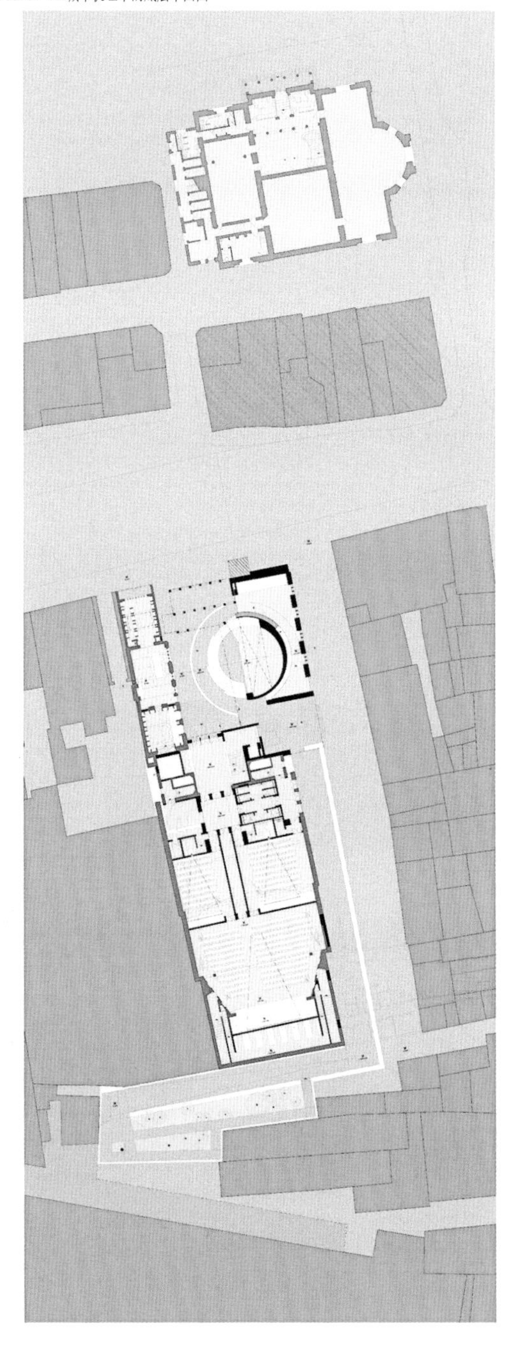

TUTOR - 指导教师: *Darren Deane*

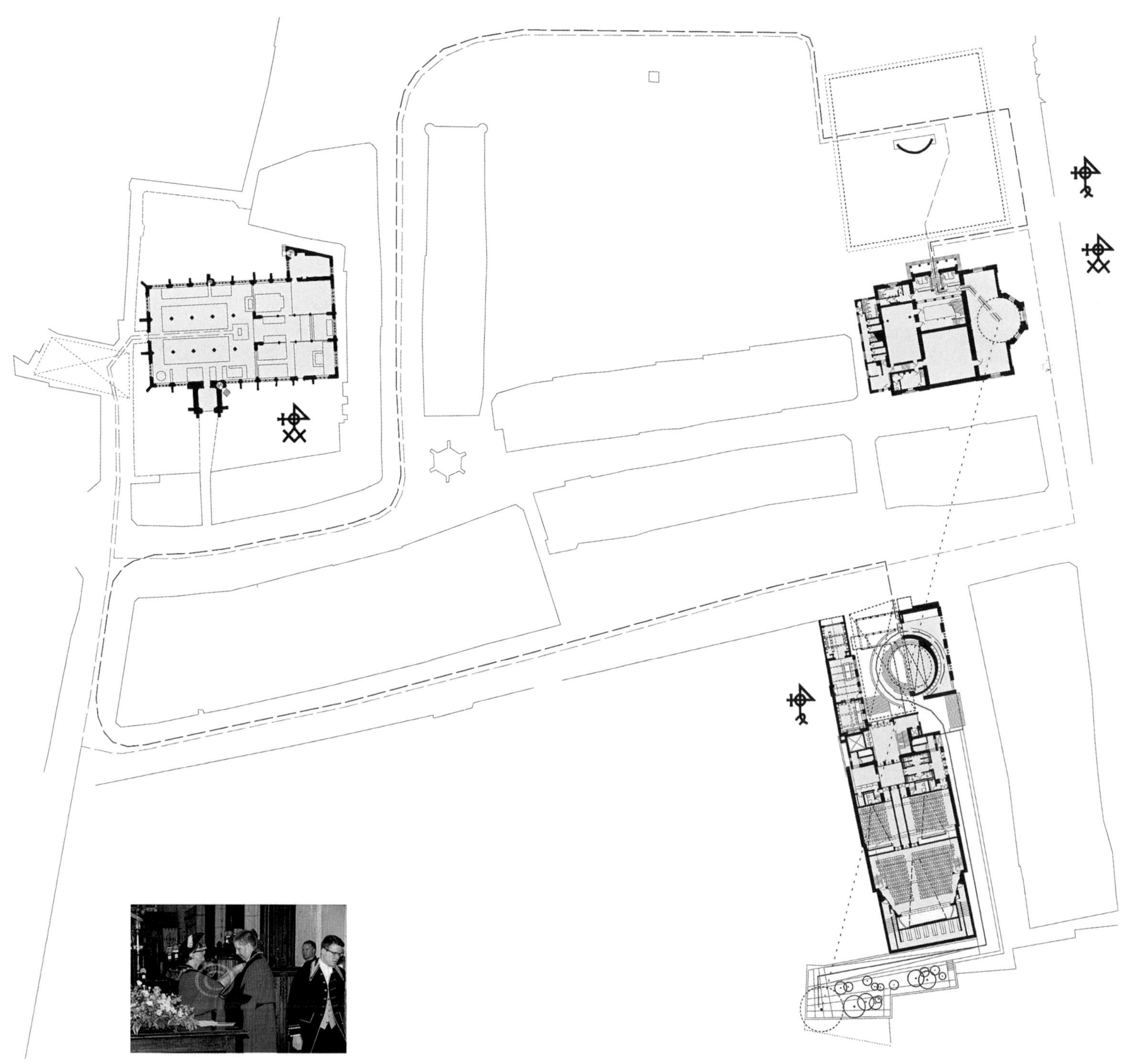

▲ - Processional Order of the City 城市秩序过程研究

▼ - Old-New Section 新旧结合处剖面图

Sectional Perspective through Chamber & The Halle of John Halle

DESIGNER-设计人: William Main

Urban Landscape Alterations

The city of Salisbury historically enjoyed an inextricable association with its surrounding hinterland. The distinction between city and landscape was blurred by a network of watercourses running through the city's streets, and water meadows populating the surrounding hinterland. The micro-programme of this proposal is a series of lido and hydrotherapy pools situated in a public garden trapped between urban and rural cultures. This scale of inhabitation forms an interface between two contrasting 'spatialities' – an immersive experience of landscape, and a contained, and anchored, civic institutional presence. The macro-programme creates a new route and threshold into the western perimeter of the Cathedral Close. This underpins a subtle reorientation of Salisbury's urban order, whilst extending the latent processional order into the wider landscape. The orchestrated movement of people along this route becomes an everyday procession which intertwines the experience of landscape and city, affording a daily reading of space.

城市景观变更

索尔斯伯里城在历史上本与它周边的内陆景观密不可分。由于沿着城市路网而蔓延开的运河网络的存在，人工景观与自然景致之间的区别变得模糊起来，而浅沼泽地则在周边到处都是。从微观角度讲，本方案意在创造一系列的露天游泳池与水疗池。它们均被设置在一处位于城市与乡村之间的公共花园内，进而在两种相互对立的“空间化”之间形成一处界面——一个是令人陶醉的景观体验，而另一个则是包容的，锚固的，市民化的机构体现。从宏观角度将，本方案旨在创造一条通往天主教堂庭院西隅的途径，一处门户。它对索尔斯伯里的城市秩序进行着微妙的重组，将潜在的过程秩序引介到更加宽广的周边景观之中。与对景观与城市的体验交织在一起，这一精心“演奏”的人民交响乐将在此成为当地日常风俗，从而使阅读空间成为一种常态。

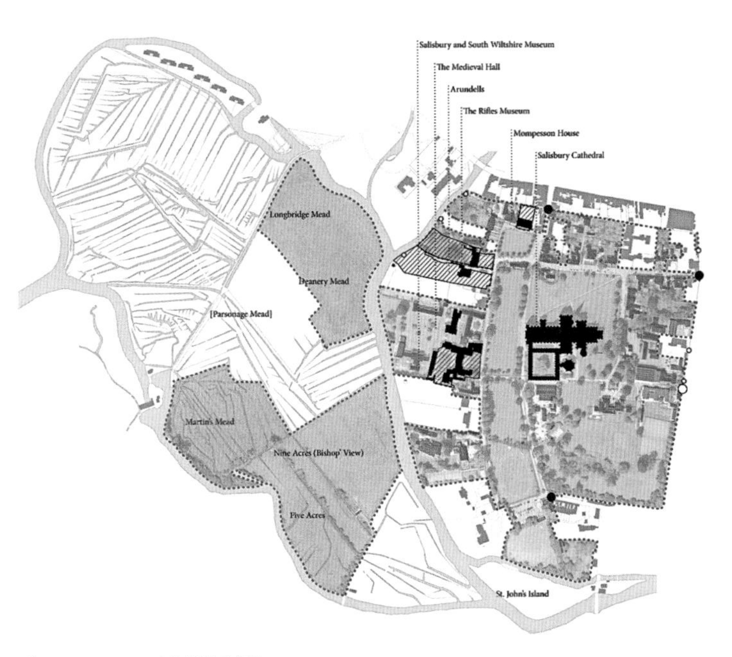

- Land Ownership 土地所有权示意图

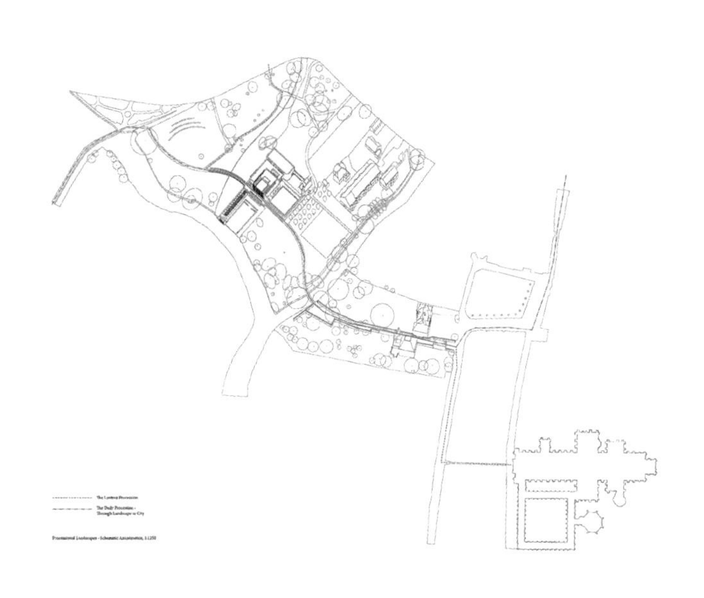

- New Western Threshold 新西隅门户空间

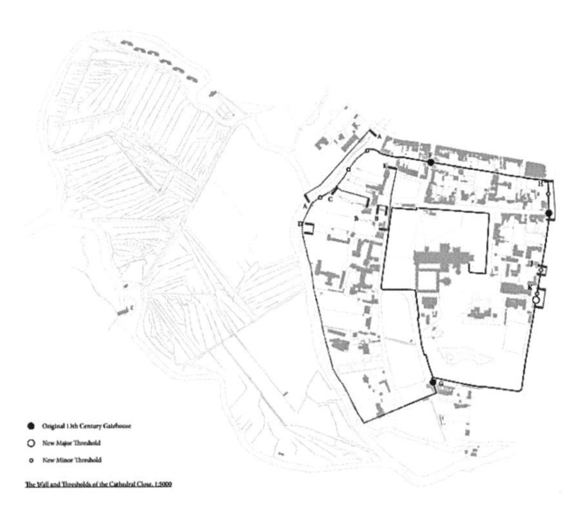

- Threshold Mapping 门户地图

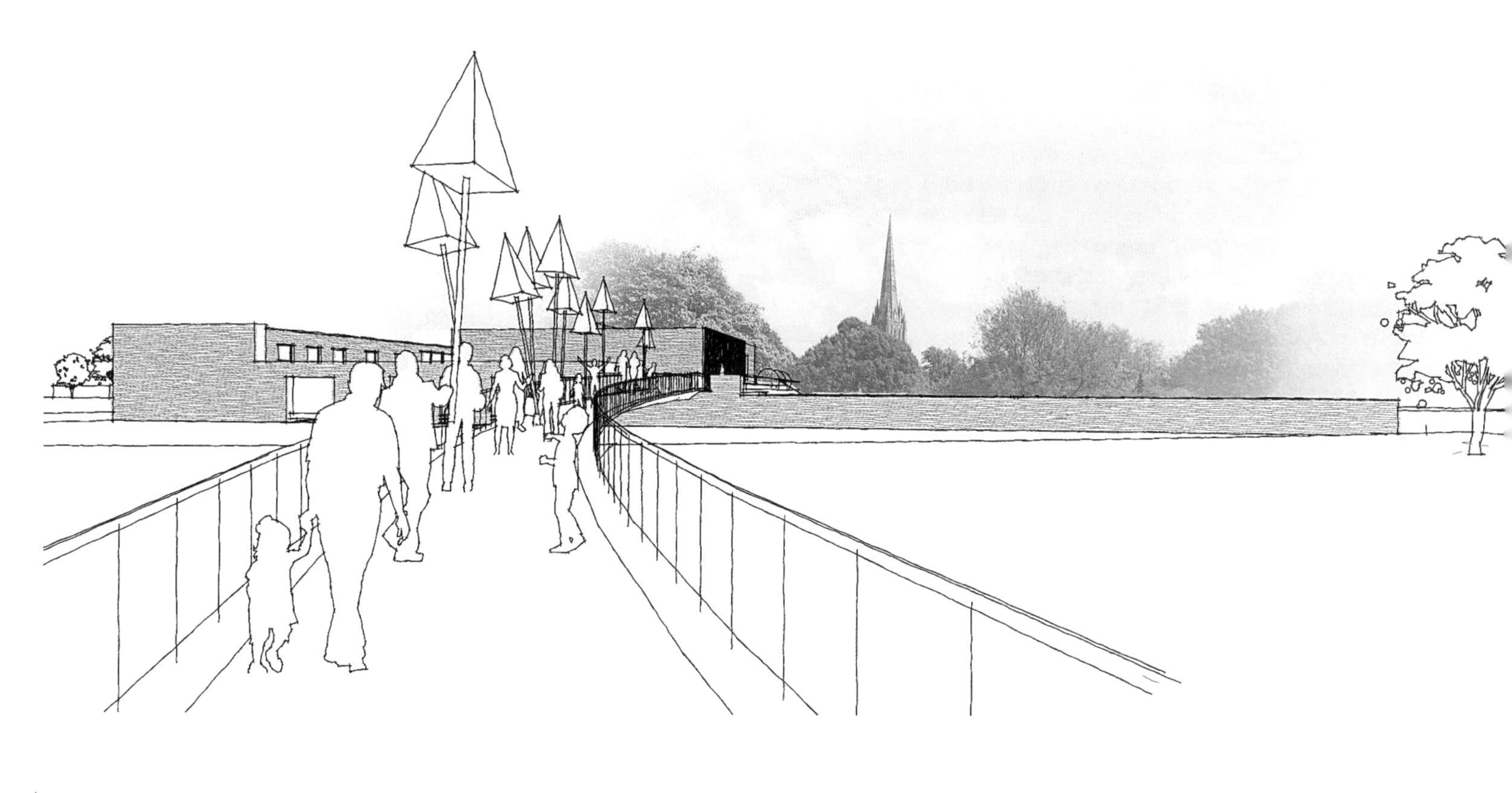

- Lantern Procession 路灯设计及透视图

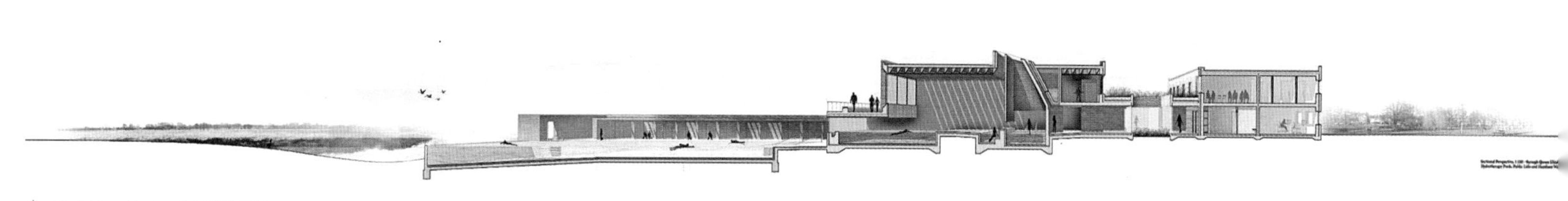

- Liquid Cross Section 亲水设施剖面图

- Threshold Photographic Studies 门户区域实景照片分析

040

DESIGNER-设计人: Nichola Finch

Theatrical Foyer

The rich, underlying processional order implicit in the foundation of the medieval City of Salisbury lies in stark contrast to the detached, isolated territory around the western fringes of the city, where, over two centuries, a collection of object buildings has accumulated. These buildings, which include a theatre and a music performance venue, fail to engage with the historic material hierarchy which once bound this Western edge into the central market place. In response, the proposal modifies existing urban and architectural structures by re-ordering the western termination of Salisbury. A brick boundary defines a new edge to the block, offering a new gateway from the railway station to the west. Relined elements adjust the perimeter of a shared box office between the two theatrical institutions, both of which are resituated on a new, unified brick surface. A ribbon of development between the existing fabric provides a theatrical foyer, a territory whose ground is shared with the city. Rather than demolish and rebuild, the proposal rearticulates meaningful relationships that create a charged, theatrical place.

剧场门厅

在那充满中世纪风情的索尔斯伯里城里，模糊地存在着一种丰富的，潜在的过程化秩序。它与围绕着城市西缘的那些相互分离的，孤立的区域形成了鲜明对比。而这些区域在以往的两个世纪中，则集中了一群拥有改造潜力的建筑。这些建筑，包括一座剧院与一所音乐演奏场，在材料方面未能与当地历史文脉相融合，从而无法将城市西边缘与中心市场联系起来。因此，本方案力图对现有城市与建筑结构进行修改，重新组织索尔斯伯里的西翼。一堵砖墙明确了街区的新边界，并提供了一处从火车站通往西区的新门户。重新组织起来的元素在两座剧院之间形成了一处共享售票厅，而两座剧院则均以新的，统一的砖墙作为外表皮加以重塑。在现有机理之间，新设计如纽带一般形成了一处剧院门厅，一处与城市共其领域的区域。并没有大肆拆毁与重建，本方案重在重新阐明了一种意义深刻的关系，从而创造出一处活力四射的戏剧化（隐喻剧院）场所。

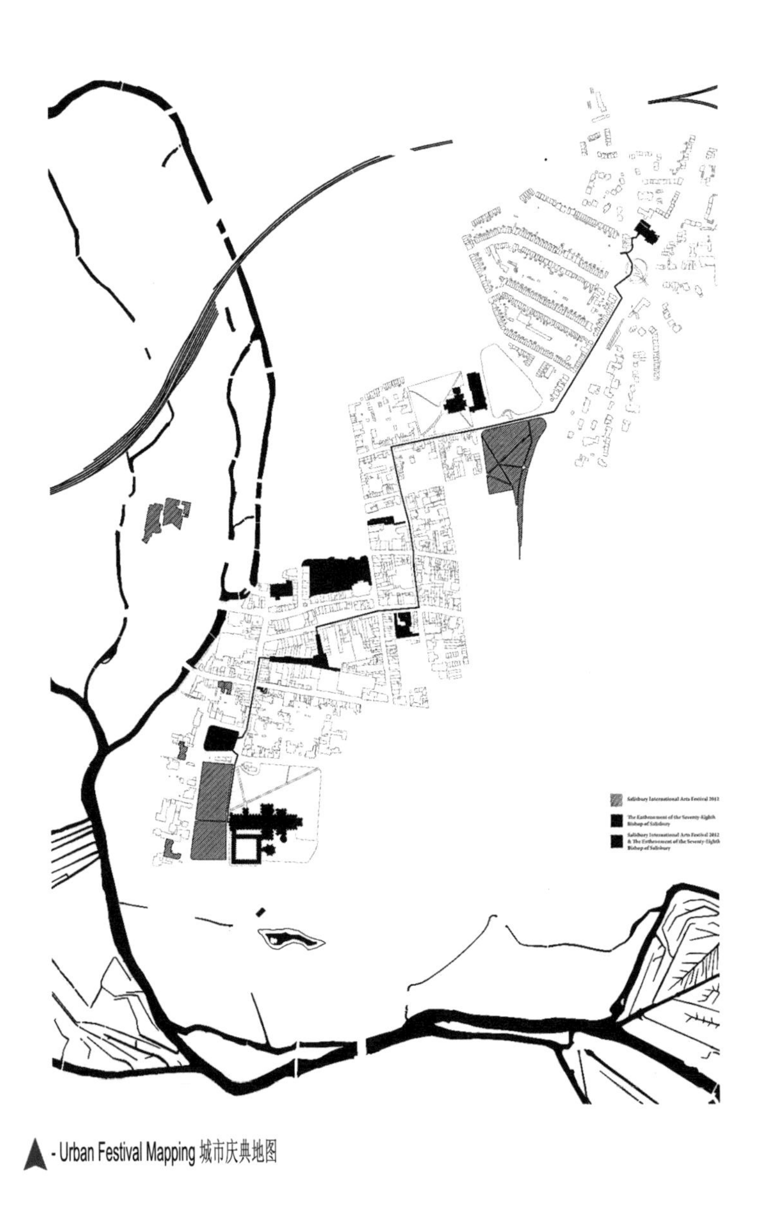

▲ - Urban Festival Mapping 城市庆典地图

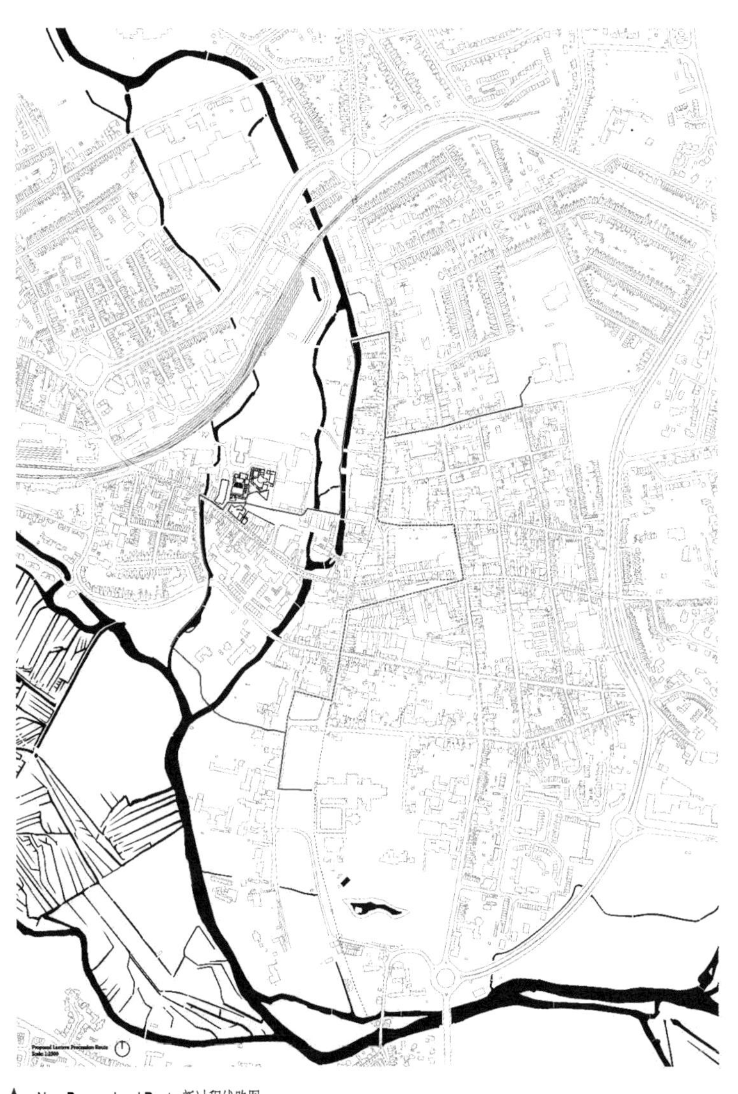

▲ - New Processional Route 新过程线路图

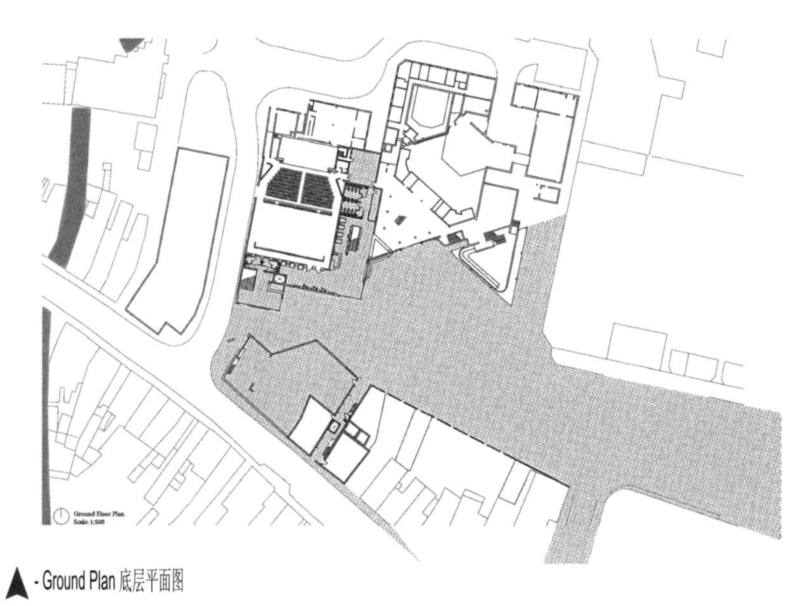

- Ground Plan 底层平面图

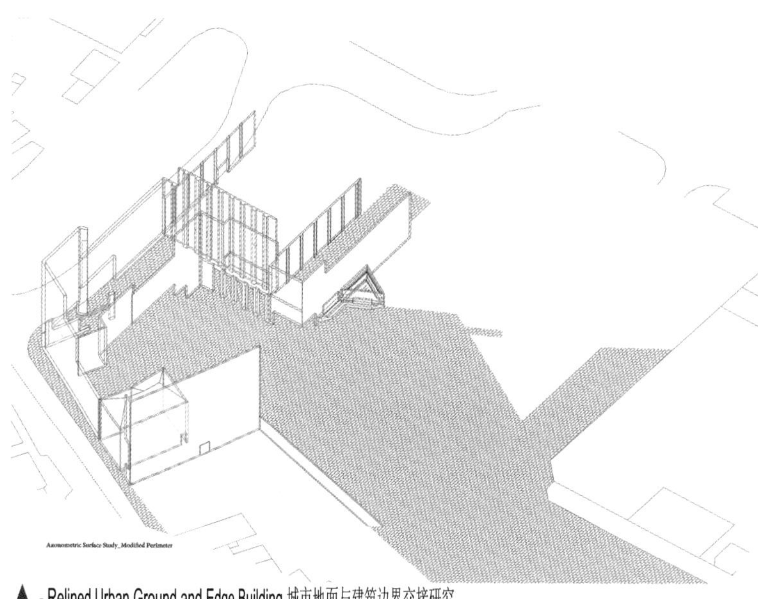

- Relined Urban Ground and Edge Building 城市地面与建筑边界交接研究

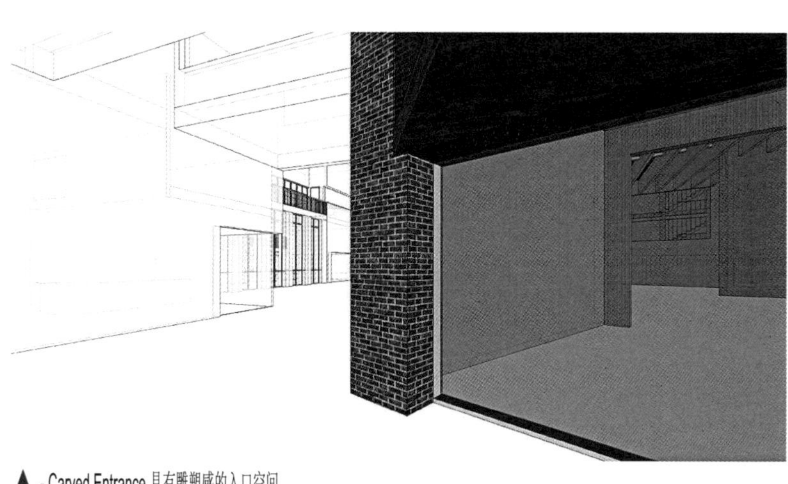

- Carved Entrance 具有雕塑感的入口空间

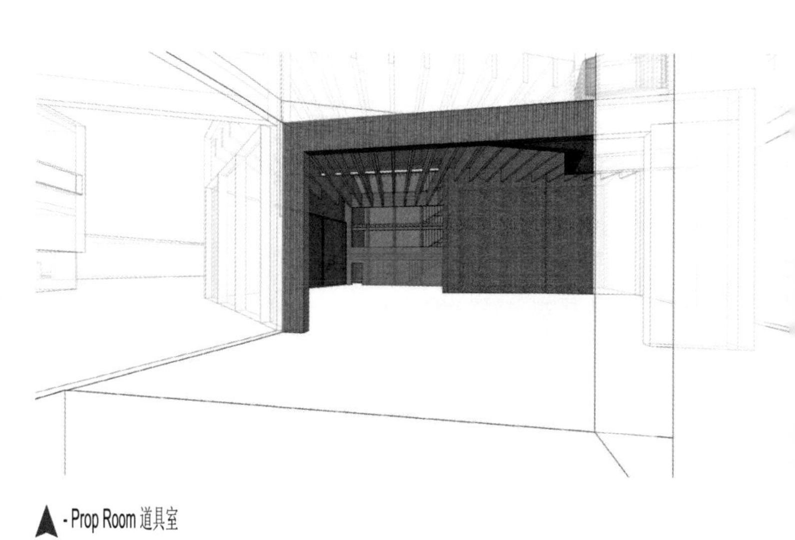

- Prop Room 道具室

- Foyer Section 门厅剖面图

DESIGNER-设计人: Toby Martin

Salisbury Cathedral Treasury

Situated within the Salisbury Cathedral, a new public intervention draws inspiration both from the Cathedral's existing figurative additions and from the processional transformation of the Cathedral within the City. By mediating between the material conditions of the stone masonry department, and the social conditions manifested in the symbolic transformation of the everyday City, the project aims to reveal the relationships between Cathedral construction, formal liturgical processions, and everyday visits. Those relationships, manifested through primary rooms such as the 'tracery' embedded above the entrance of a new formal figure, create spatial hinges between formal and informal conditions. A sense of continuity with the existing medieval fabric is created by reconstructing a space adjacent to the Cathedral's library and the cloister. This second primary room, new 'lapidarium', exhibits stone fragments accumulated over the course of the Cathedral's history. By concretising the continuity of construction within the cathedral, a new civic middle ground, between masons yard and Cathedral is established.

索尔斯伯里天主教堂宝库

本方案意图在索尔斯伯里天主教堂中嵌入一处新的公共空间。它的设计灵感既来源于教堂那富有隐喻的现有加建部分，也受到了教堂本身在城市中经历过程性转变的启发。通过中和石工部门所具有的材料特性与每天发生在城市中的象征性转变所带有的社会属性，本方案旨在展示教堂构造，正式礼拜仪式过程与每天普通游客参观三者之间的联系。而这些联系，正如镶嵌在一处新入口上方的雕花窗饰一般，存在于一些重要的房间内，进而创造出一种位于正式与非正式之间的纽带。在现有的中世纪建筑机理之中，通过对与教堂相连的图书馆与修道院围廊的改造，我们可以创造出一种连续性。而这正是另一处重要房间，一座新的“拉皮达瑞”（一种专门展示石造建筑构件与遗存的展廊），展示着在教堂漫长历史中积淀下来的各种石制建筑片段。通过在教堂中实现这一构造连续性，一种新的，属于市民的中间地带则将在石料场与教堂之间建立起来。

▼- Master Plan 总平面图

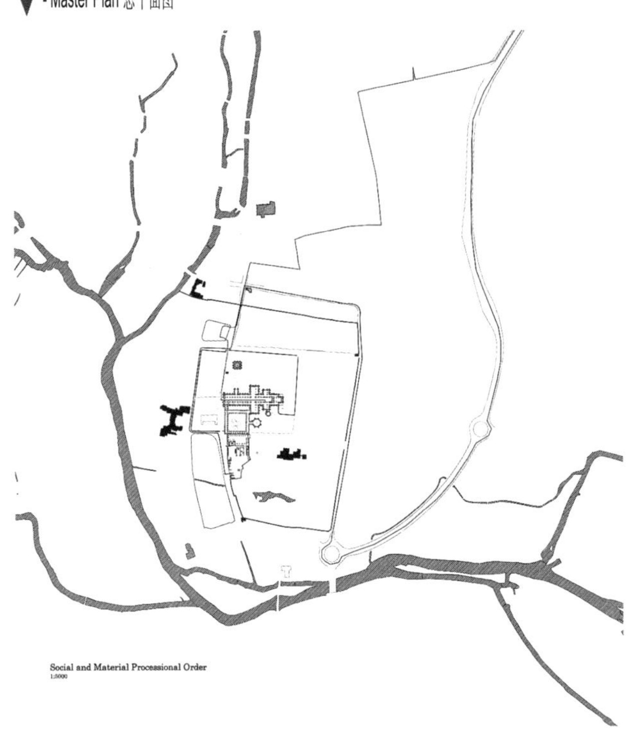

▲- Entrance 入口

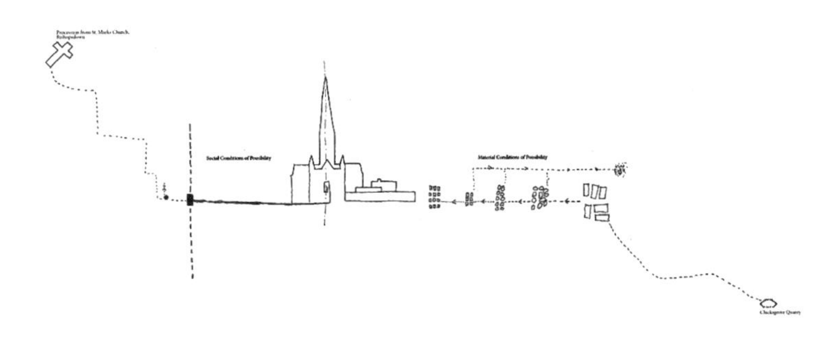

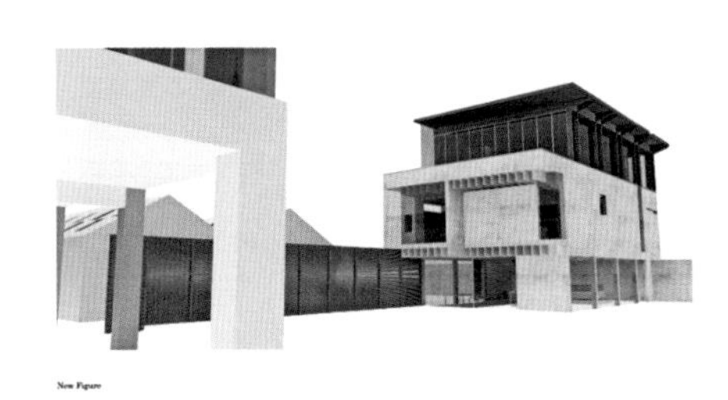

▲- Tracery Block “雕花窗饰”楼

Social and Material Conditions of Possibility
High Street and Cathedral Close Elevation
Scale 1:500

▲- Cathedral Street Section 教堂街景剖面图

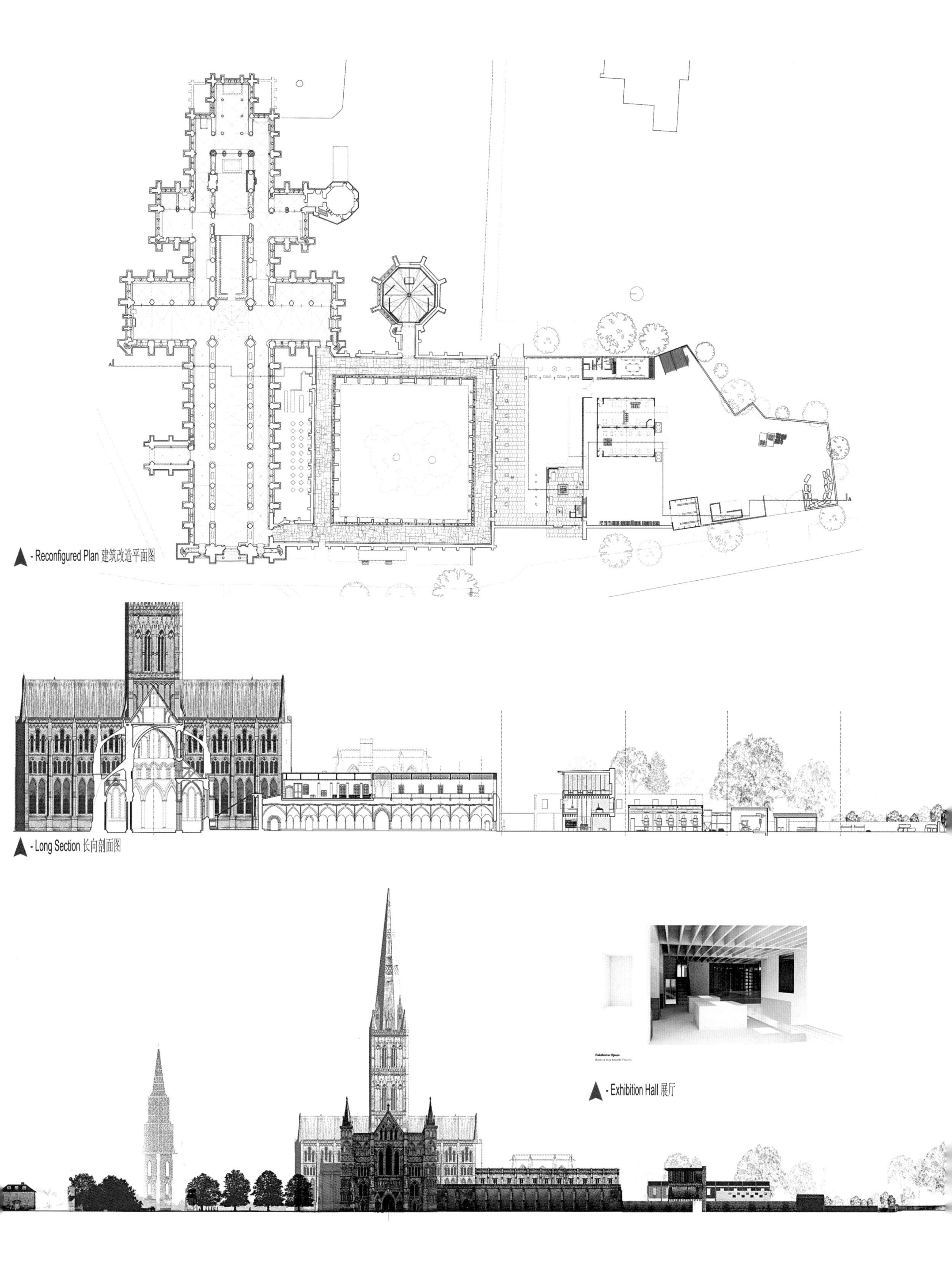

- Reconfigured Plan 建筑改造平面图

- Long Section 长向剖面图

- Exhibition Hall 展厅

—— ARCHITECTURAL ENVIRONMENTAL DESIGN
—— 建筑环境设计篇

In the last few years, the integration of sustainable and environmental issues in architecture is one of the technical aspects that have been increasingly receiving attention due to the current climate crisis. The Department of Architecture and Built Environment at the University of Nottingham has held a strong sustainability ethos, and has established itself particularly for its innovative work in the area. Teaching sustainability is only possible through a deeply embedded and widespread awareness of 'green' issues, which the department is keen to promote and attain, in the staff and student cohorts.

Architectural design is an intricate and dynamic process that must consider holistically social, economical, environmental, cultural and historical aspects. The optimisation of one facet of a project often means to compromise another, adding to the complexity of the job. Therefore the work of an architect is an interdisciplinary field drawing upon social sciences, mathematics, art, sciences, technology, politics, history and philosophy. The architect is not only expected to have an understanding of those and be able to fully and efficiently integrate them but also be able to use the same language as other project contributors such as the structural engineer or the environmental consultant.

Before Edison invented the electric light bulb and Carrier developed mechanical air conditioning, internal environmental control was mainly achieved by manipulation of the form and fabric of the building, the relationship of one space to another and the distribution of openings to allow light and air into the building. However, over the last 80 - 100 years, architects have increasingly required the environmental engineer to make good by 'brute force' what they have failed to achieve by design. Nevertheless, given a choice, most people prefer naturally ventilated and day lit buildings to mechanically ventilated and artificially lit buildings. If we add to this the environmental impact of fossil fuel dependency in buildings, there is a compelling reason to explore alternatives to mechanical conditioning, or at least to reduce our dependency. Architects and engineers must work together to regain an understanding of how building form and fabric can help to moderate the internal environment, to minimise reliance on mechanical intervention, and to create a higher quality and delightful built environment.

The work developed at our department aims to inform the sustainable practice of architecture and enhance the quality of the built environment through teaching and research in environmental science and design. The work is largely related to mitigating and adapting to the impacts of climate change, and reducing carbon emissions through appropriate building design, including testing of innovative solutions. The teaching in studio is very hands-on enabling students to develop the skills required to produce architectural designs with low environmental impact, gain an understanding of the application of technical knowledge to design, and to explore how theory and practice may inform each other. The studio design themes are kept as close to practice as possible, and frequently involve real-life projects and clients. This is the case in most of the projects featured in this section, such as the Sixth Form Centre for a Comprehensive School in Derbyshire, and the Energy Technologies Building (ETRI).

在过去几年中，受当前气候危机的影响，将可持续发展与环境因素整合入建筑设计之中的技术理念已经得到了越来越多的关注。在诺丁汉大学建筑与建造环境学院，可持续性发展理念业已蔚然成风，许多创新性工作在这里落地开花。传授可持续发展理念唯有深入地，广泛地宣传‘绿色’意识，而这也正是学院希望在教职工和学生中推广与维持的方面。

建筑设计是复杂而充满变化的过程，必须全面地考虑社会、经济、环境、文化和历史等诸多方面。在一个项目中，对某一方面的优化强调往往意味着对其他方面的妥协，而这只能使得整个工作变得更为复杂。因此，一位建筑师的工作通常是跨领域的，包含了社会科学、数学、艺术、科学、技术、政治、历史，以及哲学。我们希望建筑师不仅仅要理解并有效地整合这些学科，而且能与项目的其他工种负责人，如结构工程师或环境咨询顾问等，进行有效的交流与沟通。

在爱迪生发明灯泡，卡里尔改进机械式空调之前，建筑的室内环境通常由建筑形式和材质来进行控制，例如通过设计空间之间的相互联系与开口的布置来引入光线和空气。然后，在过去的80至100年中，建筑师开始不断地请求环境工程师通过其“蛮力”来弥补那些建筑设计无法达到的目标。可是，如果可以选择的话，相对于机械通风和人工照明而言，大多数使用者还是更喜欢自然通风与自然采光。除此之外，如果我们再考虑到因为建筑对化石燃料的依赖而产生的影响，那就足以证明探寻现有机械式环境调节系统之替代者的紧迫性，或者，至少我们应该试图去降低对化石能源的依赖程度。建筑师和工程师因此必须合作，重新专注于建筑形式和材质，探索如何借助它们来调节室内环境，进而减少对机械的依赖，创造出更加高质，更加赏心悦目的建筑环境。

我们学院所开展的工作主要通过针对环境科学和设计方面的教学与研究，传授建筑的可持续性实践，提高建筑环境的质量。这些工作，在很大程度上与三方面相关，即减轻并适应气候变化所产生的影响，通过合理的建筑设计减少碳排放，以及测试创新的设计方法。设计组的教学具有很强的实践性，要求学生们去掌握设计低环境影响的建筑，理解技术在设计中的运用，探索理论与实践的结合。设计组的题目则会尽可能的贴近实践，经常会与实际的项目和客户挂钩。本篇中的大多数项目即反映了这种情况，例如德比郡一所中学的第六阶年级（高三）中心，以及能源技术大楼（ETRI）等等。

—— **Dr. Lucelia Rodrigues**
卢塞利亚·罗德里格斯 博士

The Head of the Architectural Environmental Design Research Group
建筑环境设计研究组主任

DESIGNER-设计人: Thomas Bennett

Marine Archaeology & Sericulture

Located on a temporal coastal site, the project aims to preserve St Andrews church, Covehithe, which is threatened by the approaching coastline. Strategically, the intervention makes use of long-shore drift, to preserve land and re-configure the site over a number of decades. Wave-energy generation is built into the fabric of the infrastructure; so that as land is lost, energy is recovered. The program combines a nostalgic land-based activity (sericulture) with an explorative marine aspect (marine archaeology), at the interface between land and sea. Public exhibition spaces and galleries bring the public into the building, weaving and reconciling the two key aspects of the program. Exposing the programmatic activities to public interaction and making them accessible, this hybrid-institution takes on a lateral character. The intervention preserves and re-appropriates the existing church buildings and spaces. Elements of the existing fabric are projected into a new realm below and at the periphery of the church.

海洋考古学与养蚕业

该项目坐落在科夫海斯的一处暂时稳定的海岸线之上，它旨在保护正在受到不断逼近的海岸线威胁的圣安德鲁教堂。在设计策略上，新加建的部分将在未来数十年间，利用沿岸的冲积层来保护土地，重塑周边环境。潮汐能发电机被嵌入地层构造之中，因此随着土地的流失，能量可被源源不断的制造出来。在岸与海相交接的地方，该方案计划将怀旧的，依附大地的活动（如养蚕业）和海洋探索（如海洋考古学）联系起来。而公众则可进入建筑，来展览空间与画廊参观相关的展览，从而将计划中的这两个关键方面编织缠绕在一起。于是，借助将这些计划中的建筑行为与公众之间实现简单可行的互动，这一综合性研究中心具备了横向联系特点。这一加建项目保护并重新利用了现有教堂与空间，同时，老机理之上的元素细节也被如实“投影”在了位于教堂下部与外边缘的新区域之内。

▼ - Massing of Existing Ruin 现有遗址体块研究

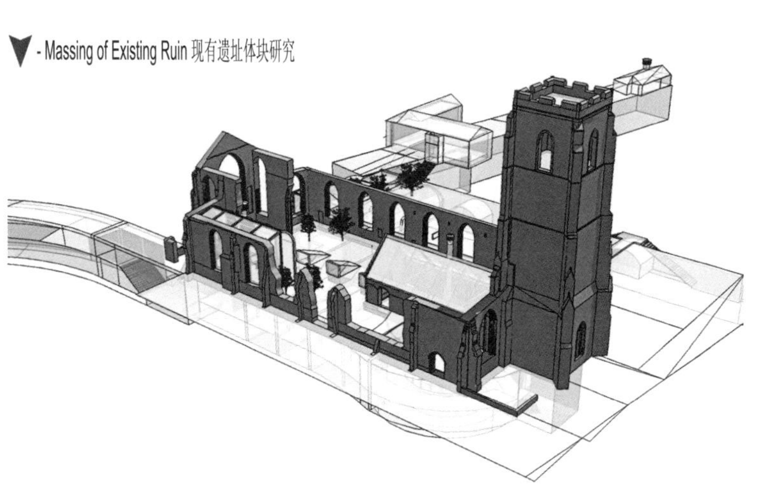

▼ - Coastal Strategy - "Archipelago" 海岸线分析 - "群岛概念"

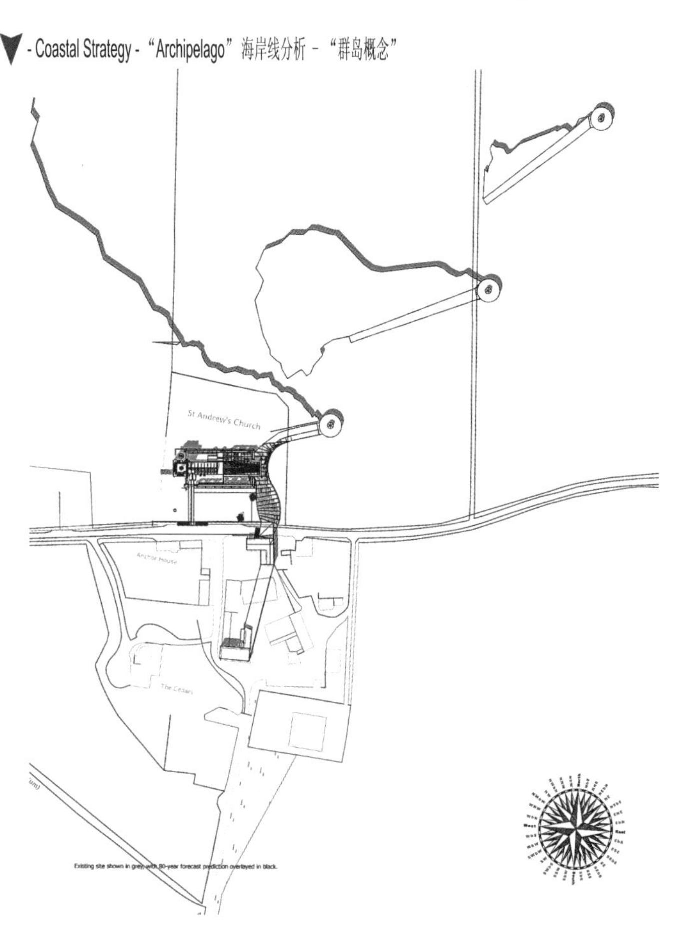

▲ - Tectonics 构造技术策略分析图

- Layers 层概念

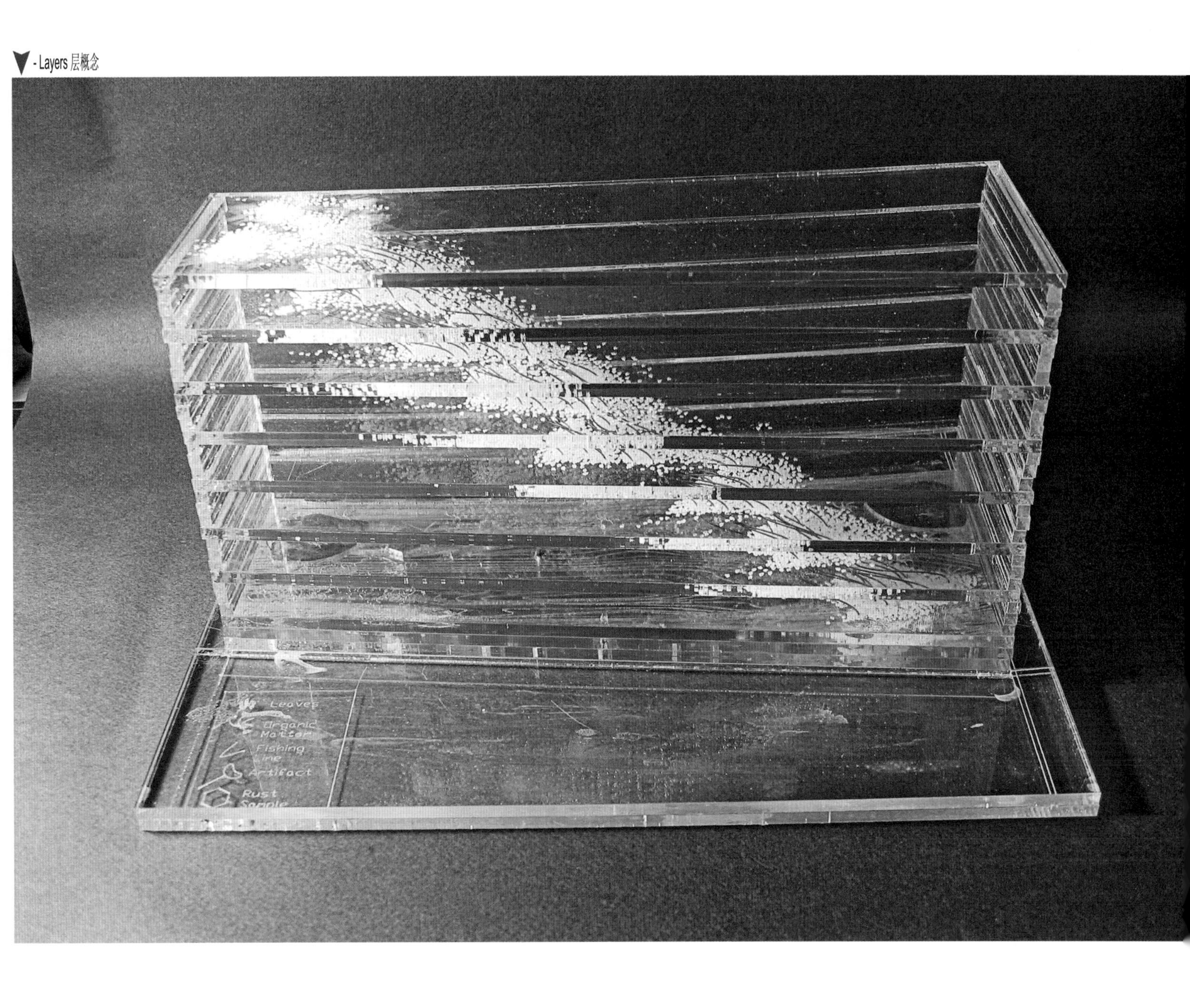

- Sectional Temporal Study of Water 模型剖面 - 示意即时水体研究

- Material Study 材料分析

043

DESIGNER-设计人: Adam Chambers, Alex Dale-Jones

Urban Wind Farm, London

Located in London's Canary Wharf, the initial concept for this design is to lean the building forward to the south, creating an element of self-shading, and thus reducing the incidence of unwanted solar gain at the hottest times of the year, when the sun is at its highest. At the tower's summit and along its corners – where wind speeds are greatest – sits a vast wind farm consisting of an array of Combined Augmented Technology Turbines (CATT) within a lightweight structure, reminiscent of early proposals for the New York Freedom Tower. These, it is hoped, would supply the vast majority of the building's energy needs. The scheme convinces with some fantastic technical studies in evaporative façade cooling (utilising water from the adjacent Heron Quay), structure and wind turbine strategies – in particular proposals for a turbine maintenance robot, used to safely access and maintain the wind farm.

城市风力发电厂，伦敦

位于伦敦的金丝雀码头，该设计的最初概念是将建筑朝南向倾斜，形成基本的自身遮阳，从而在全年最热时期，当太阳高度角最大时，减少不需要的太阳能。在塔楼的最高点和四个角——即风速最快的地方——坐落着一个巨型的，由一系列称为CATT的小型风力发电系统组成的风力发电矩阵，整个矩阵由轻钢结构支撑，整体形象让人联想起纽约自由塔的一个早期方案。我们希望这个发电厂能够提供这栋楼大部分的用电需要。这个方案也得到了来自一些极富特色的技术研究方面的支持，例如建筑立面的蒸发冷却系统（利用邻近苍鹭码头的水资源），结构和风力发电机组相结合的策略——这是一项关于涡轮机维修的机器人的特别设计，用来安全地进入风力发电厂，并维修机组。

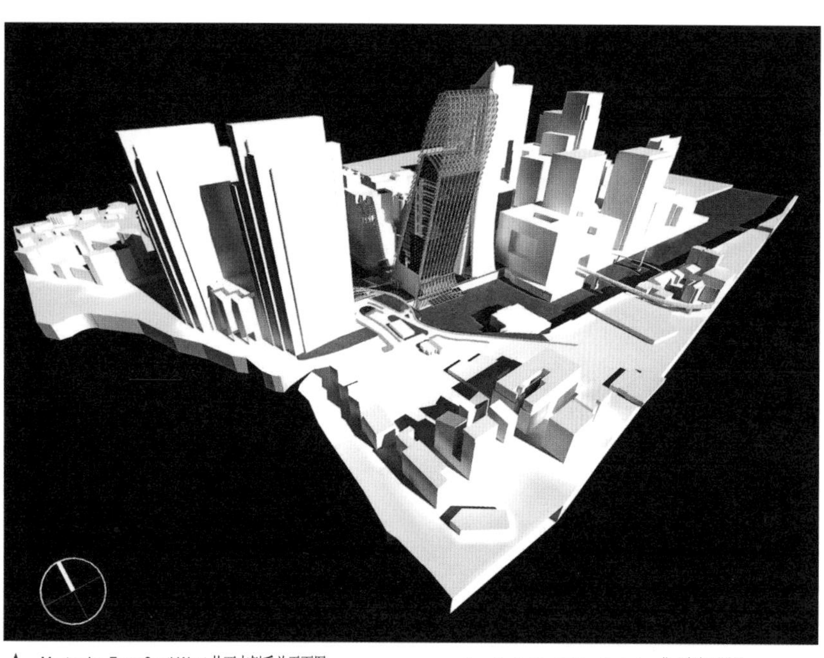

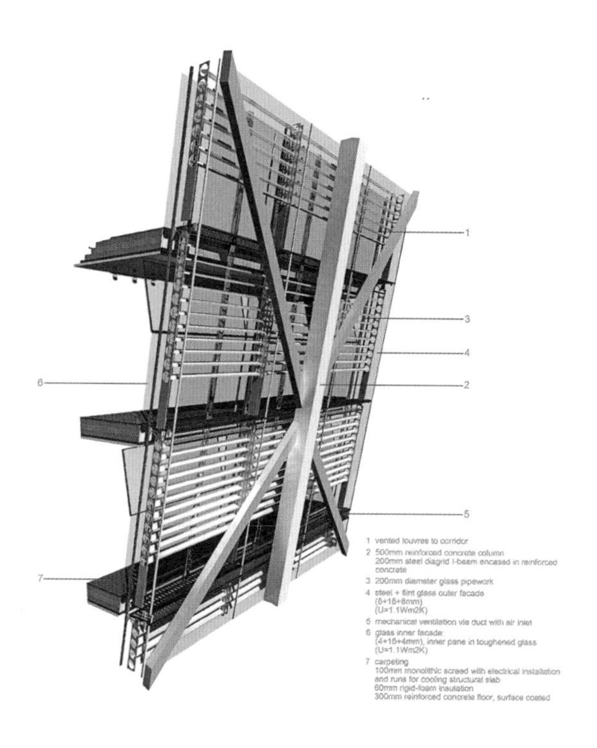

▲ - Masterplan From SouthWest 从西南侧看总平面图

◀ - North / South Facade design 北/南立面设计

▼ - Structural Strategy 结构设计分析图

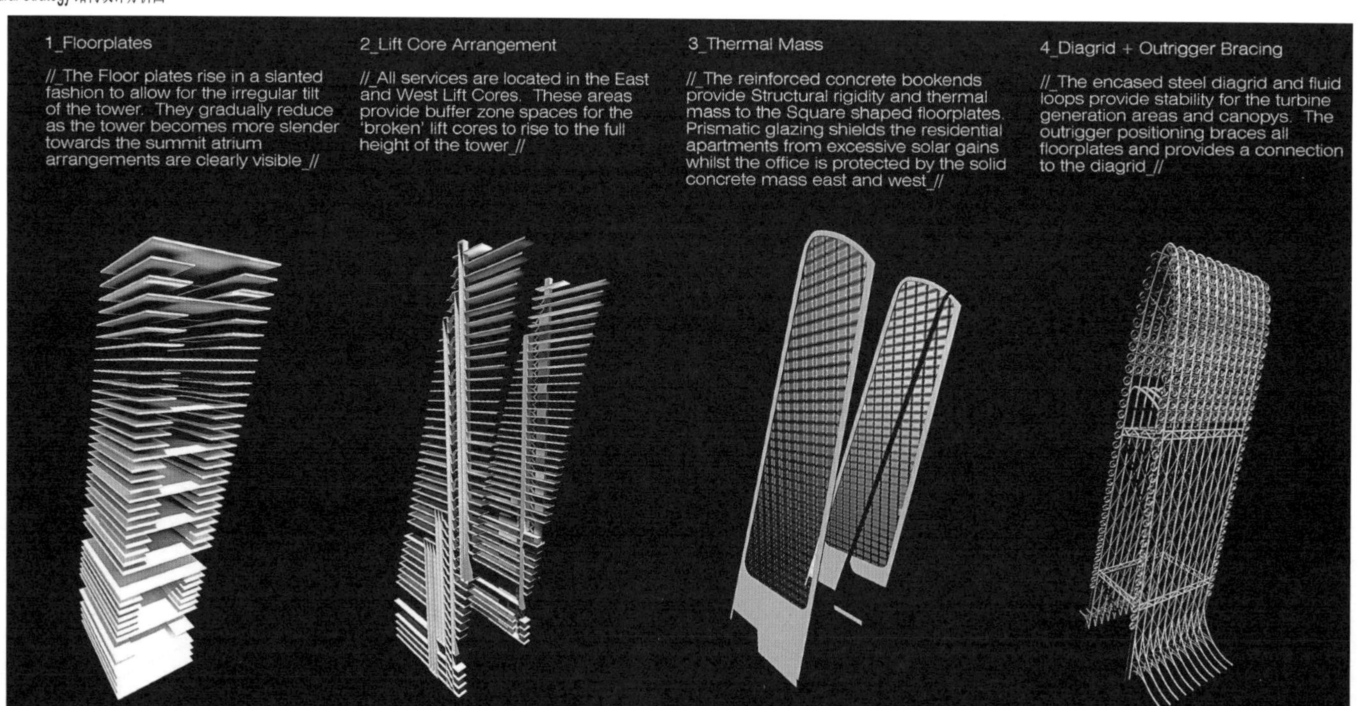

▲ - South Elevation 南立面

◢ - Bird View from South 南侧鸟瞰图

➤ - Interior Perspective of Wind Turbine Maintance Robot 风力发电机维护机器人内部透视图

▼ - Bird View from West 西侧鸟瞰图

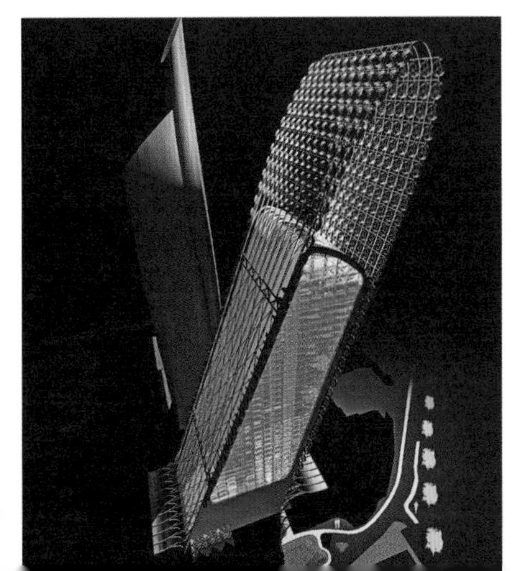

DESIGNER-设计人: Minh Ngoc Phan

Stacked Courtyards, Abu Dhabi

The Courtyard House is a key residential typology in the Middle Eastern region, allowing occupants access to a semi-open space that is shaded from the harsh desert sun and wind, whilst also maintaining their privacy. Acknowledging the environmental and cultural importance of this building type, this project aims to reinterpret the Courtyard House within the high-rise realm. The result is a series of stacked six-storey villages, each centred around a semi-private courtyard. Large Mashrabiya screens shade these spaces, but still allow the cooling breeze from the adjacent Persian Gulf to penetrate. As the building rises, the courtyards twist towards the north, maintaining the best views of the Corniche. Wrapped around the courtyard spaces are residential apartments and these too are inspired by local living patterns – many are planned for extended families with each also having access to a smaller, private courtyard space. Externally the building façade emphasises the courtyards, with glazing restricted to continuous narrow slots for shading purposes, resulting in opacity rather than transparency dominating.

堆叠式庭院，阿布扎比

庭院式住宅是中东地区主要的住宅类型，这种院落所形成的半开放空间，既可以让住户避免沙漠中的烈日狂风，又可以保持其私密性。基于对这种建筑类型的环境与文化重要性的尊重，该方案旨在在高层领域对庭院式住宅进行新的解读。最终方案是一系列的六层叠落式村落，每个中心都有一个半私密的庭院。巨大的马什拉比亚格栅窗不仅可以为这些空间进行遮阳，也欢迎来自附近波斯湾的习习凉风。随着建筑的升高，庭院转到北向，保持欣赏海滨的最佳视角。庭院周围的空间是居住单元，这些套房也是受当地生活模式的启发——许多是为大家庭设计的，并同时也为家族中的个体小家都提供了更小的，私密的空间。建筑的外表面也起到强化庭院的作用，为了遮阳，窗户玻璃部分被限制为连续的窄条，并且使用了乳白色半透明的材料，不是全透明的普通玻璃面。

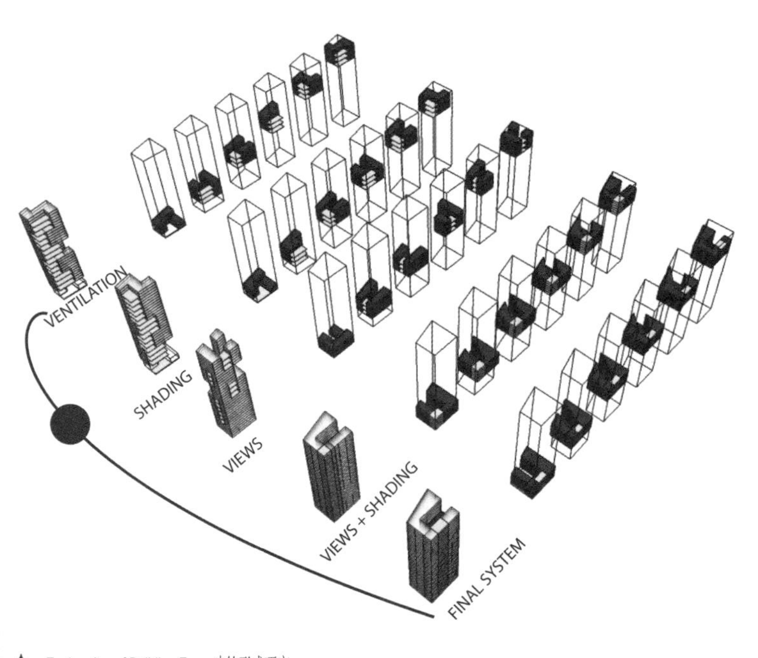

▲ - Exploration of Building Form 建筑形式研究

► - External Perspective 外部效果图

▼ - Typical Plan 标准层平面图

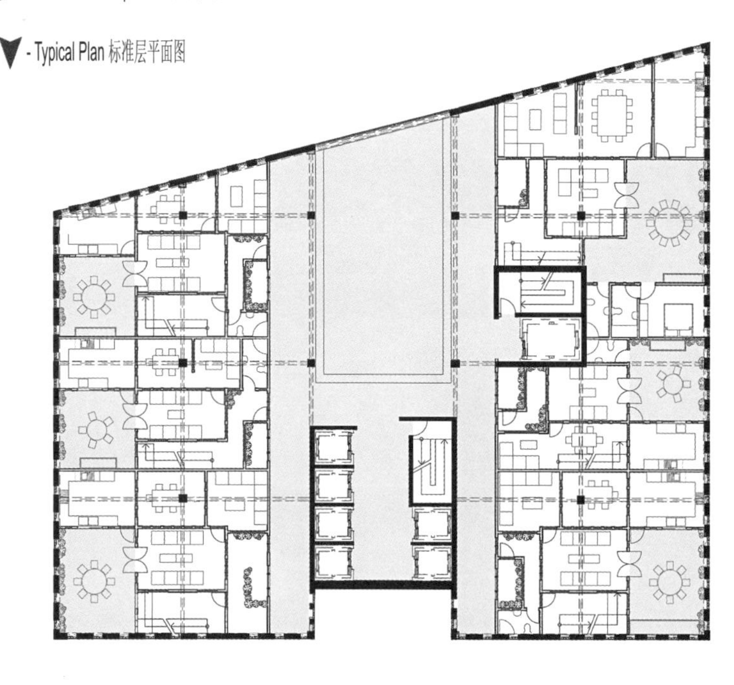

▲ & ▼ - Communal Courtyard 公共庭院

DESIGNERS-设计人: Matthew Bryant, David Calder, Amrita Chowdhury

Leamouth Masterplan and Skybridges, London

The 2010 'Vertical Communities' studio envisages a new urban vision for Leamouth in East London consisting of eight mixed-use tall buildings linked together by a groundscraper at the base and a skybridge network at height. The groundscraper is conceived as a green extension to the Lea Valley 'eco corridor', tying the site into its surroundings. A new public realm is created on a site-wide inhabitable green roof, with all building services and parking located below. A skybridge network is created at height between the towers, lifting the public domain into the sky and connecting all the key public facilities within the buildings. Each tower has its own vertical public transportation linking the ground floor interface with the skybridge network above, through shuttle elevators, escalators or ramped green pathways. The skybridges are designed as modular units that can be easily assembled on site, with prefabricated 'retail pods' suspended along their length. The northern-most skybridge includes a vast green roof, creating a public skypark some 200 metres above the city.

利茅斯总体规划和天桥，伦敦

2010年的“垂直社区”设计组设想了一个全新的城市景象，那就是将位于东伦敦利茅斯的八座综合大楼在底部用裙楼，顶部用天桥相连。底部的裙楼可以作为利亚谷（Lea Valley）“生态走廊”的一个延伸，将基地与周围紧密联系起来。一处面积如基地一般大的怡人绿色屋顶变为了新的公共空间，其下面设有各种建筑辅助设施和停车场。天桥网络位于每个塔楼的上部，将公共场所提升到空中，并将每个建筑重要的公共设施联系了起来。每个塔楼都有单独的垂直交通系统，通过竖直电梯、自动扶梯，或是绿色坡道，将地面与天桥网络相连。为了简便地在基地中组建，天桥遵循了模块化设计，而一些预制的“豆荚小商店”则沿着长桥悬在空中。最北边的天桥包括一个巨大的绿色屋顶，形成了一个离地约200米的空中花园。

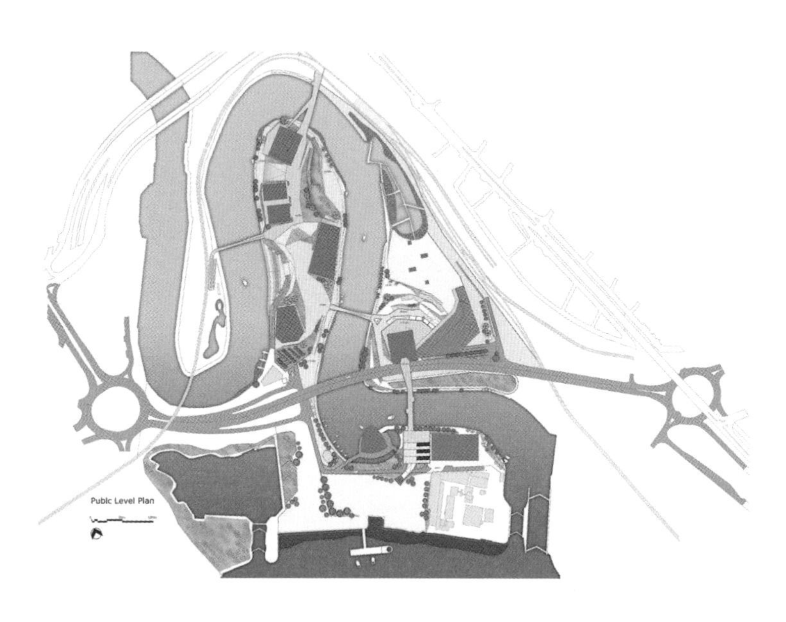

▲ - Model of Building Facade and Skybridge 建筑立面与天桥模型

▲ - Masterplan, Consisting of Eight Interlinked Tall Buildings 总平面图，包括8栋相互联系的高层建筑

► - Model of Public Skypark 200 Metres Above Ground 距地面200米高的公共空中花园模型

▼ - Skybridge Network and Vertical Transportation Systems 天桥网络与垂直交通系统

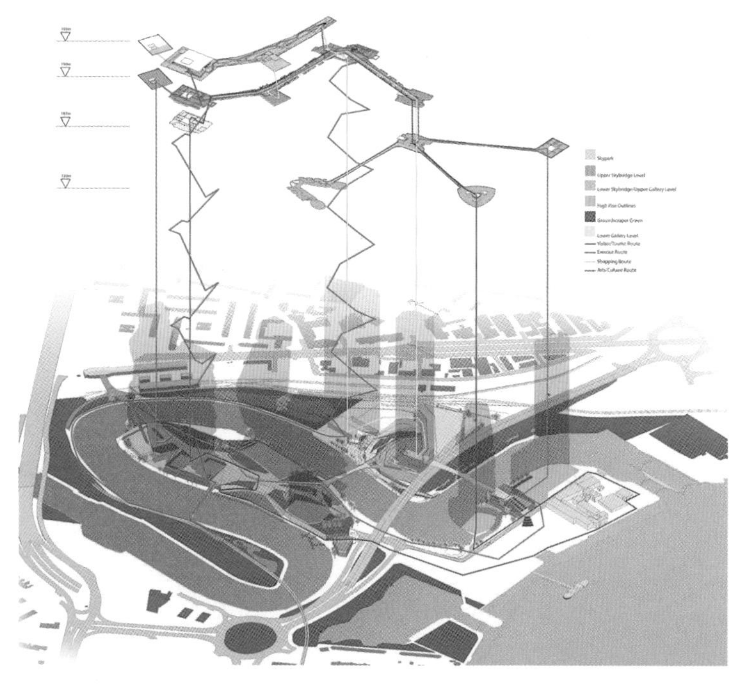

YEAR 5 & MASTER STUDENTS PROJECT 2009-2010 —— 五年级学生与硕士研究生作品 2009-2010

TUTORS - 指导教师: *Philip Oldfield, David Nicholson-cole*

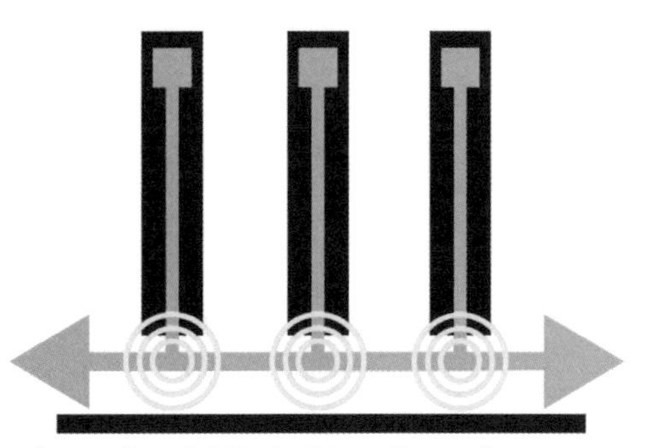

Standard High Rise Development

- Unlinked group of isolated towers
- Ground Level Stimuli
- Movement flow blocked at ground level
- Wayfinding ability reduced
- Lack of social integration between towers
- Lack of maintenance in dilapidation spiral
- Unrest and miscontent creates crime
- Low property values

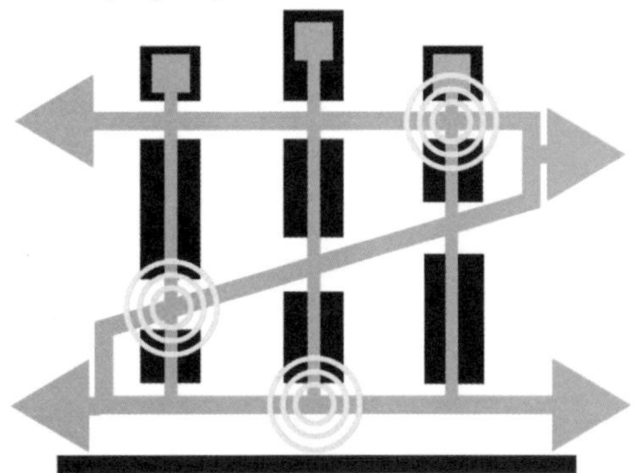

Proposed High Rise Community

- Linked towers at multiple levels
- Multi Level Stimuli
- Interdimensional movement
- Wayfinding ability increased
- Integration of residents and visitors
- Social interaction across development
- Maintenance of environment
- Low crime
- High property values

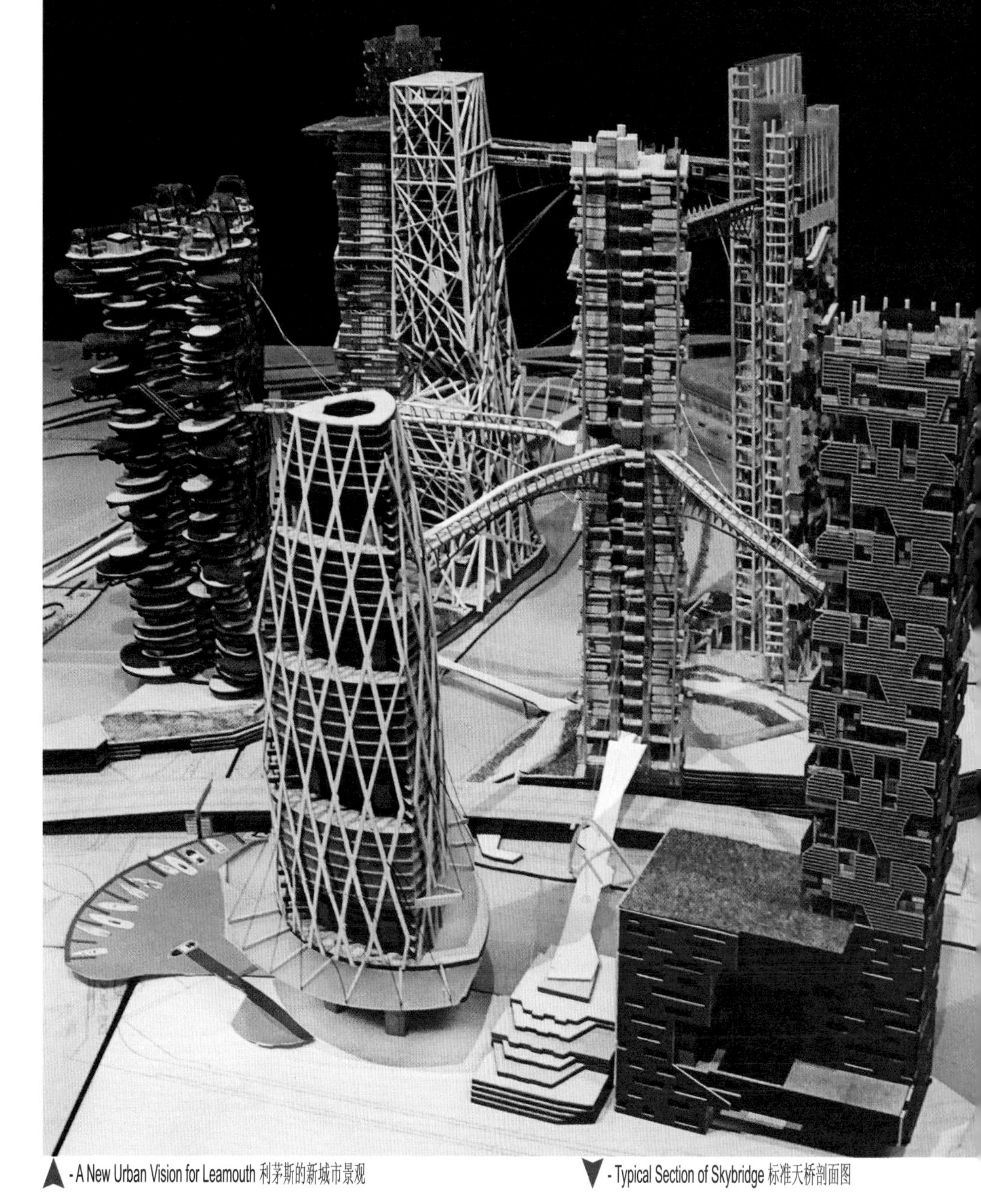

▲ - A New Urban Vision for Leamouth 利茅斯的新城市景观

▼ - Typical Section of Skybridge 标准天桥剖面图

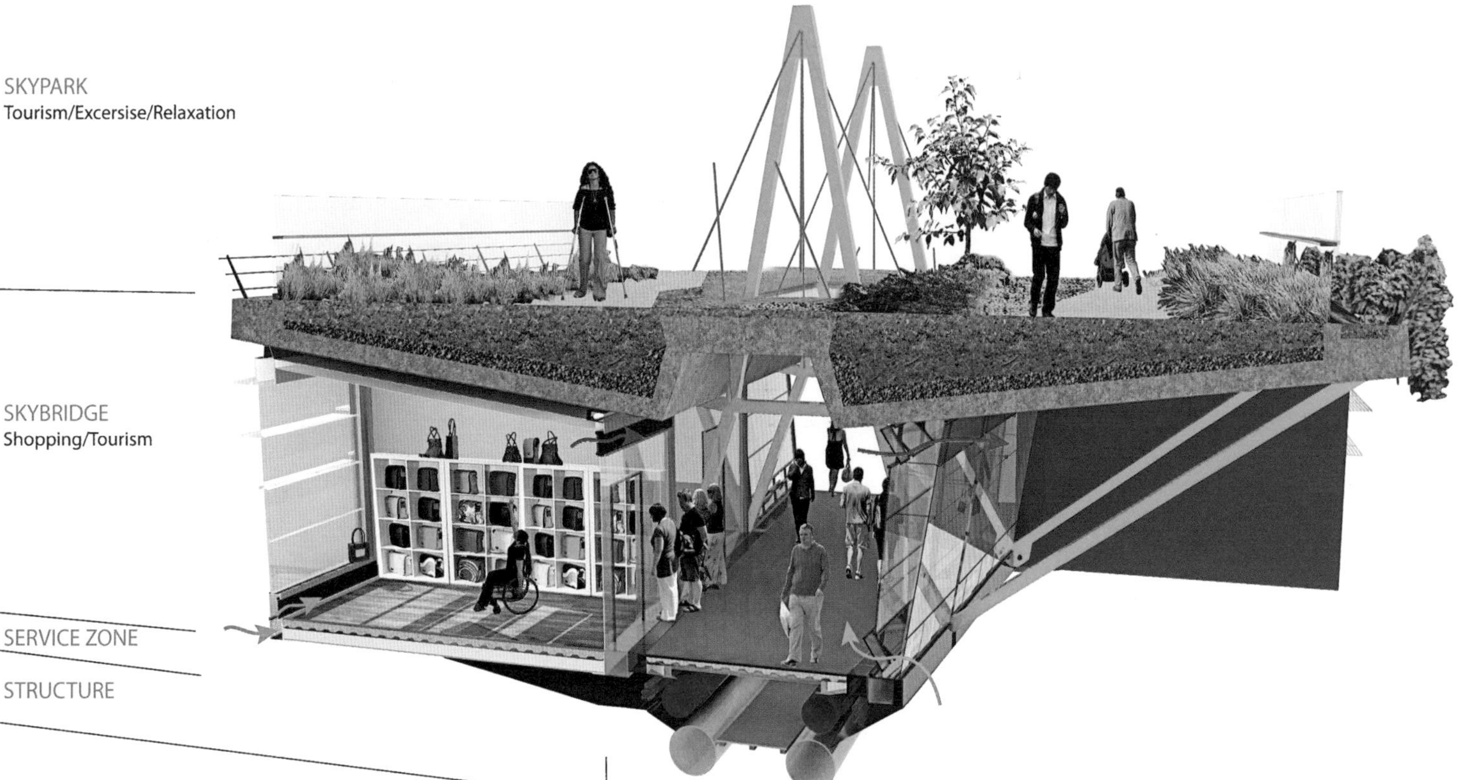

DESIGNERS-设计人: Chandni Chadha, Arham Daoudi, Elnaz Eidinejad, Akshay Sethi

Colour My Thames, London

This design aims to create a vertical residential community bound together by aspects of colour and art. Rather than restrict the tower to the visual monotony and environmental failings of typical curtain-walling, the design aims to integrate colour and vibrancy into the building in a meaningful way, beyond just the aesthetic. This is achieved by 'mapping' the colours of the surroundings onto each façade. Thus the south façade takes on the primary colours of the adjacent Container City, the blues and greys of the Thames, the white of the O2 Dome and the reds, browns and greens of residential communities beyond. In this way, each façade reflects the context which it faces in an abstract manner. Timber louvers on the façade provide shading and privacy to the residents, but also serve to create a dynamic, ever changing elevation as blocks of colour are exposed and hidden as the louvers are opened and closed throughout the day and night.

为我的泰晤士添彩，伦敦

这个设计希望打造一个充满色彩和艺术的垂直住宅社区。为了避免出现空洞乏味的塔楼，破坏环境的幕墙，我们的设计会通过一种有意义的方式把色彩和活力整合到整个建筑中去，而不仅仅是出于美学考虑。实现这个想法的途径是把周围的颜色"贴图"到每块立面表皮上。比如南立面主要由邻近集装箱城的主色调构成，还有泰晤士河的蓝色和灰色，02运动场穹顶的白色，以及周围居住小区的红色，棕色和绿色。通过这样的方法，每个立面都抽象的反映了它所面对的语境。木质的百叶窗不仅为住户提供了遮阳和私密性，而且产生了一种不断变化的立面，因为在白天和夜晚，随着百叶窗的打开和关闭，会相应地显露或隐藏不同的色块。

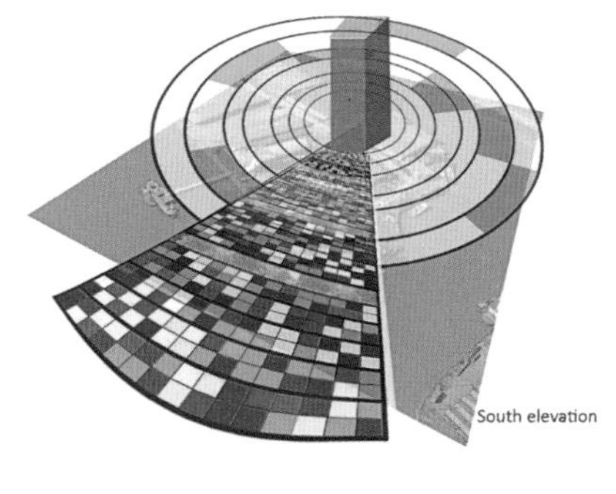

▲▼ - Sections 剖面图

➤ - Colour Mapping of the South Facade 南立面色彩分析图

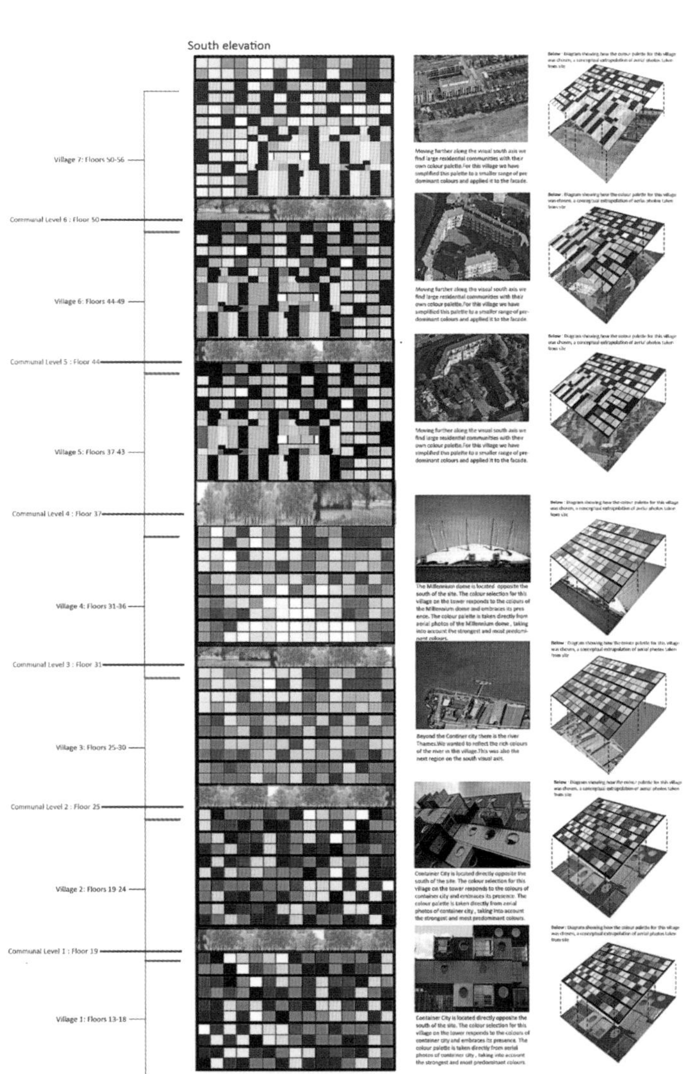

TUTORS - 指导教师: Philip Oldfield, David Nicholson-cole

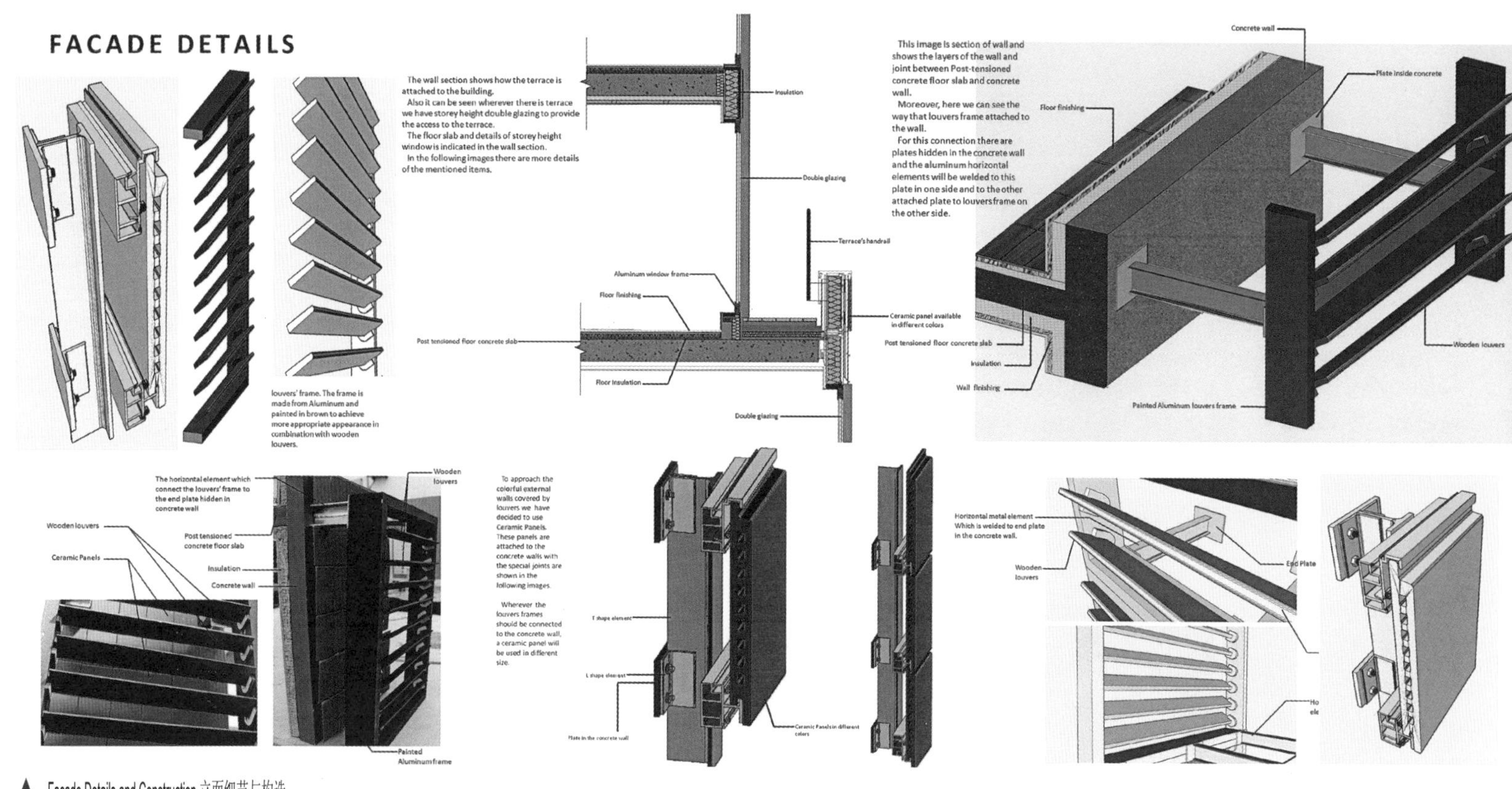

- Facade Details and Construction 立面细节与构造

- Facade Detail 立面细节

- Physical Model 实体模型

047

DESIGNERS-设计人: Kok Keong Tew, Linshou Wang

Adaptable Tower, Rotterdam

This project is influenced by a number of factors. The first is the desire to create a tall building that is adaptable over its lifetime, such that it can change function, density and even height if required. The second, inspired by the city's vast container port, is the use of modular and prefabricated technologies to facilitate this adaptability. Finally, the project aims to acknowledge the unique aesthetic of the Dutch De Stijl movement. The resultant design consists of a steel mega-frame which acts a structural framework within which prefabricated and part-fabricated accommodation spaces are supported. These can be removed, extended and rearranged to cater for future scenarios that we may not be aware of yet, such as changes in market forces, changes in demographics, etc. It also allows for occupants to extend their apartments into vacant grids as their family grows, rather than having to move to a larger house out in the suburbs.

可变之塔，鹿特丹

这个项目受到许多因素的影响。首先，我们希望这个高层在使用期间是可以进行改变的，例如在需要的时候，它的功能，密度，甚至是高度都可以变化。其次，受到城市中巨大的集装箱码头的启发，模数和预制技术的优点得到充分应用来实现高层建筑的可变性。最后，这个项目也想表达出荷兰风格派运动的独特美学观点。整个项目将一座巨型钢构架作为基本结构框架，以此支持位于其中的预制和半预制的套房空间。这些都可以根据我们所不能准确预见的未知情况进行移除，扩展和重新组合，例如根据经济变量，社会人口结构改变等等。它同时也允许使用者将套房的面积扩大到空置的网格中，来应对家庭成员的增加，而不必被迫搬到郊区的一处大房子中。

▲ - Masterplan 总平面图

◄ - Section Perspective 剖透视图

▼ - Typical Plan 标准层平面图

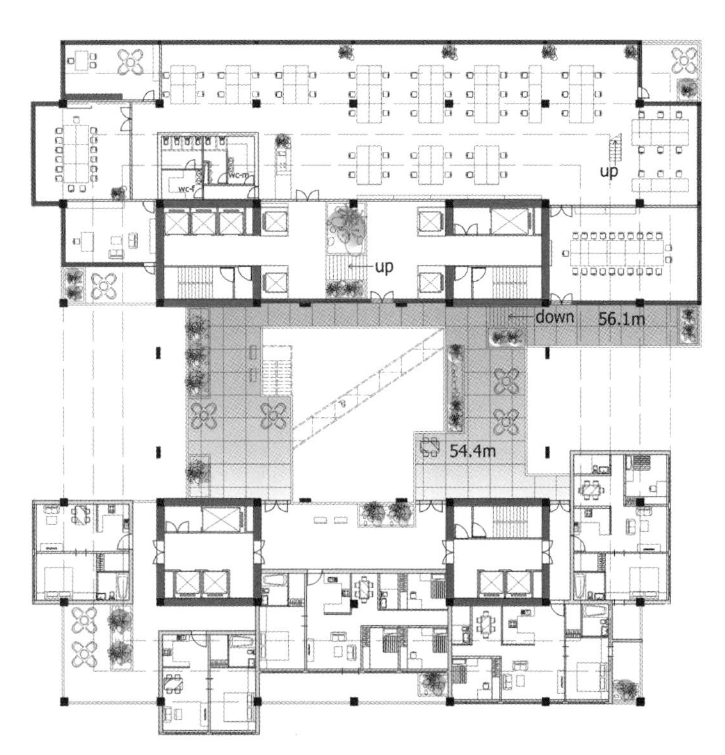

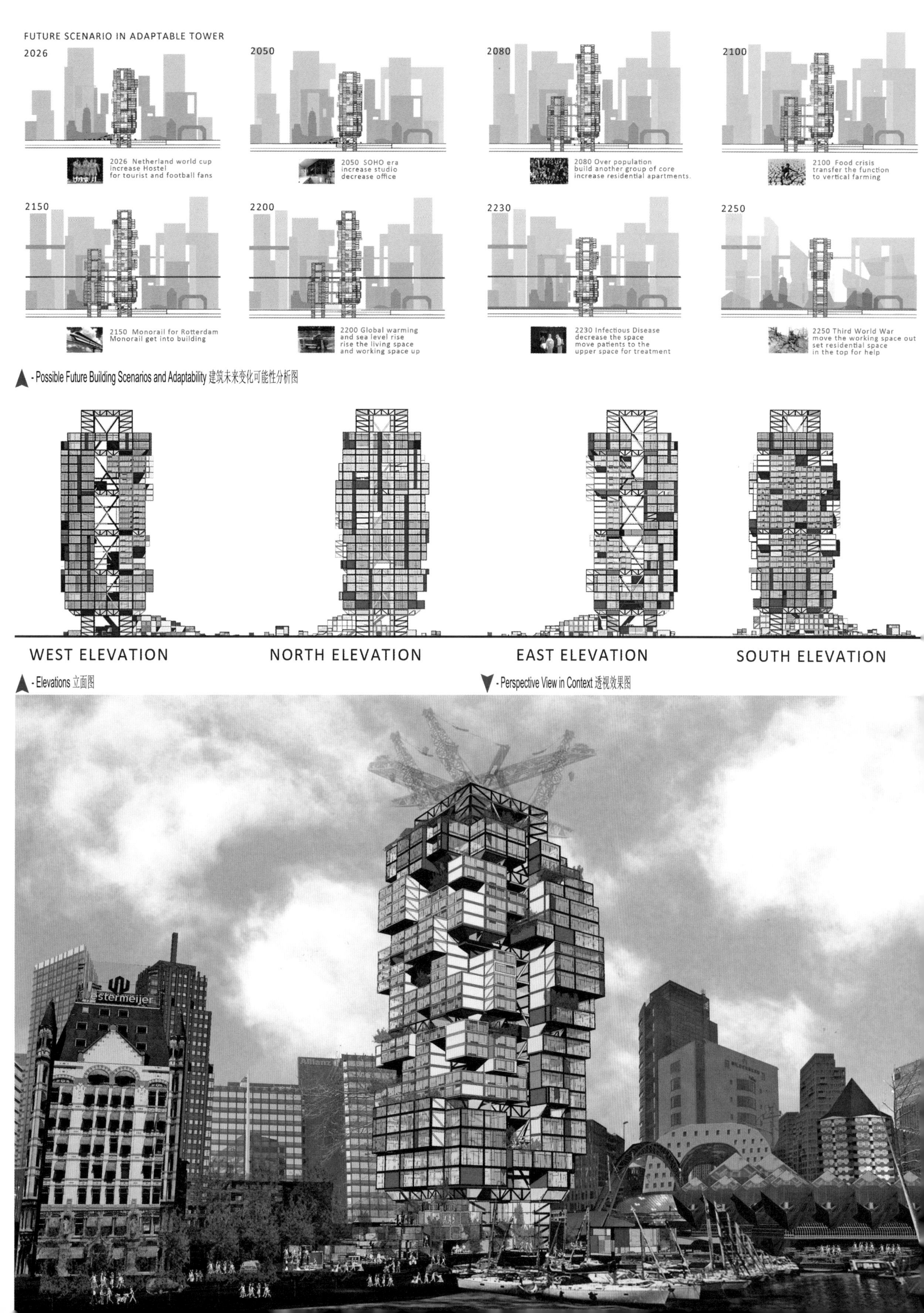

- Possible Future Building Scenarios and Adaptability 建筑未来变化可能性分析图

- Elevations 立面图

- Perspective View in Context 透视效果图

DESIGNER-设计人: Samuel Holt

Vertical Kampong, Singapore

The Kampong House is a vernacular architectural typology, typical of South East Asia, and designed to respond to the social and environmental characteristics of the region. Once common in Singapore, the 20th century has seen the typology reduced to just one village in the whole of the country. Inspired by its response to local climate and culture, this scheme aims to reinterpret Kampong living in a contemporary and vertical manner. The result is a design which whilst modern, embodies many of the principles of the Kampong. Firstly, the building is orientated with its short sides facing east – west, where the sun is hottest, and incorporates large balconies and overhangs to minimise unwanted solar gain. In section it is broken down into a series of vertical villages, each with its own communal / recreational spaces such as gardens, child care facilities, exhibition and performance spaces, sport spaces, etc. Apartments are designed with flexibility and comfort in mind with sliding doors and folding shutters allowing for units to be opened up, whilst also promoting cross ventilation.

垂直村落，新加坡

底层架空的杆阑式建筑是一种在东南亚相当普遍的传统建筑类型，它非常好地适应了当地社会和环境特点。尽管也曾经在新加坡随处可见，但到了20世纪，整个国家就只剩下一个村子依然保留有这种类型的建筑了。受其适应当地气候和文化的启发，本方案计划用当代的，垂直的建筑方式来重新诠释杆阑屋生活模式。最后完成的塔楼虽然体现了现代风格，却包含了许多杆阑屋的核心理念。首先，建筑的短边朝向东西两面来减少东西晒的影响，并结合宽大的阳台和出挑来尽量遮挡不需要的光照。在剖面上，它分散成一系列竖向的村落，每个村落都有自己的社区和休闲空间，例如花园、儿童保育设施、展览和表演空间、运动场所等。公寓设计尽量考虑了灵活性和舒适性，布置了滑动门和折叠式百叶窗，使得每个单元均可以自由打开，使得凉风可以穿堂而过。

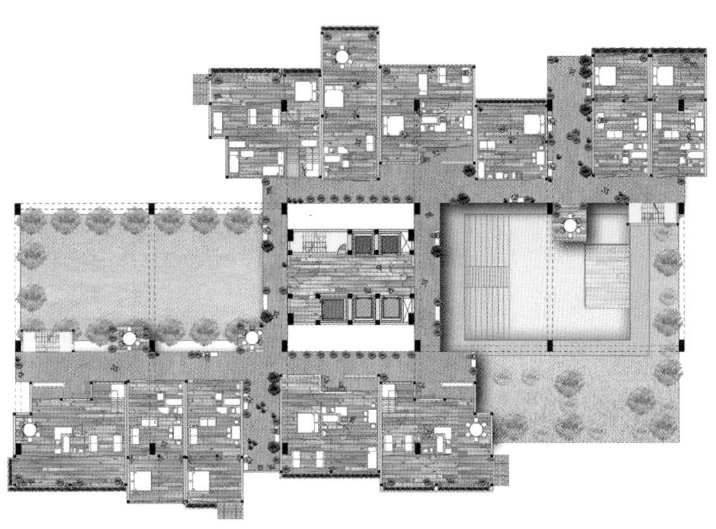

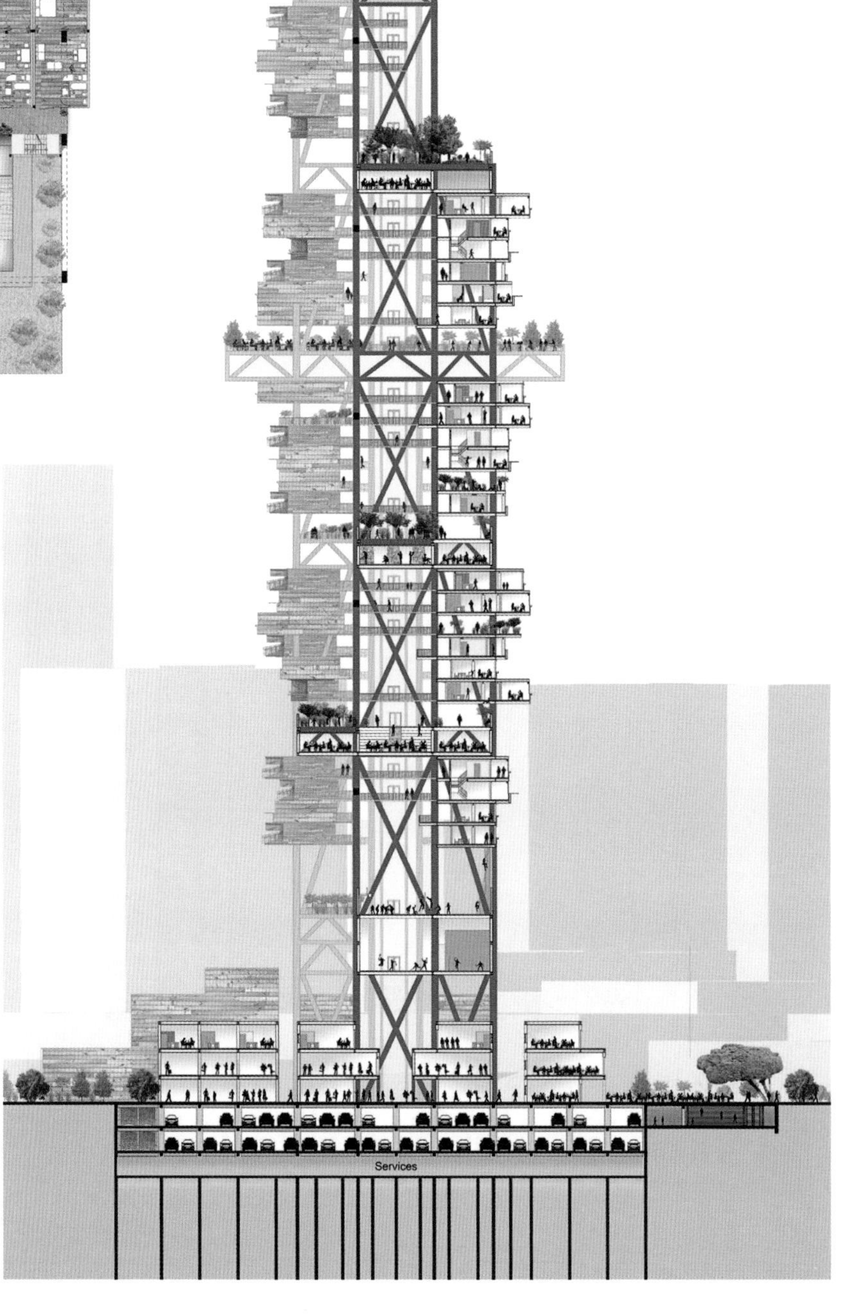

▲ - Typical Floor Plan 标准层平面图

► - Section 剖面图

▼ - Typical Apartment Design 标准公寓设计

Typical flat: sliding doors closed

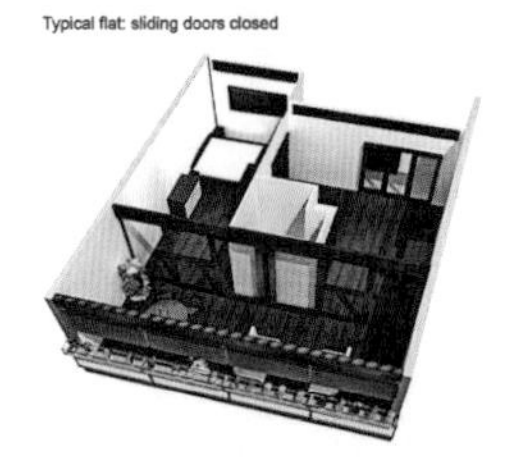

Typical flat: sliding doors open

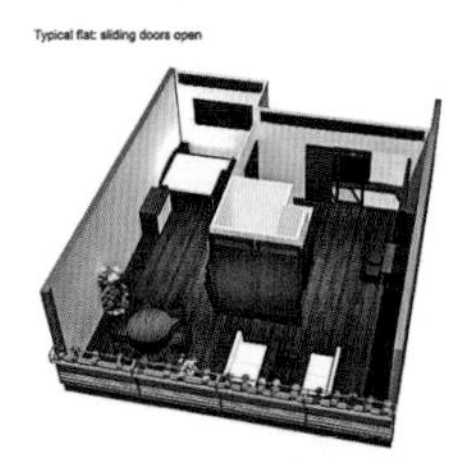

▲▲ - Village Internal View 村落内景透视图

▲ - Physical Model 实体模型

➤ - External View in Context 外景透视图

049

DESIGNER-设计人: Fei Qian

Solar Parking Tower, Abu Dhabi

This design aims to address the problem of excessive use of private fossil-fuel powered cars. It envisions an Abu Dhabi of the future, where residents move about the city in all-electric autonomous pod cars which are stored in huge vertical car pools. The tower itself acts as one such car pool, accommodating 200 apartments and 750 vehicles which are parked using automated platforms. This vertical parking strategy would provide maximum parking density for the city, freeing up the hundreds of ground-level car parks that currently litter Abu Dhabi for future building development, thus creating a denser more concentrated city. In addition, the tower also provides the energy for all the cars through the integration of 19,000m² of photovoltaic panels on the south, east and west facades, making the most of the high solar incidence of the city. These panels also serve to shade the north-facing apartments and the dramatic full-height central atrium where residents, scientists and visitors alike would watch the automated car-parking process occur.

太阳能停车塔，阿布扎比

本设计旨在解决当下过度使用燃烧化石能源（石油）的私家车所产生的各种问题。它设想在未来的阿布扎比，居民们将使用纯电动车在城市中活动，而平时这些车则存放在一个巨大的竖向停车场里。塔楼本身就是一个停车场，有200户公寓和可供750辆交通工具停放的自动平台。这个竖向停车场计划可以为城市提供最大密度的停车空间，从而可使数百个正在阻碍阿布扎比未来发展的地面停车场得以解放，并最终建立起一个更紧凑的城市。另外，这个塔可以通过整合在南、东、西三个方向的立面上的19000平方米太阳能电池板为所有的机车提供电力，进而成为这个城市太阳能利用率最高的地方。这些太阳能板同时可被用作遮阳板，为北向公寓和中间那处惊人的全通高巨大中庭提供阴凉。而在中庭里，住户，科学家和游客则可观看自动停车场的工作，运行过程。

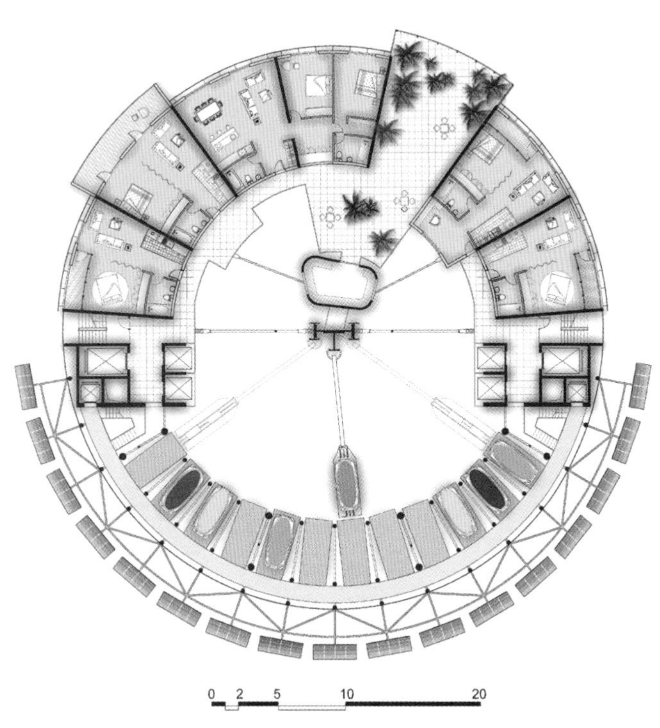

▲ - Typical Plan 标准层平面图

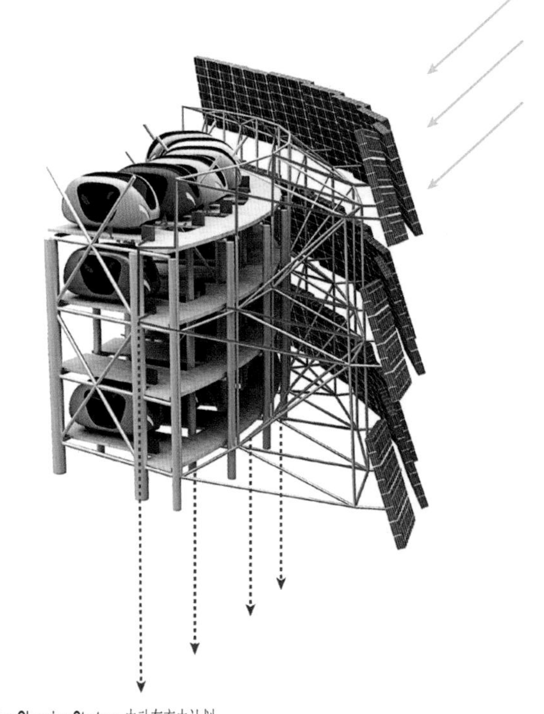

▲ - Electric Car-Charging Strategy 电动车充电计划

▼ - Exploded Diagram of Building Layers 建筑分层分析图

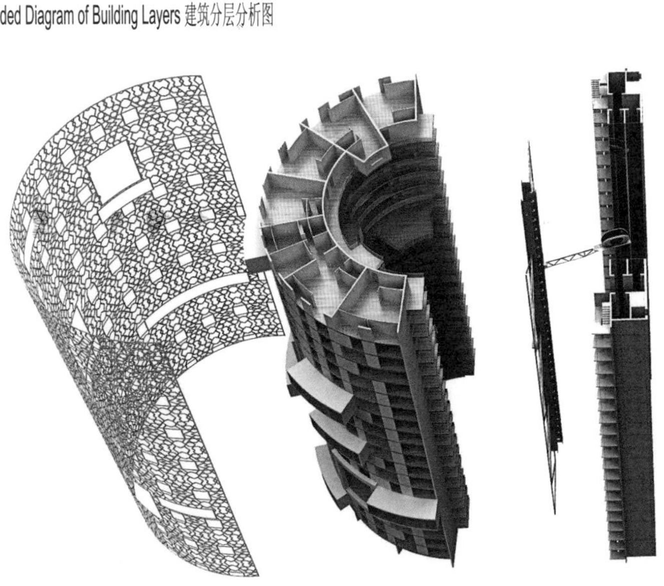

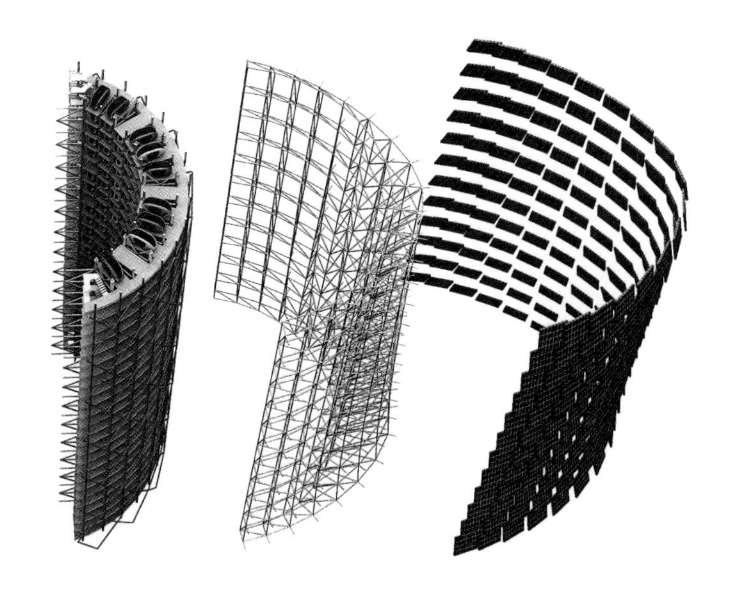

▲ - A New Urban Vision for Abu Dhabi, With Solar Parking Towers
太阳能停车塔为阿布扎比天际线增添新景观

◀ - View in Context 城市景观透视图

▶ - Section 剖面图

▼ - View up the Parking Atrium 仰视停车塔中庭

DESIGNERS-设计人: Samuel Holt, Shreela Sharan, Willie Yogatama

Manhattan Sky Podium, New York

This project aims to lift the vibrancy and activities of the ground into the sky via flying escalators and skybridges linking together existing building rooftops. Where these links meet, some 115 metres up, a vast 'podium in the sky' is created, accommodating a host of public facilities typically found on the ground – a skatepark, outdoor restaurants, gardens, an ice skating ring in the winter and more – but with the added benefit of spectacular views over the city and the Statue of Liberty beyond. The sky podium spans three main towers (two of which are merely structural legs), with a long, slender residential block rising above, orientated south to maximise passive solar gain and for the best views. In the floors beneath the podium a student hotel and fashion hotel are accommodated, which benefit from south-facing rooms with large atria for deep daylight penetration. The office block is located at the lowest level, shaded by existing buildings, but with good access to indirect day-lighting from the north.

曼哈顿空中平台，纽约

设计者试图利用空中自动扶梯和天桥来连接原有高层建筑的屋顶，来把地面的生气和活动转移到空中。在115米高空，这些相互交汇于一处巨大空中平台，这里不但容纳了地面通常所拥有的公共设施——滑板运动场、室外餐厅、花园、冬天的滑冰场等等——而且还可以眺望整个城市和旁边的自由女神像。这个天空平台跨越了三个主塔楼（其中两个仅仅是结构支撑），平台上还承托一个细长的住宅楼，它面朝南向来获得最大的太阳能和最佳视角。平台底下还设有一间学生旅店和一间时尚酒店，宽大中庭与围绕其布置的南向房间使得阳光可以深入建筑内部。办公区位于最低的楼层，这些区域不但有原有建筑为其提供遮阳，而且可以获得舒适的北向天光。

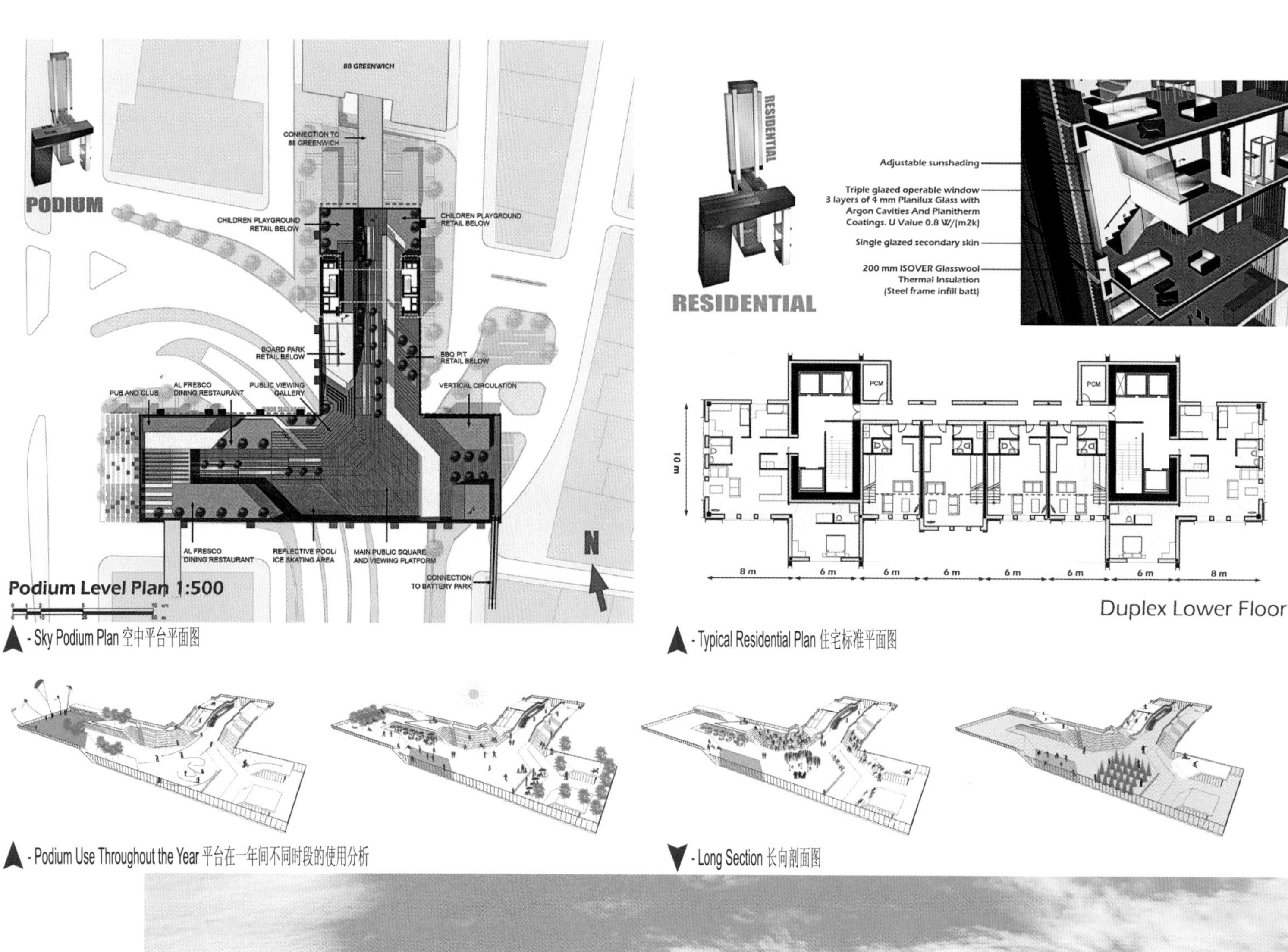

- Sky Podium Plan 空中平台平面图

- Typical Residential Plan 住宅标准平面图

- Podium Use Throughout the Year 平台在一年间不同时段的使用分析

- Long Section 长向剖面图

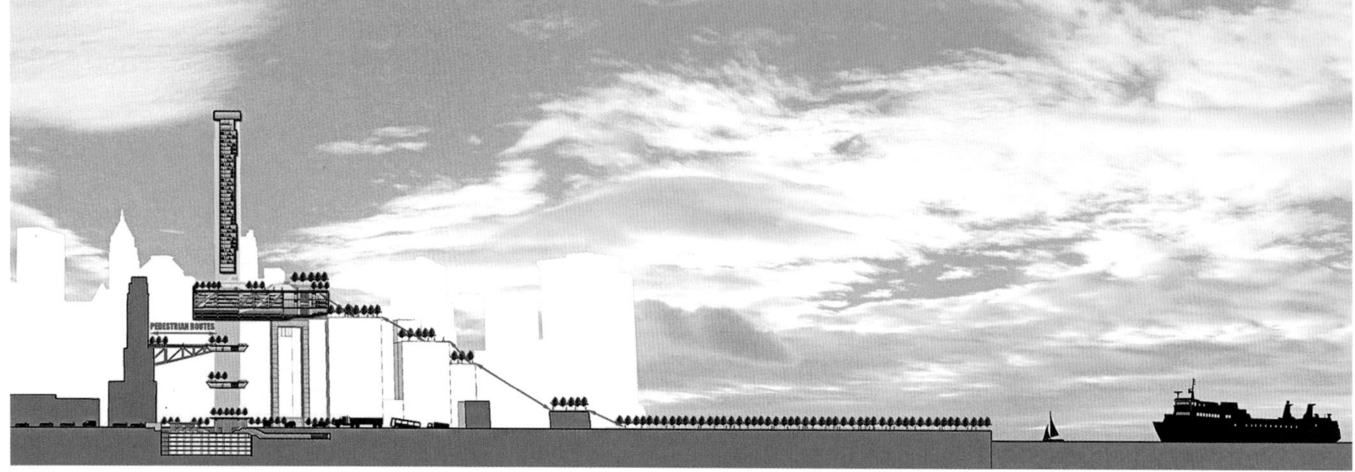

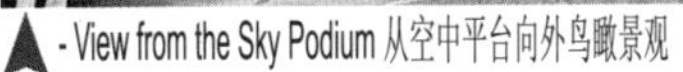

- View from the Sky Podium 从空中平台向外鸟瞰景观

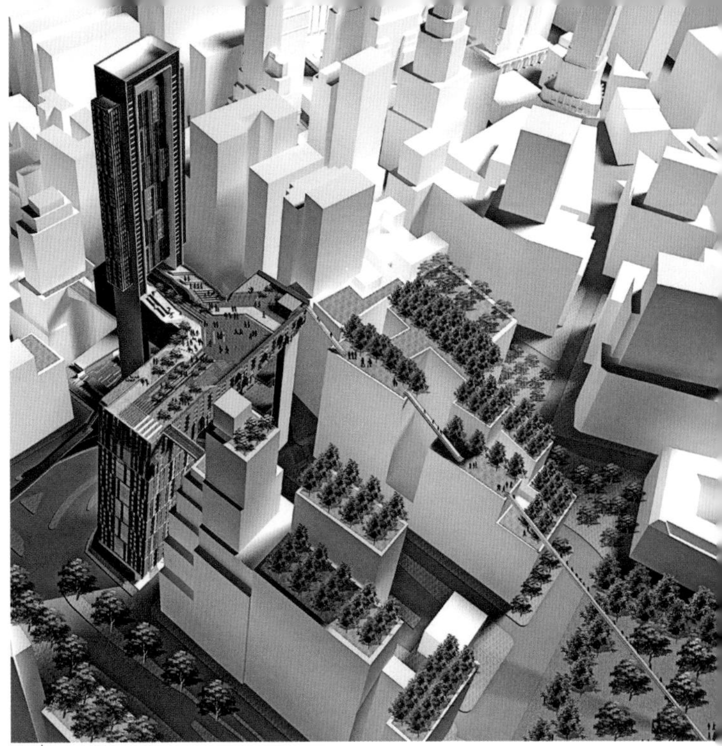

- Aerial View 鸟瞰图

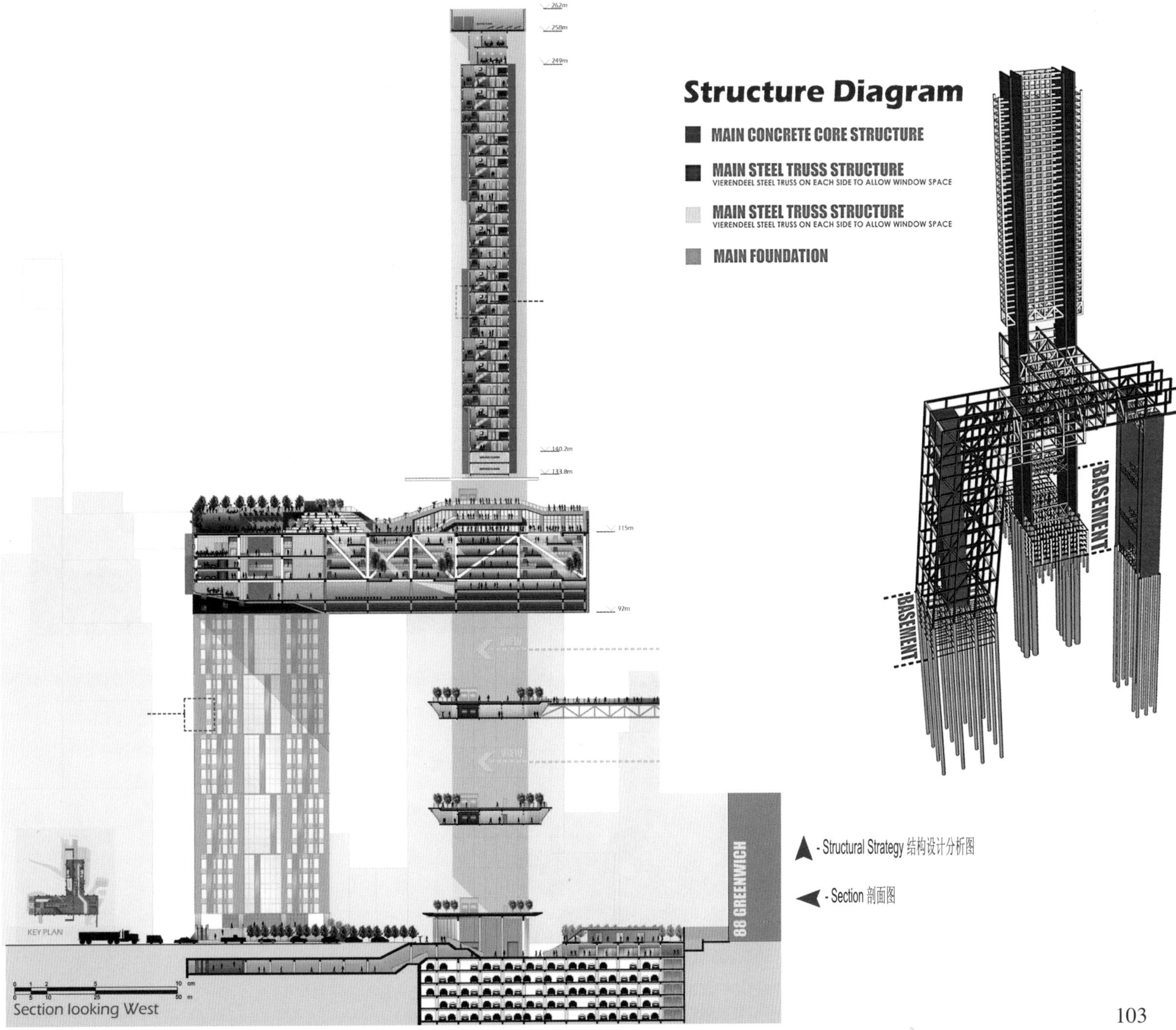

- Structural Strategy 结构设计分析图

- Section 剖面图

051

DESIGNERS-设计人: Venu Madhav Chippa, Avinash Davidson, Ranjit Shekhar

The Green Ramp, New York

This project is informed by a desire to better integrate Lower Manhattan's green spaces into the city fabric, and in particular, to create a link between Battery Park and the tower. The design solution is a building form which ramps from Battery Park all the way up to the site, where it culminates in a 'PassivHaus' skyscraper. Above the ramp a vast inhabitable green roof is proposed, acting as an extension to the park and creating an exciting recreational space within the city. The Green Ramp is envisaged as a 'vertical eco-corridor', with continuous vegetation linking ground and sky, reducing the heat island effect and allowing for the migration of plant species throughout the building, and by extension, the city. Planting is integrated into the tower's southern façade through deciduous vines which grow along cables to provide summer shading, but die away in the winter when increased solar heat gain is more useful.

绿色坡道，纽约

这个项目着重于更好地将曼哈顿下城绿化空间整合到城市肌理中，特别是建立起巴特里公园与塔楼之间的联系。所设计出的建筑形式是一条大坡道，从巴特里公园一直上升延伸到基地上空，并在此陡然向上成为一栋"被动式建筑"类型的摩天大楼。在坡道之上是一个宽敞舒适的绿色屋顶，作为公园和城市娱乐空间的扩展。绿色坡道被设想为一条"垂直的生态走廊"，植被像纽带一样连接着地面和天空，不仅可以减少城市热岛效应，还使得各种植物物种可以在建筑中，以致整个城市中蔓延。植被被集中布置在塔楼的南向立面，落叶藤本植物沿着绳索生长来提供夏日遮阳，而当冬天到来时，叶子的飘落则可以使塔楼获得更多的太阳光。

▲ - Physical Model, View from Battery Park 实体模型，从巴特里公园方向看

◀◀ - Ground Floor Plan 一层平面图

◀ - Inhabitable Green Roof Plan 绿色屋顶平面图

▲ - Section 剖面图

▼ - Aerial View 鸟瞰图

DESIGNERS-设计人: Chuyu Qiu, Ankur Modi, Suruchi Modi

Green Canyons, New York

This project consists of a mixed-use tall building in Lower Manhattan, designed to meet 'PassivHaus' performance requirements. From an urban standpoint, the design aims to reconnect the site with its surroundings. It thus proposes a series of green tendrils, spreading out and creating physical connections with key nodes in the urban fabric such as Battery Park and the new World Trade Center. These connections are made in the horizontal and vertical realm, allowing the public to circulate up above the city, lifting the vibrancy and activities of the ground into the sky. Where they meet a 'green wall' is created; this huge central space hosts the tower's social and communal activities such as parks, gardens, sports, conference centres and more. Located either side of the green wall are compact, super-insulated blocks of accommodation, organised as per their ideal PassivHaus orientation with hotel and residential spaces on the south for maximum solar gain and the best views, and offices to the north for indirect day-lighting.

翠谷，纽约

该项目位于曼哈顿下城，我们试图设计出一个满足“被动式建筑”要求的综合性大楼。从城市角度出发，这栋大楼旨在重新连接基地和周围的环境。因此我们打算将一系列绿色触须蔓延向四周，以此与城市肌理中的关键节点产生实质性联系，例如巴特里公园和新的世界贸易中心。这些连接点分布于水平与垂直空间内，使得公众交通系统延伸至城市上方，将地面的生气和活动带向空中。人们会在这里看到一堵“绿墙”，这个巨大的中央空间承载了塔楼的社会和社区活动，例如公园、花园、运动、会议中心等等。在绿墙的两边是紧凑的，超级保暖的公寓模块，根据被动式建筑的概念，酒店和居住空间为朝南向，来取得最大的日照和最好视角，而办公室则布置在北向，主要利用间接采光。

- Ground Floor Plan 一层平面图

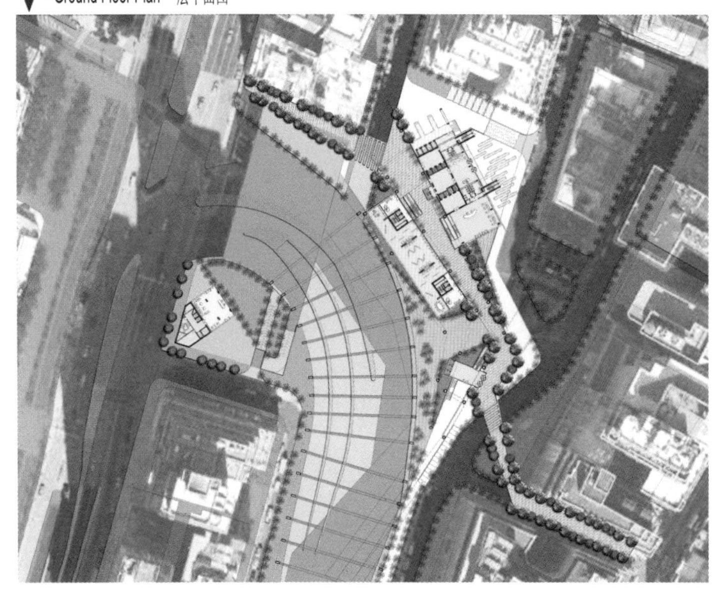

- Typical Floor Plan 标准层平面图

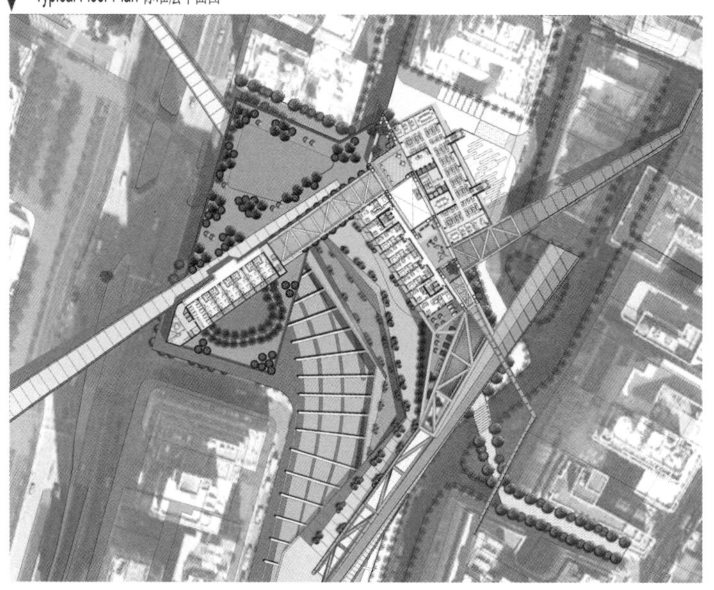

- Detailed Section and Elevation of the Residential Facade 剖面详图与公寓立面图

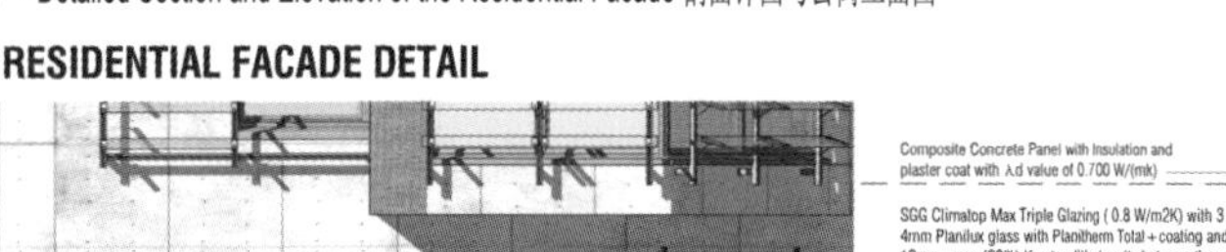

- Interior View of the 'Green Wall' “绿墙”内景透视图

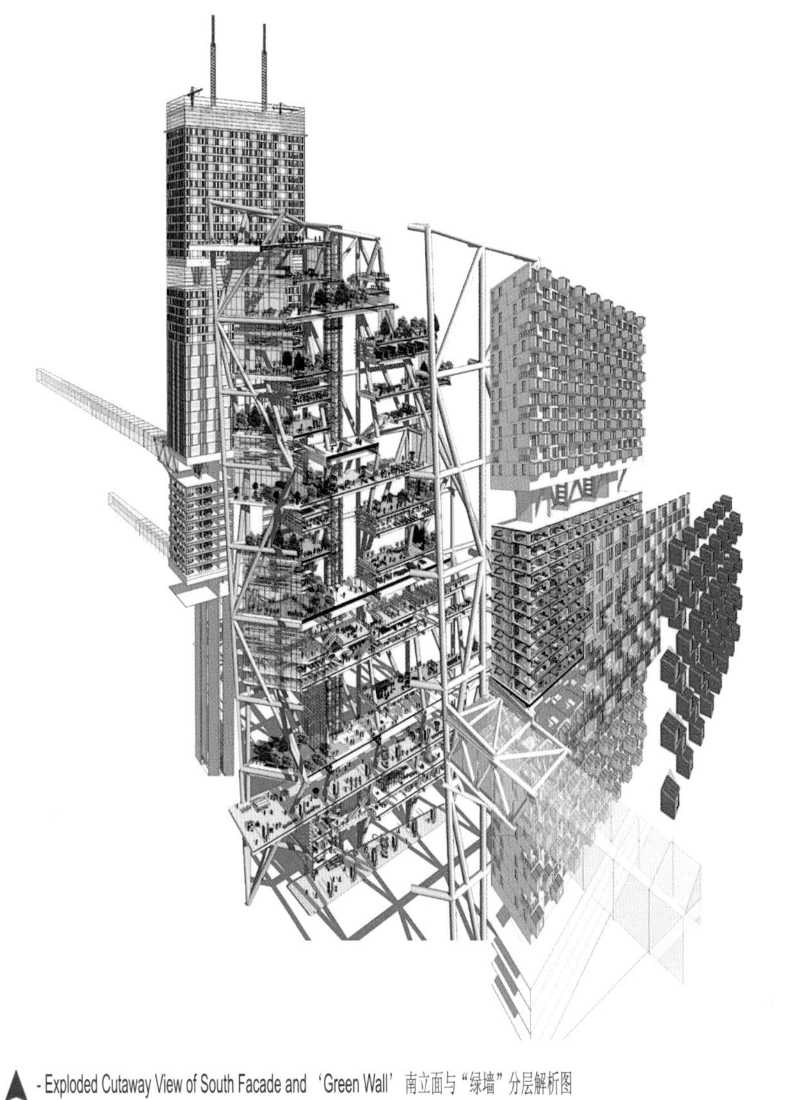

▲ - Exploded Cutaway View of South Facade and 'Green Wall' 南立面与“绿墙”分层解析图

► - Overall View 总体鸟瞰图

▼ - Heating/Cooling and Ventilation Strategy 采暖/制冷与通风设计

053

DESIGNERS-设计人: Najla Gunnur, Soha Hirbod, Fahimeh Soltani

A New Shibam, Abu Dhabi

This design for a desert skyscraper in Abu Dhabi drew inspiration from a vernacular precedent - the 500-year-old towers of Shibam, Yemen. Known often as the 'Manhattan of the Desert', Shibam is a magnificent walled city of mud towers, some rising up to 11 storeys high, with labyrinth-like alleyways and shaded courtyards below. Influenced by Shibam, and its response to the harsh local climate, the students were challenged to reinterpret this historic design in a modern and contemporary manner. The result is a series of solid, slender towers, tightly grouped together and stepping up across the site to visually tie together the skyline from the Chamber of Commerce at the east to the ADIA Tower at the west. The towers are clustered around open, yet shaded courtyards, with alleyways providing circulation from a series of shared cores. Carved into the mass of the buildings are skygardens, orientated towards the Gulf for the best views and to harness sea breezes. The building's structural strategy is a based on large concrete shear walls – a modern take on Shibam's construction with the same environmental benefits of thermal mass.

新希巴姆，阿布扎比

这是一座位于阿布扎比沙漠中的高层建筑，其设计灵感源自一处传统建筑先例——有500历史的希巴姆塔楼群（也门）。希巴姆被誉为“沙漠中的曼哈顿”，围绕全城的城墙后有着壮观的泥造塔楼，有一些甚至达到11层高，其下方是迷宫般的小巷和阴凉庭院。受希巴姆应对当地严酷气候方式的启发，学生们试图用一种当代手法来重新诠释这一古老的设计。他们将一系列表面封闭的细长塔楼紧密地组合在一起，从远处看基地，从东边的商会到西边的ADIA塔楼，整个天际线仿佛都被捆绑在了一起。塔楼群由开放的阴凉庭院串联起来，其间的小巷则将每个共享中心联系了起来。一处空中花园切入整块建筑实体之中，朝向海湾的最佳视角，并可以享受凉爽的海风。高层的结构主要依靠大面积的混凝土剪力墙，这是一种现代版的希巴姆结构，也拥有优异的热惰性，可适应当地的环境。

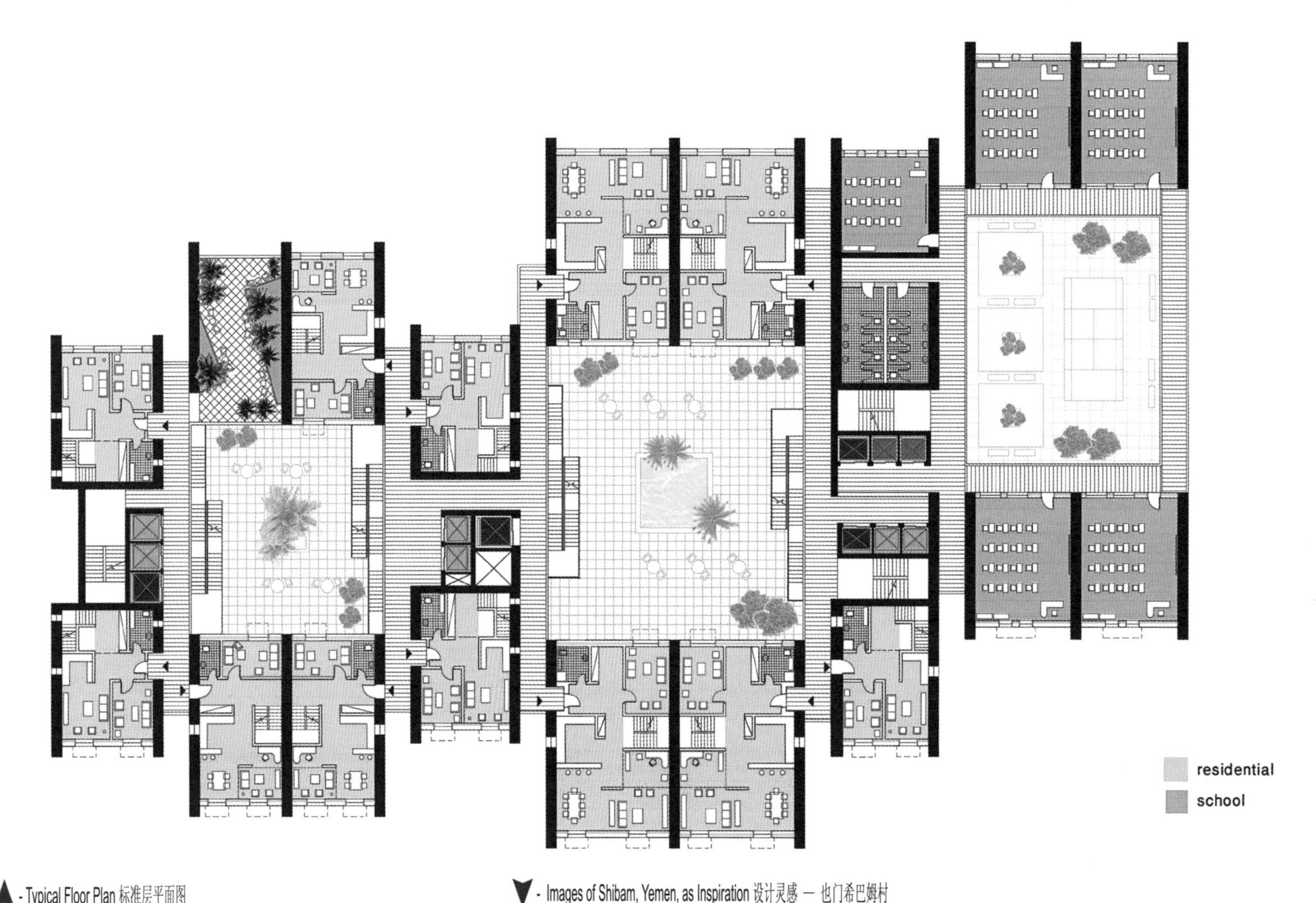

▲ - Typical Floor Plan 标准层平面图

▼ - Images of Shibam, Yemen, as Inspiration 设计灵感 — 也门希巴姆村

TUTORS - 指导教师: *Philip Oldfield, David Nicholson-cole*

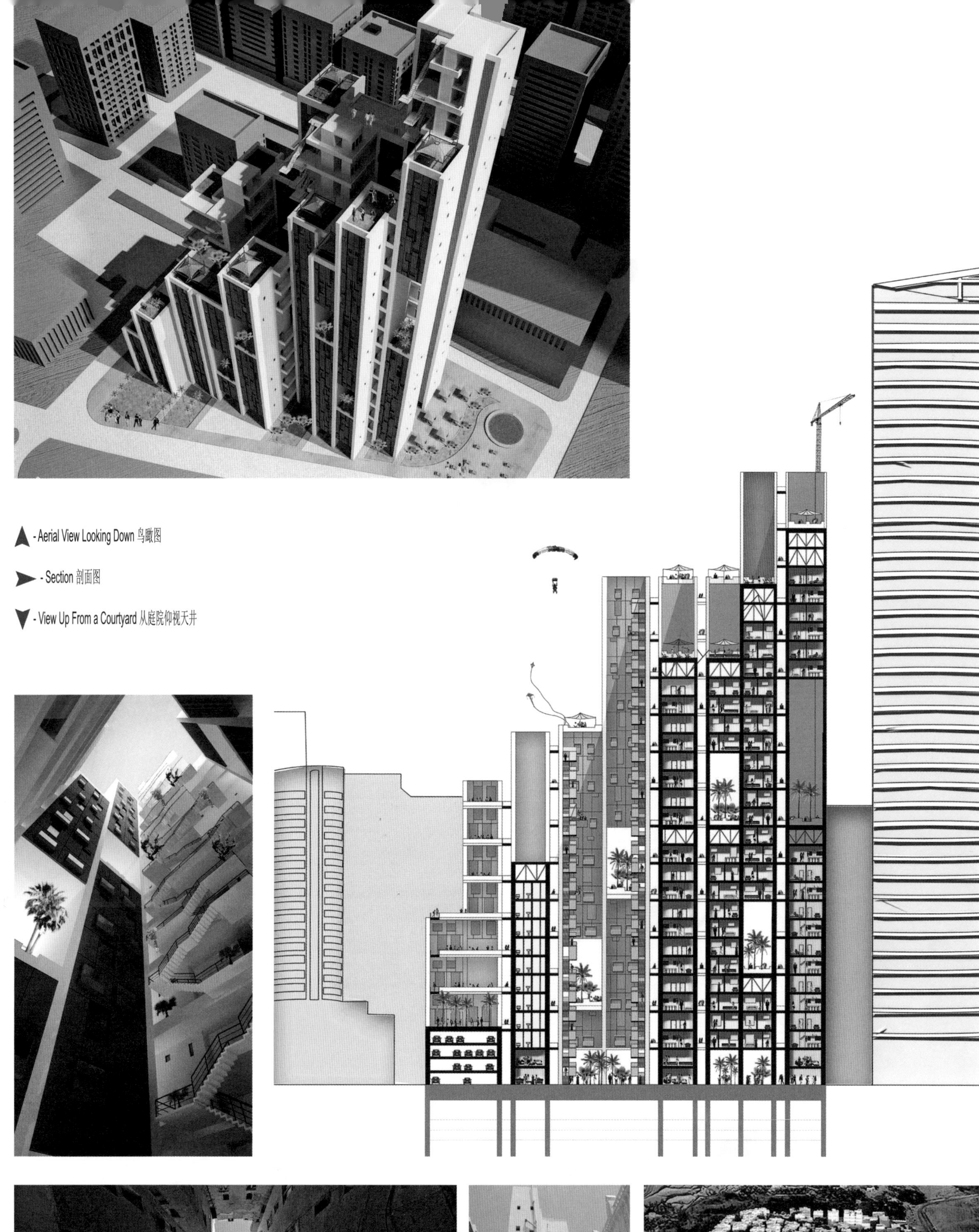

▲ - Aerial View Looking Down 鸟瞰图

➤ - Section 剖面图

▼ - View Up From a Courtyard 从庭院仰视天井

DESIGNERS-设计人: Ankur Modi, Suruchi Modi

Festival Tower, Singapore

This design aims to create a vertical residential community bound together by aspects of celebration and festival. Site research identified a series of important processional routes used for religious and national events through and around the site, and these have been strengthened and tied together with the creation of a spiralling processional pathway up the building. The pathway culminates in a significant celebration space at height within the scheme, elevating the excitement, colour and vibrancy of festivals that usually occur on the ground up into the sky. Located above and around the processional route are four slender residential towers, split to maintain and enhance a key urban axis across the site. Hung between the towers are a series of vertical gardens, providing a place of community and recreation and mitigating Singapore's urban heat island effect. Influenced by the Malaya House, the apartments are designed with flexibility and adaptability in mind - open terraces can be built up to allow families to grow and change over time, creating homes for a lifetime.

庆典之塔，新加坡

本设计旨在创造一处可以用于节日庆典的垂直社区。对基地的研究确认了一系列用于宗教和国家庆典的重要活动路线。它们现在均穿过或绕过基地，但在设计中这些线路均得到了加强，汇合在一起，沿着一条螺旋状的游行路径延伸到建筑之上。路径在高处汇聚为一处引人注目的欢庆空间，将活动的喜庆感，节日的颜色与活力也从地面升腾到空中。在游行路径的周围是四个细长的住宅塔楼，其分散的布局可以延续和加强穿过基地的城市轴线。悬浮在塔楼之间的是一组竖向花园，为居民提供了社交和休闲场所，也同时减轻了新加坡的城市热岛效应。受马来西亚民居的影响，设计时尽量考虑了公寓套房的灵活性和适应性——开放的平台使每个家庭得居住面积可根据需要而扩展和改变，为家族的长时间聚居提供了可能性。

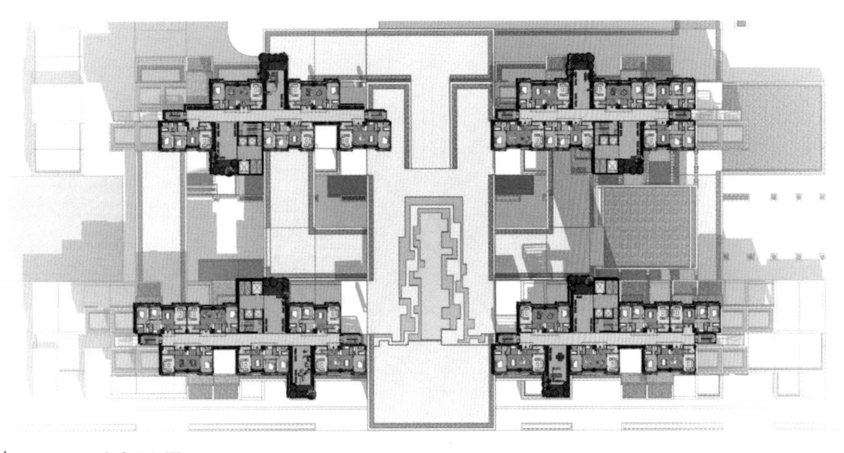

▲ - Typical Plan 标准层平面图

▲ - Aerial View 鸟瞰图

▲ - Concept 概念

► - Section 剖面图

▲ - View of Vertical Gardens with Celebrations Below 竖向花园透视图，其下方正在进行庆典

▼ - 3D Cutaway View 三维剖透视图

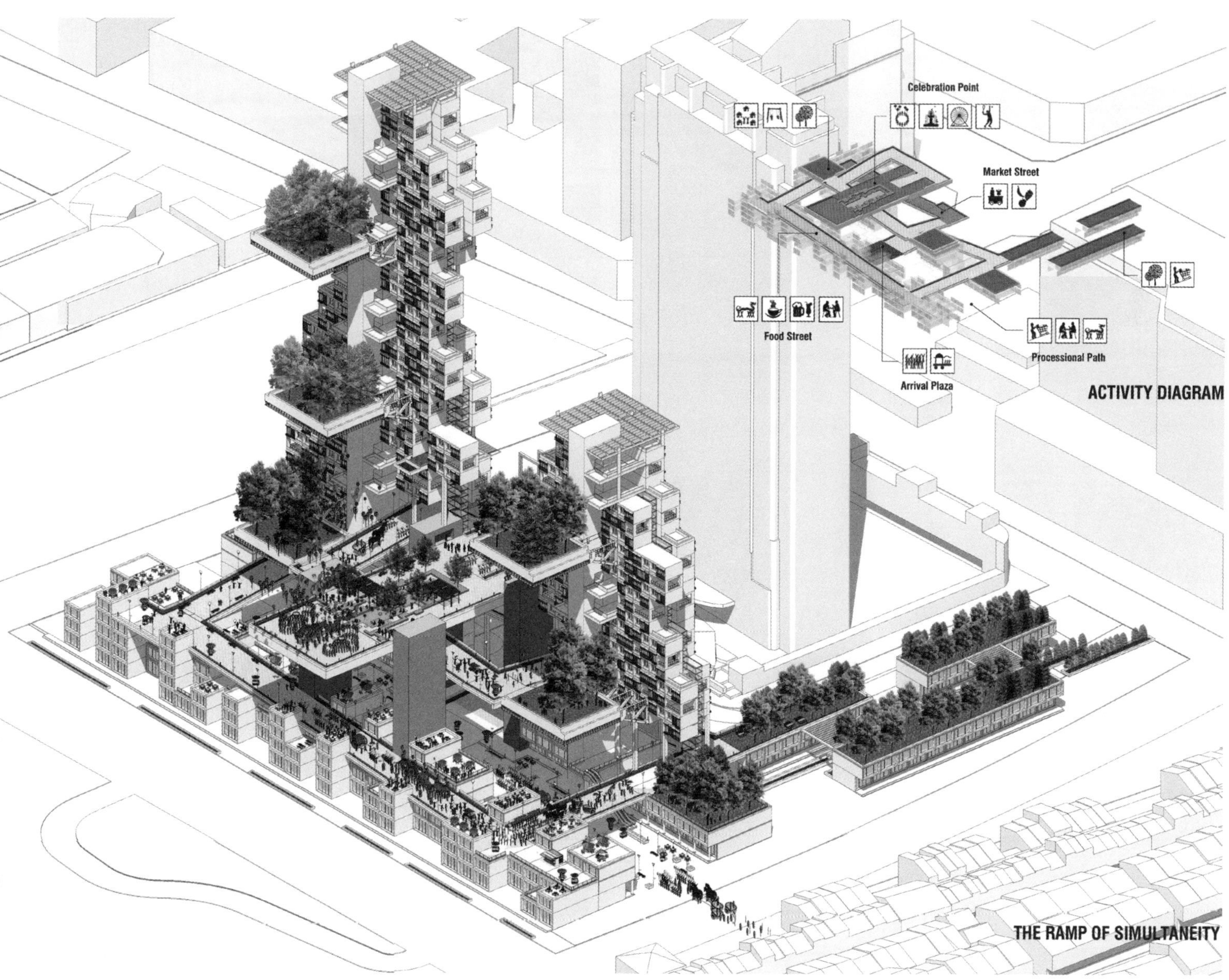

DESIGNER-设计人: Akshay Sethi

JO_2 Tower

The first stage of project was to produce a facade which improves the quality of living by removing air pollution. The aim of the proposed facade was to create a system that is highly efficient at ambient temperatures, renewable and as efficient as natural systems. Meanwhile, the system implements the beauty of nature. By function, the facade uses methods of pollution removal found in nature. Aesthetically, the skin of the façade takes many forms as it responds uniquely to various pollutants. Taking inspiration from nature, the JO_2 Tower is a closed loop system that provides various attractions for car-users. Attractions such as a car boot sale, a car-wash and a drive-through, have been reimagined and provide a completely unique experience. The location was chosen close to a congested route in London, to maximize the productivity of the façade. The JO_2 Tower is essentially an organism with each function working in collaboration. The building breathes through its skin; it grows; it senses and expresses the state of the environment and it creates unique experiences for the user.

JO_2塔

本方案的首要目的是设计一种立面表皮系统，使其能够消除空气污染，进而提高生活质量。进行这一设计主要旨在建立一套对周围温度高度敏感，可自我更新，且与相应自然系统一样有效的人工系统。此外，自然之美也将在上面得以体现。从角度功能看，该表皮系统借鉴了自然界自我净化的方法。从美观角度看，该表皮系统可针对不同的污染物而变换很多特定的形式。J02塔的灵感源自自然，它形成一种闭合的环状系统，为汽车使用者提供了不同的服务。在这里，例如汽车零件零售，洗车服务或仅仅驾驶汽车穿过塔楼本身等，所有这些功能都被重新定义，进而可为使用者提供一种全新的体验。方案基地被选择在伦敦市内，靠近一条拥堵的道路，从而可以最大程度上发挥表皮系统的功能。JO_2塔内各项功能均相互配合，就像有机体一样。建筑通过其表皮呼吸，它生长，它感觉，并显示着环境状况。这对于使用者而言无疑是一种独一无二的体验。

► - Typical Plan 标准层平面图

▲ - Carwash Experience 汽车清洗场景

▼ - Functional Modes 建筑不同功能模式

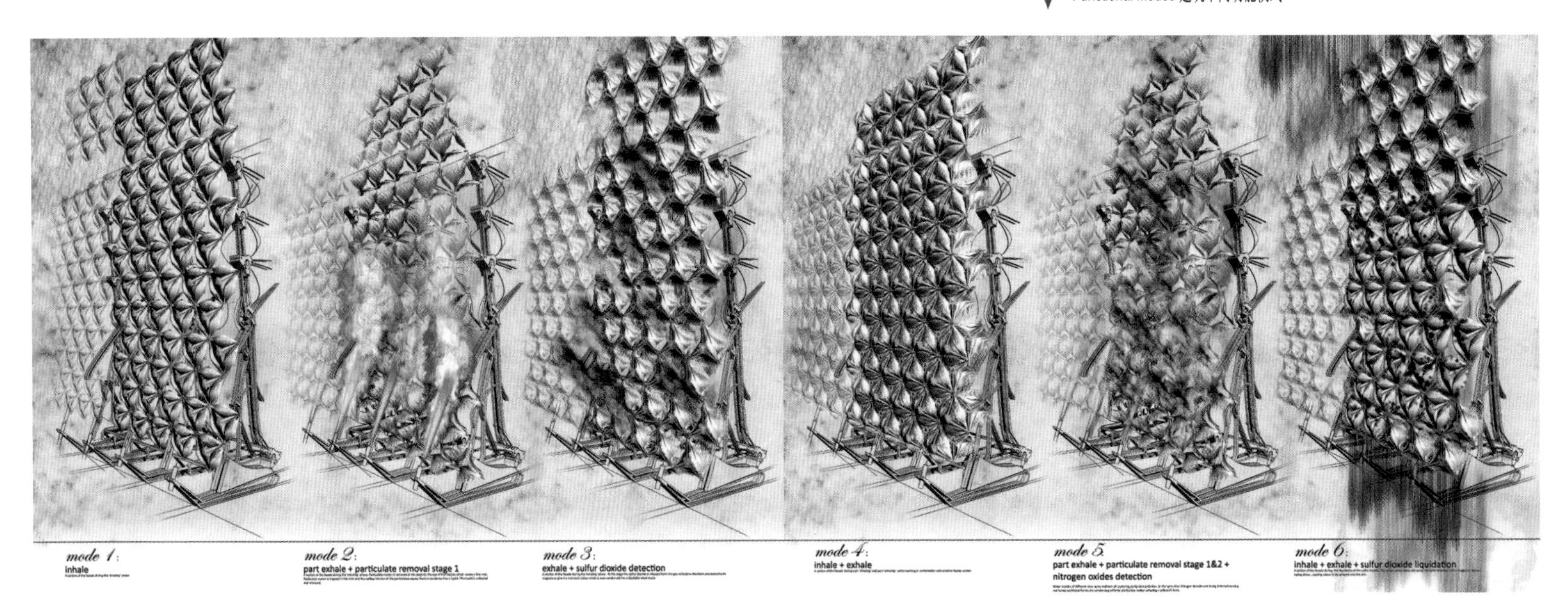

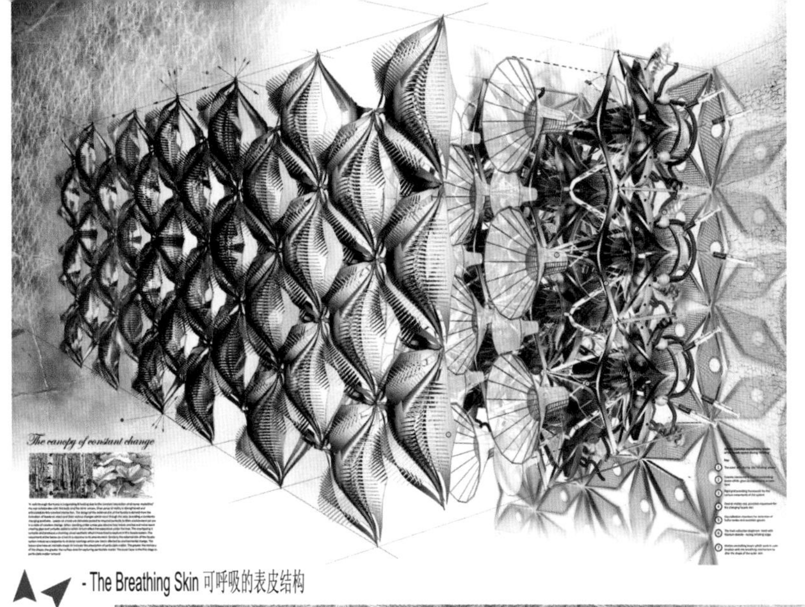

- The Breathing Skin 可呼吸的表皮结构

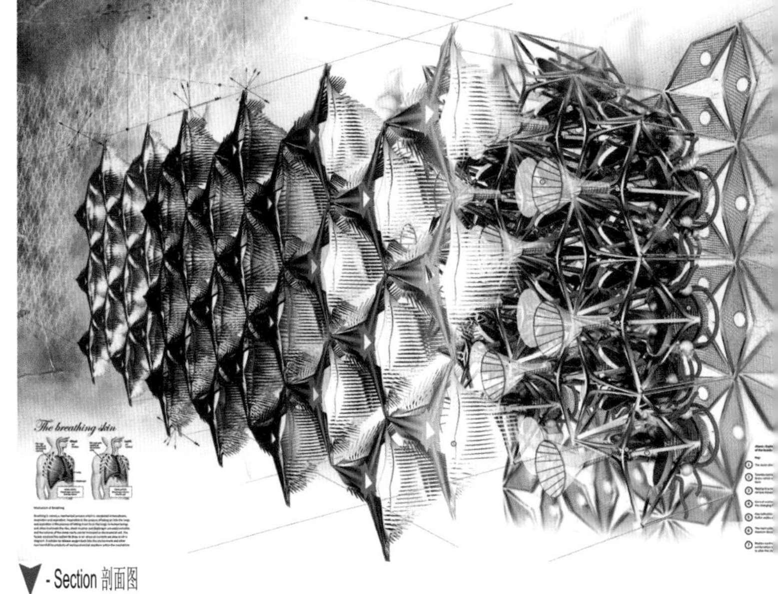

- Section 剖面图

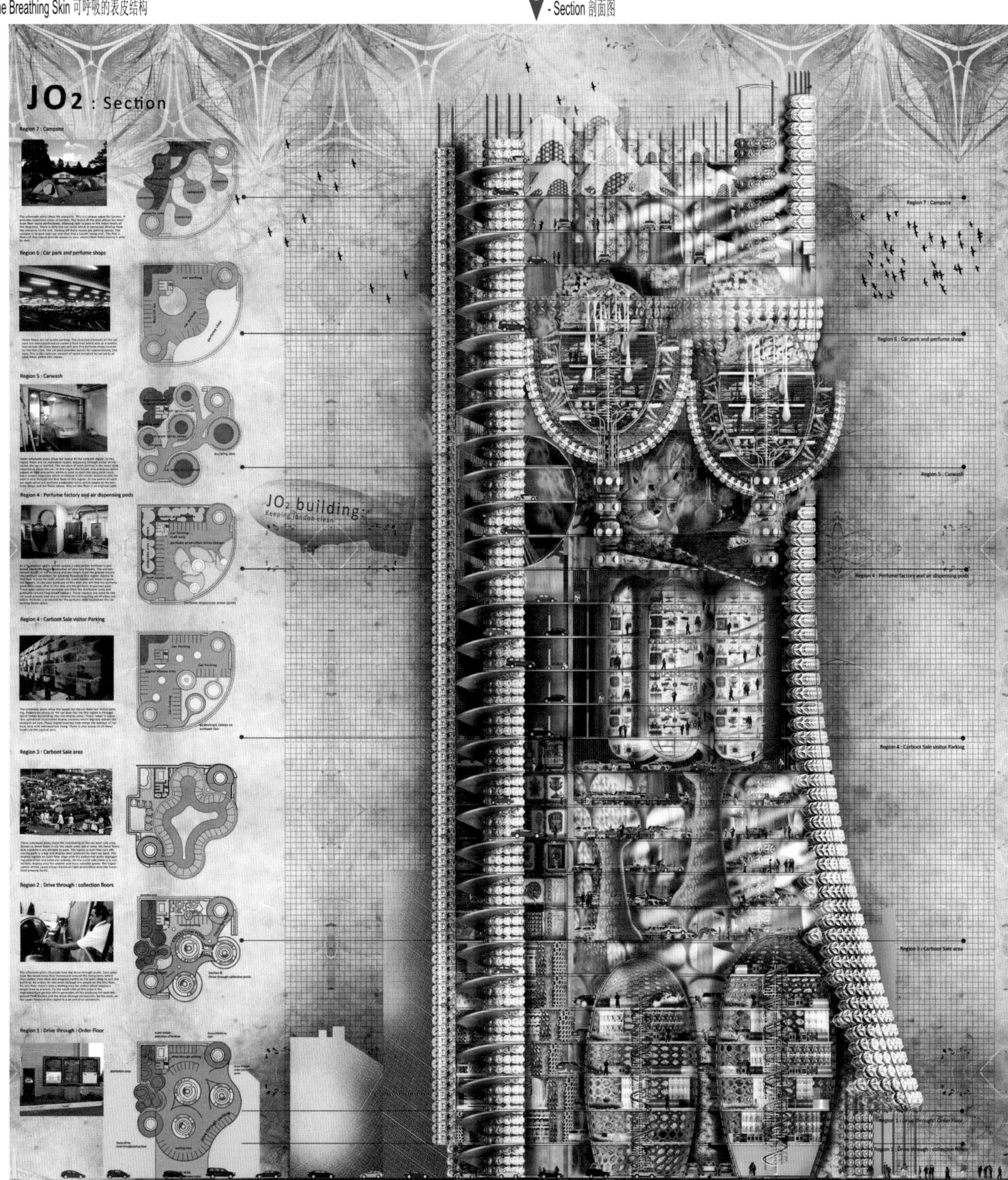

DESIGNERS－设计人: Philip C.L Etherington, Joe Bamber, Chenhao Yang

ETRI

Having decided upon using 'the University of Nottingham's Innovation Park' as a direction for the project, we began to define the parameters behind how to strength the direction in particular the following aspects: to activate the pedestrian street frontage through presence, connection, purpose and engagement to create an active spatial threshold; to reinforce views – make new and resolve existing weak public spaces, links and nodes; to enhance view points and better site definition through progressive threshold, lining urban edges and softening 'parkland' threshold; to develop a programmatic journey for visitors within the ETRI; to develop a programme of events and associated functions; to create an integrated, considered landscape strategy which fused boundaries.

能源科技研究所

在决定采用“诺丁汉大学创新产业园”作为本项目的发展方向后，我们便开始定义参数，并特意从以下几个方面来巩固发展方向：通过其存在性，连接性，目的性与参与性来达到激活步行街临街立面的目的，并创建积极的空间门户；去加强视野——创造新的，并提升现有的，薄弱的，公共场所，连接和节点；通过渐进式门户区，整合城市边缘和软化“公园”门槛，去提升视角并提供更好的场地定义；在研究所内位访问者制定程序化的路径；基于一定程序来开发多种事宜与相关功能；以及去创建一个综合的地景战略，融合各方界线，使之成为整体。

▼- Model 实体模型

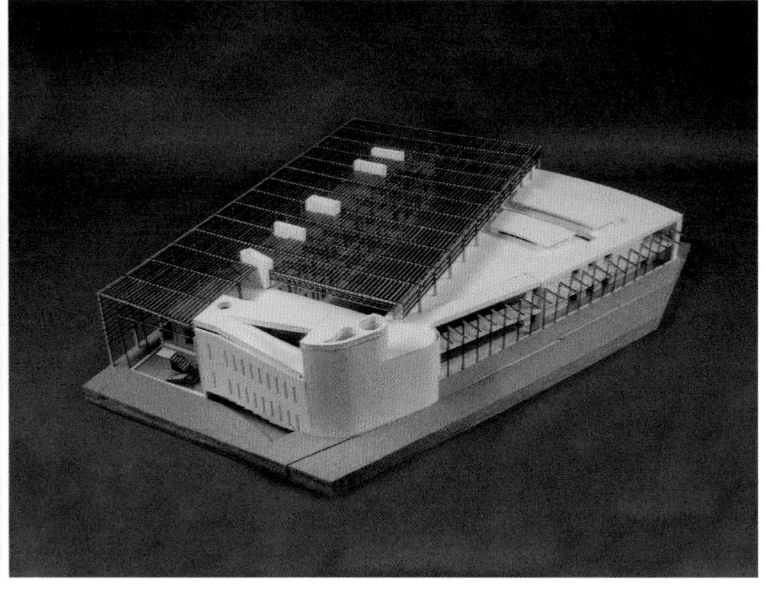

▼- Summer Day Environment Analysis 夏季室内环境分析

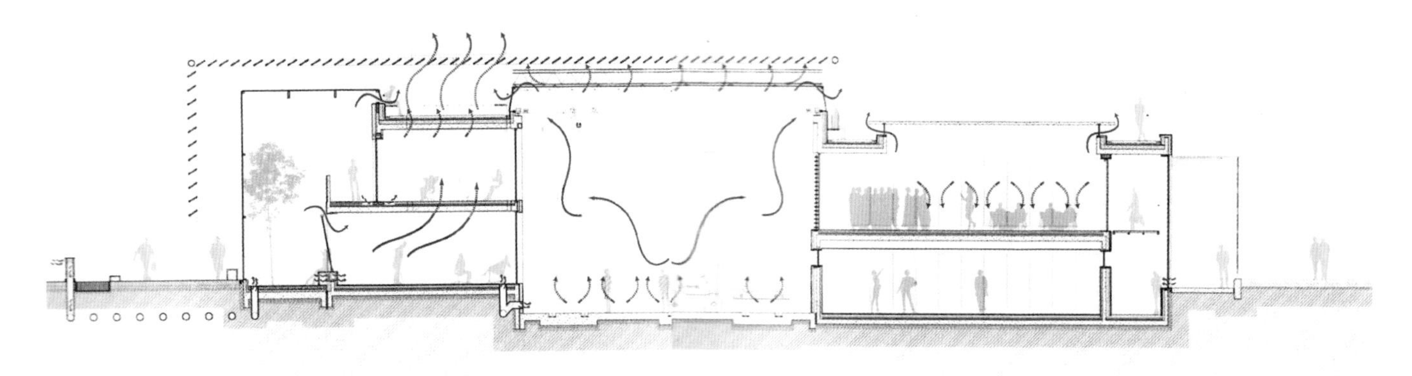

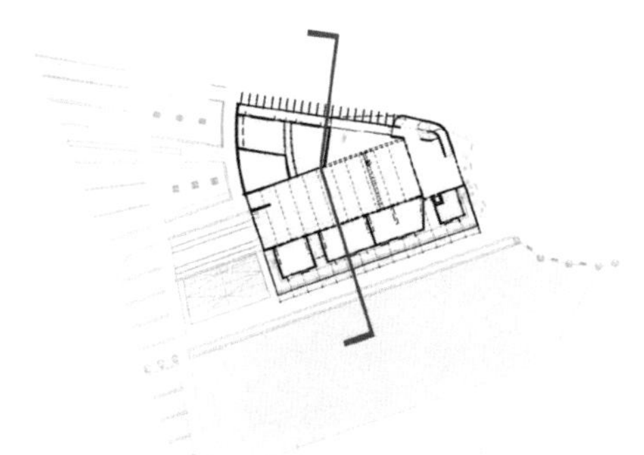

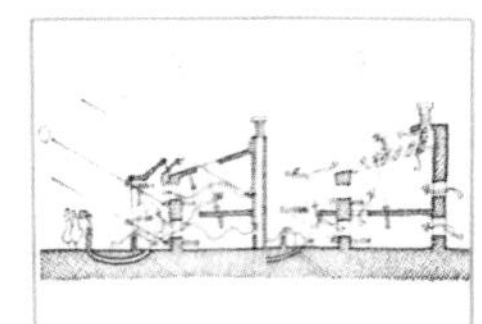

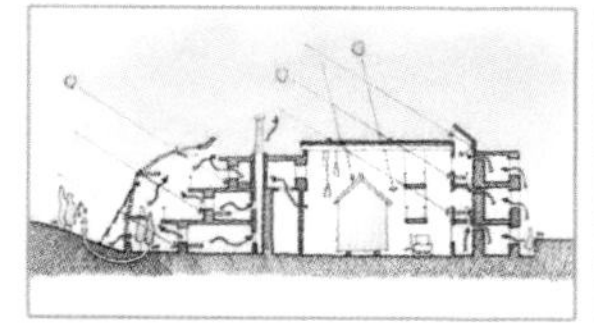

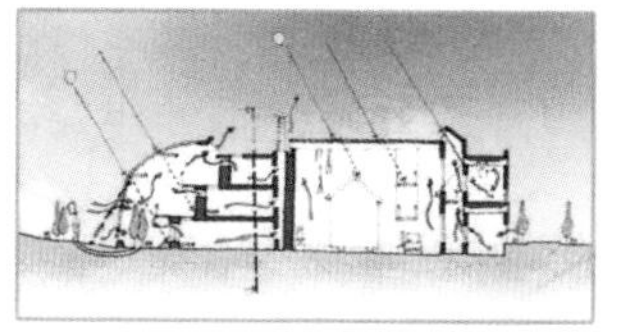

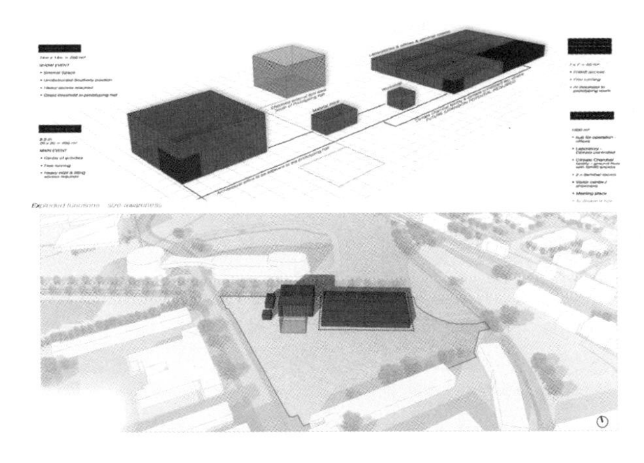

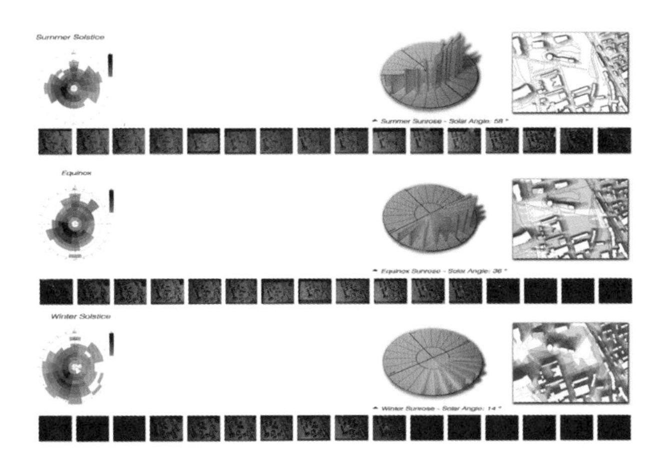

- Volumetric Analysis of Functions and Climatic Modelling 基于功能与气候模型所进行的建筑体量分析

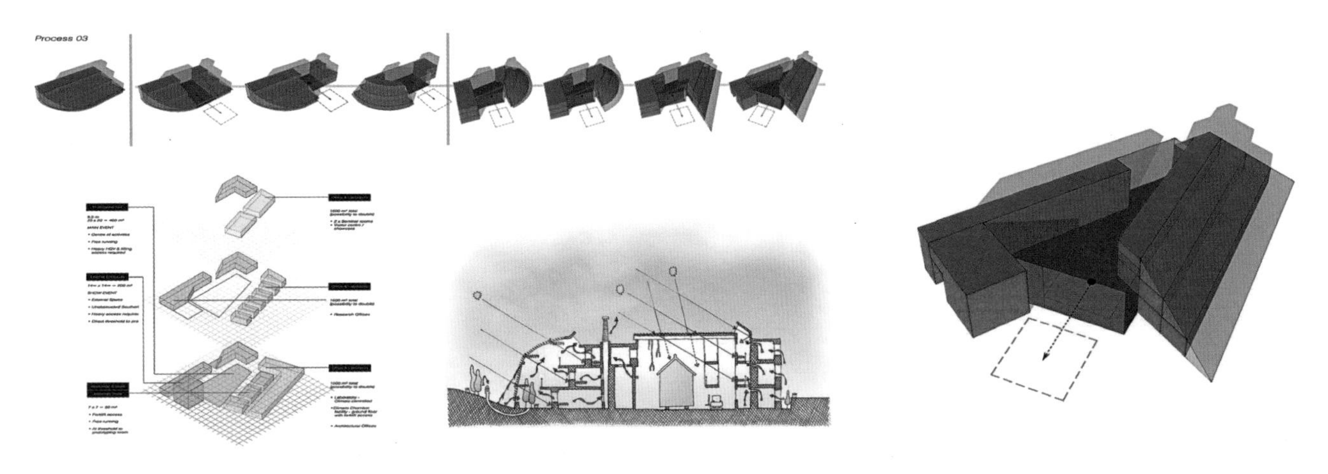

- Programme Defining 建筑整体运行计划

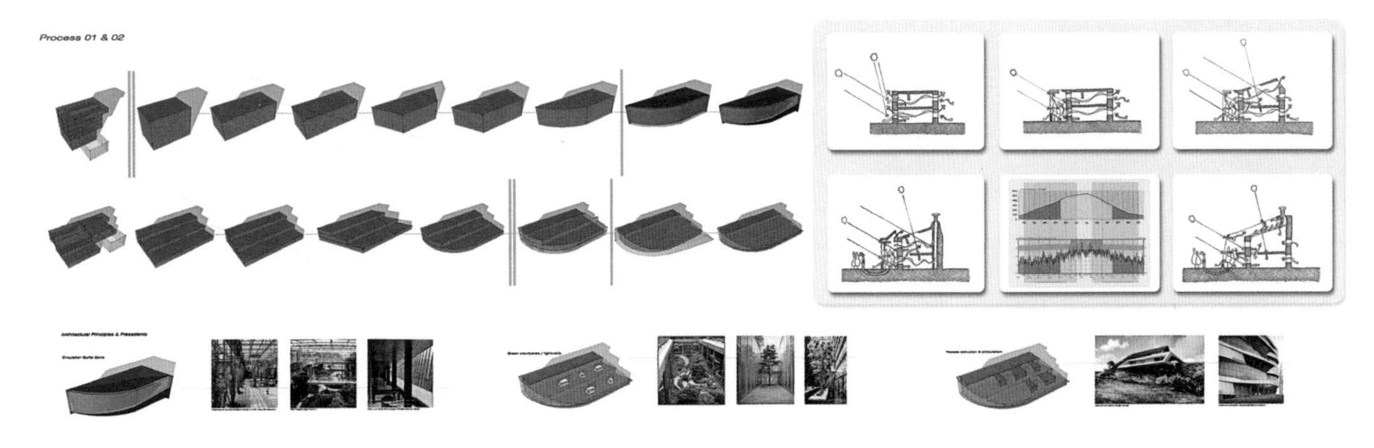

- Initial Massing and Environmental Sections 初级建筑体块分析与建筑环境剖面分析

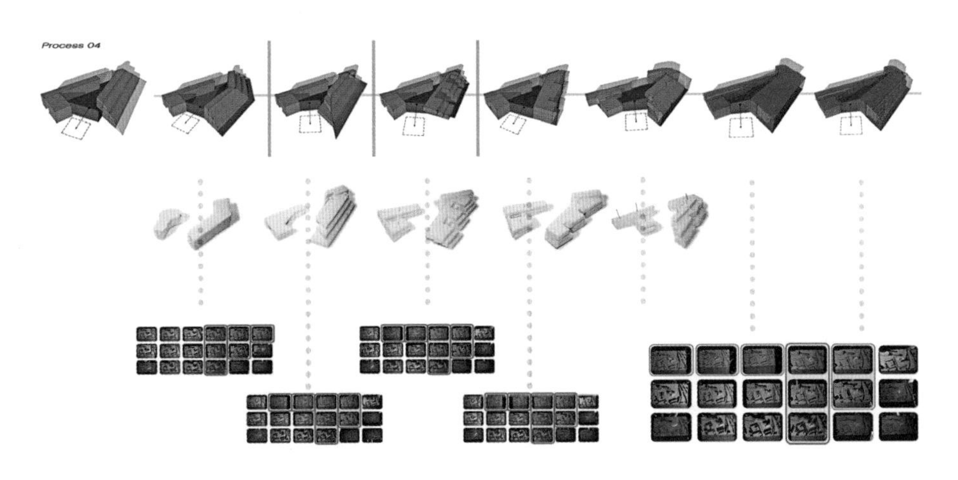

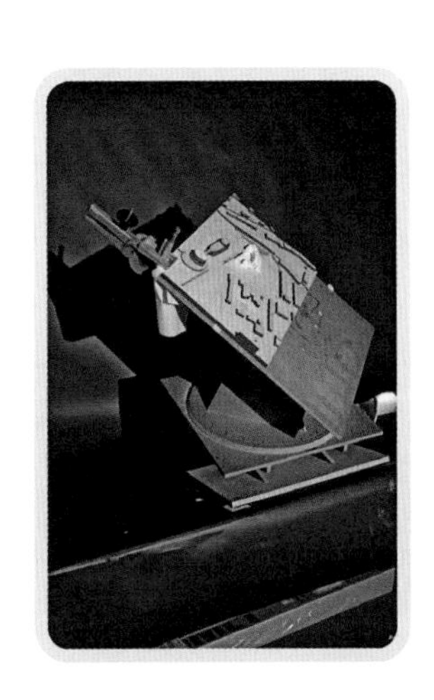

- Further Massing and Hellodon Testing 后期建筑体块分析以及实体模型借助日照分析仪进行日照分析

DESIGNERS-设计人: Richard Pulford, Matt Sandhu, Richard Bray, Ben Hyett

The Creative Construction Centre, Jubilee Campus, University of Nottingham

The ETRI building in the Jubilee Campus of University of Nottingham is designed specifically to address typical environmental issue in the local Nottingham climate. As emphasized to be the aim of the Environmental Design Studio, several questions have been asked as the basic principles of design:

Due to external climatic conditions, what issues should our passive strategies to address?

On a diurnal basis, is there a big enough range for passive strategies such as night ventilation to be considered?

Is the sky cover in Nottingham generally overcast or clear? To what extent can we utilize natural lighting?

Our outcomes are generated based on the answers of these questions all crystallized on one tangible building.

创意构建中心，朱比利校区，诺丁汉大学

能源科技研究所新楼位于诺丁汉大学朱比利校区，其设计特别专注于解决一些典型的诺丁汉本地环境问题。作为一座环境设计研究室，其运行目标揭示了一些被用作设计基本准则的问题：基于外界气候情况，我们的被动式设计策略应该解决什么问题？在日常使用的基础上，是否有一个包容性足够大的被动式设计方案战略可供参考，比如采用夜间自然通风？

在通常情况下，诺丁汉的天空是多为多云还是晴空？我们有多大的可能性来利用自然光？

我们的结果均来自于这些问题的答案，且同时结晶为一座有形的建筑。

- Plans 平面图

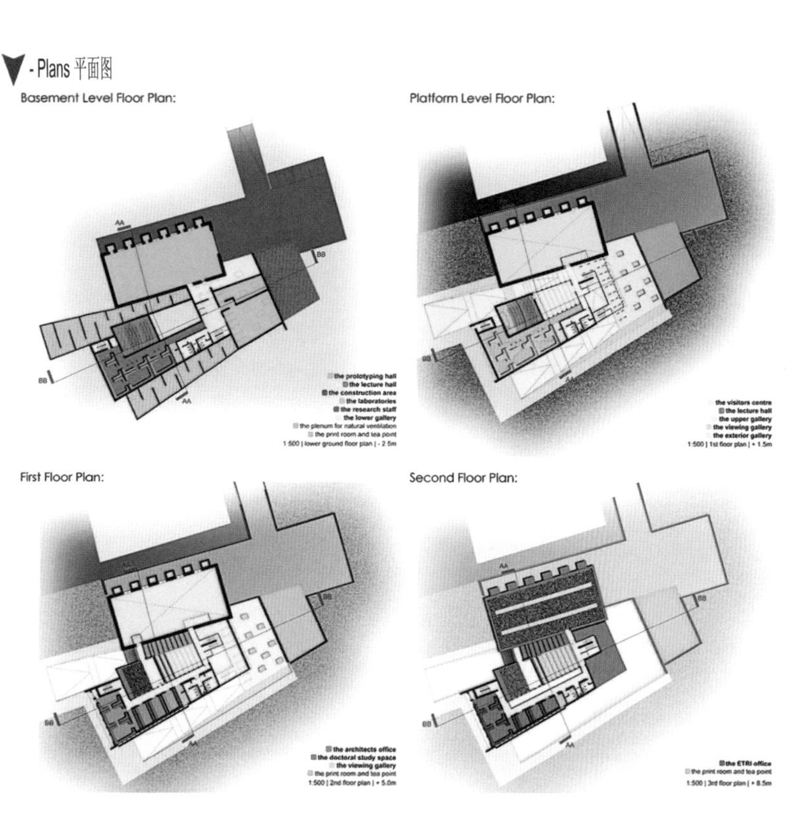

- Prototype Hall Relationship 足尺寸建筑原型测试室空间关系分析图

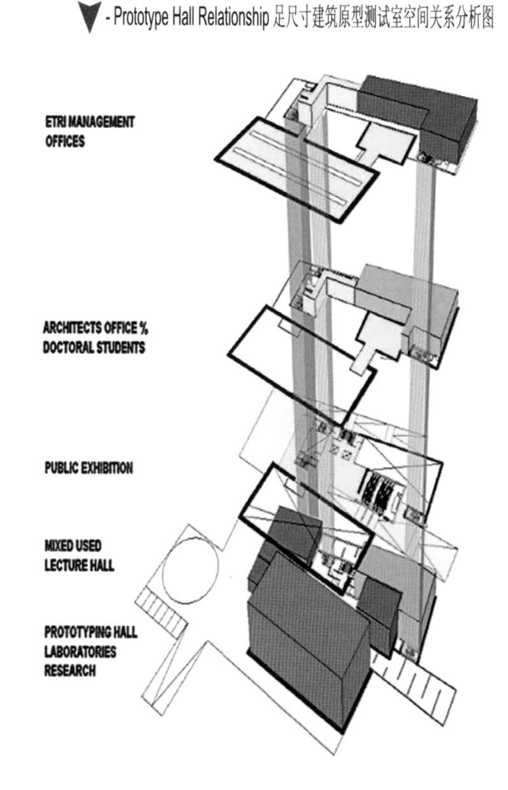

- Digital Models and Perspectives 数码模型与局部透视图

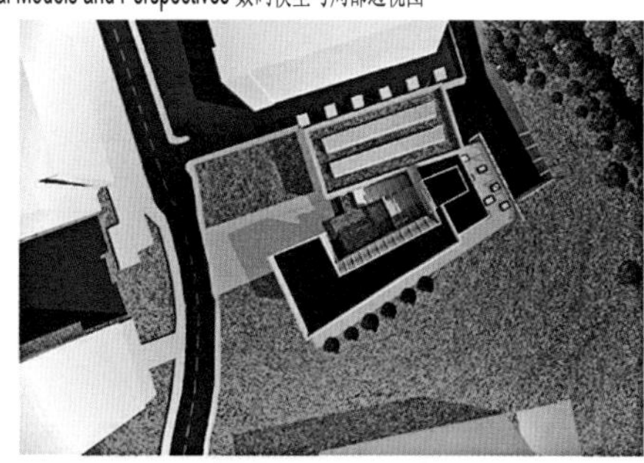

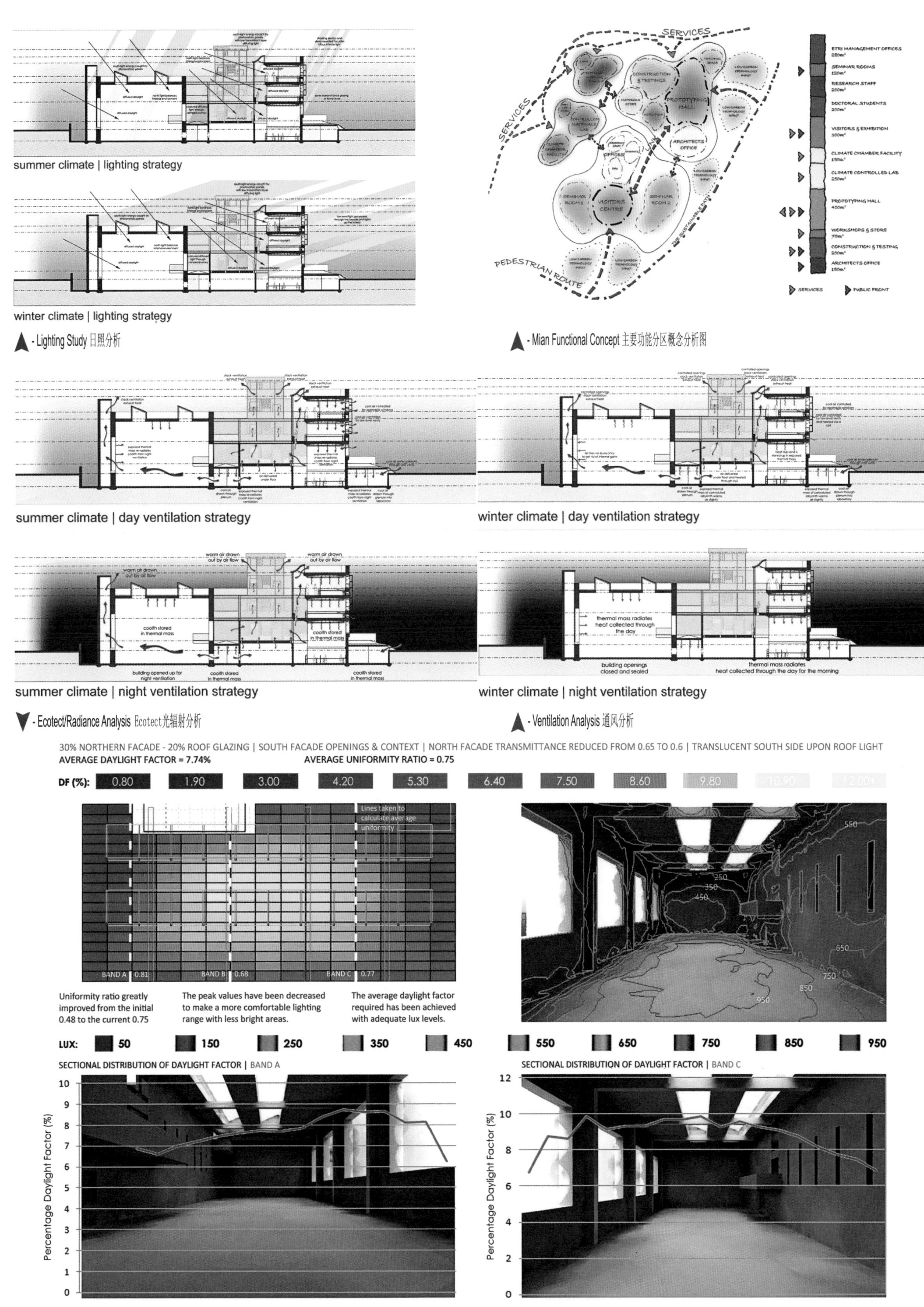
summer climate | lighting strategy
winter climate | lighting strategy
- Lighting Study 日照分析
SERVICES
SERVICES
PEDESTRIAN ROUTE
SEMINAR ROOM 1
VISITORS CENTRE
SEMINAR ROOM 2
PROTOTYPING HALL
CONSTRUCTION & TESTING
ARCHITECTS OFFICE
ETRI MANAGEMENT OFFICES
SEMINAR ROOMS
RESEARCH STAFF
DOCTORAL STUDENTS
VISITORS & EXHIBITION
CLIMATE CHAMBER FACILITY
CLIMATE CONTROLLED LAB
PROTOTYPING HALL
WORKSHOPS & STORE
CONSTRUCTION & TESTING
ARCHITECTS OFFICE
SERVICES
PUBLIC FRONT
- Mian Functional Concept 主要功能分区概念分析图
summer climate | day ventilation strategy
winter climate | day ventilation strategy
warm air drawn out by air flow
cooth stored in thermal mass
building opened up for night ventilation
thermal mass radiates heat collected through the day
building openings closed and sealed
thermal mass radiates heat collected through the day for the morning
summer climate | night ventilation strategy
winter climate | night ventilation strategy
- Ecotect/Radiance Analysis Ecotect光辐射分析
- Ventilation Analysis 通风分析
30% NORTHERN FACADE - 20% ROOF GLAZING | SOUTH FACADE OPENINGS & CONTEXT | NORTH FACADE TRANSMITTANCE REDUCED FROM 0.65 TO 0.6 | TRANSLUCENT SOUTH SIDE UPON ROOF LIGHT
AVERAGE DAYLIGHT FACTOR = 7.74%
AVERAGE UNIFORMITY RATIO = 0.75
DF (%): 0.80 1.90 3.00 4.20 5.30 6.40 7.50 8.60 9.80 10.90 12.00+
Lines taken to calculate average uniformity
BAND A 0.81
BAND B 0.68
BAND C 0.77
550 250 350 450 650 750 850 950
Uniformity ratio greatly improved from the initial 0.48 to the current 0.75
The peak values have been decreased to make a more comfortable lighting range with less bright areas.
The average daylight factor required has been achieved with adequate lux levels.
LUX: 50 150 250 350 450 550 650 750 850 950
SECTIONAL DISTRIBUTION OF DAYLIGHT FACTOR | BAND A
SECTIONAL DISTRIBUTION OF DAYLIGHT FACTOR | BAND C
Percentage Daylight Factor (%)
RANGE REDUCED TO VALUES BETWEEN 6.27 AND 8.75 % DAYLIGHT FACTOR ACCROSS BAND A |
ALL VALUES BELOW 10% DAYLIGHT FACTOR WITH A GOOD DISTRIBUTION ACHIEVED |

DESIGNERS-设计人: Yanti Chen, Mingwei Sun

ETRI, Jubilee Campus, University of Nottingham

This project aimed to design a complex for the new Energy Technology Research Institute with a Prototype Hall, located in the Jubilee Campus, the University of Nottingham. It is an exemplar low carbon building, both in construction and daily use. The original site is a meadow with a small pool, playing as a landscape as well as a habitat for wild life, thus the concept of this design was to create an efficient research centre which hiding under a 'Green Carpet' that offered a peaceful land not only for human but also for wild life. The combination between architectural design and environmental considerations enhances communication between the building and ambient conditions. Here environmental design embraces more than just the architectural techniques; it aims to create a balance among the architecture, the occupants, and the surrounding ecosystems.

能源研究中心，朱比利校区，诺丁汉大学

该项目致力于为坐落于英国诺丁汉大学朱比利校区的新能源技术研究所与原型大厅设计一综合设施体。无论是在建造过程还是在日常使用方面，该综合体被设定为一个具有典范作用的低碳建筑。其原始位置是一个作为观赏景观以及野生动物栖息地的草甸，之内另有一池塘。因此这一设计理念是建立一个隐藏于"绿地毯"下的高效科研中心，该"绿地毯" 不仅仅为人类也为野生动物提供了一片宁静的土地。建筑设计和环境因素的结合旨在增强建筑与环境之间来自不同方面的沟通。该设计所体现的是环境设计所涵盖的不仅仅是建筑技术；相反，其目的是设法实现建筑物、居住者和周围的生态系统之间的平衡。

- Plans 平面图

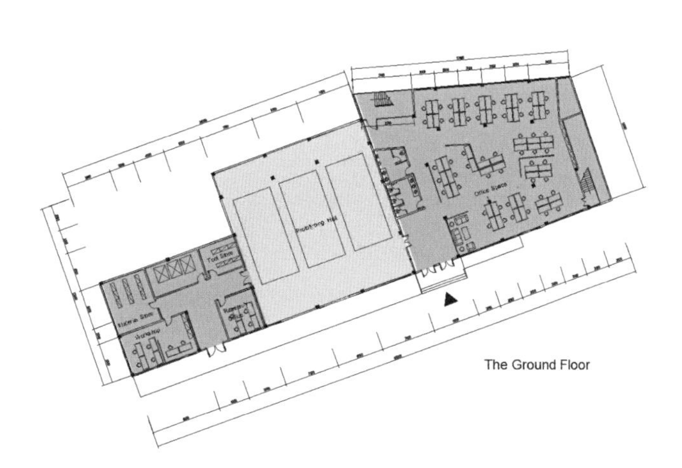

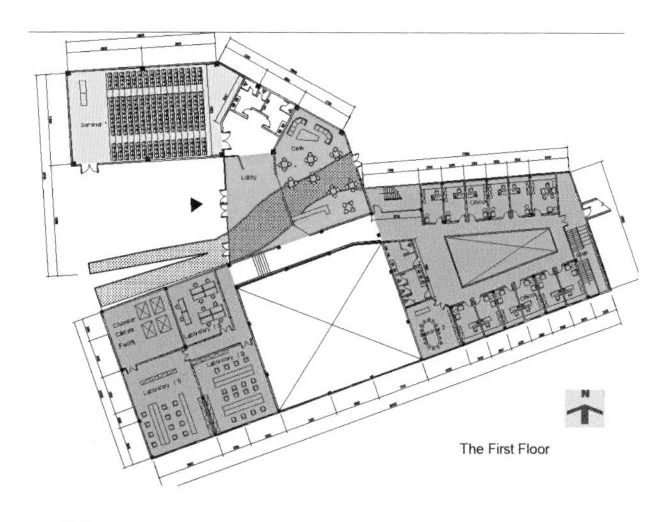

- Digital Models 数码模型

- West Elevation 西立面

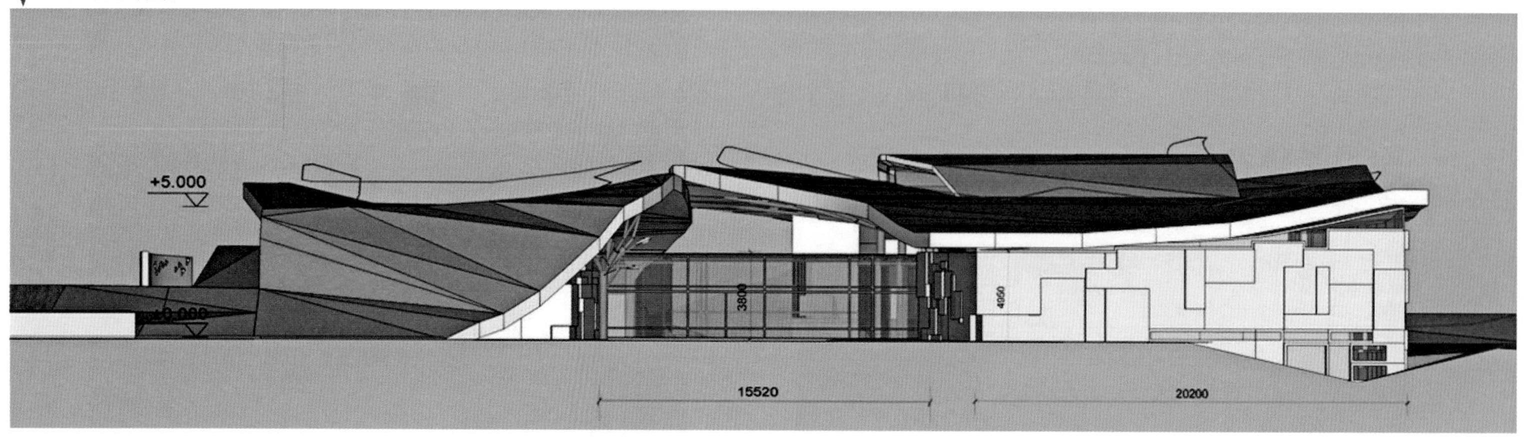

TUTORS - **指导教师**: *Brian Ford, Benson Lau, Lucelia Rodrigues*

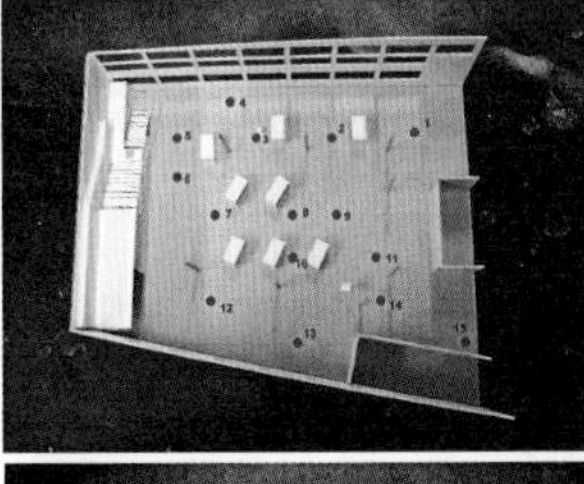

DAYLIGHT FACTOR TESTING:

We chose the office space for daylight factor testing in artificial sky and Ecotect.

After the artificial sky testing, we found that the DF was 12.9%, but the Uniformity roatio was 0.33, so we designed 3 types of window for a better result.

At the end, the best result was:

The average DF: 8.22%
The uniformity ratio: 0.61

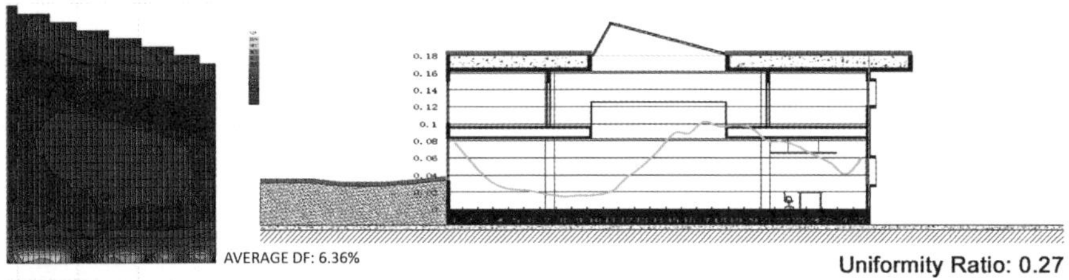

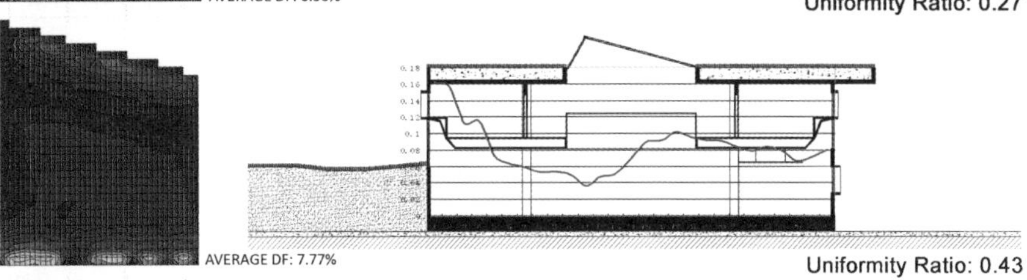

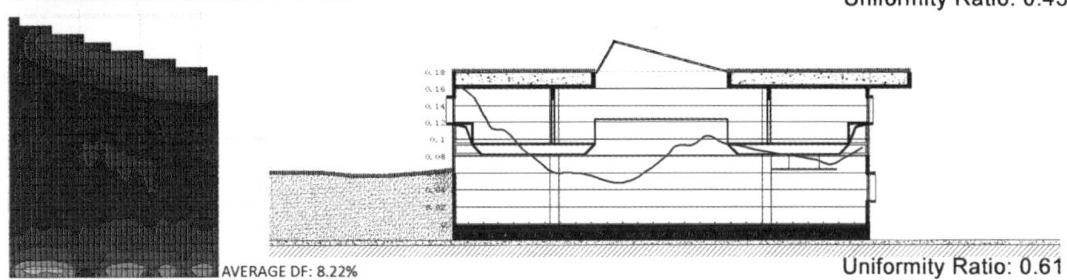

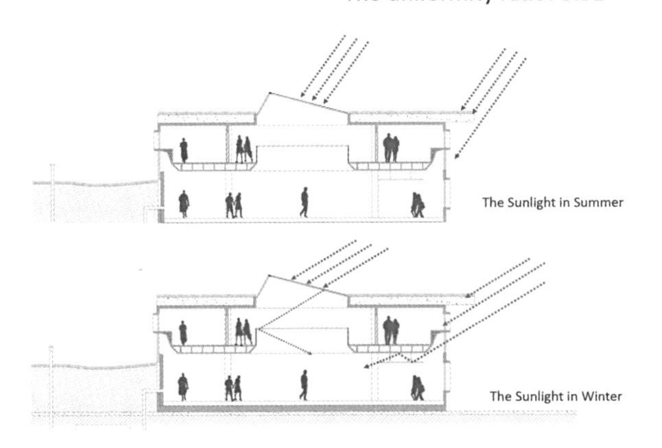

Daytime with wind 01

▲ - Light Analysis 日照分析

▼ - Ventilation Analysis 通风分析

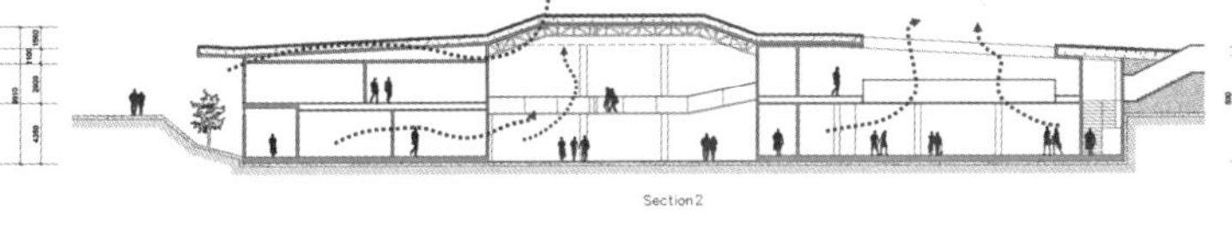

Daytime with wind 02

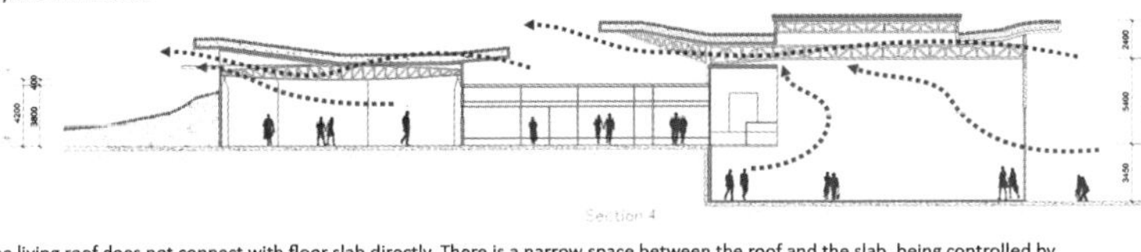

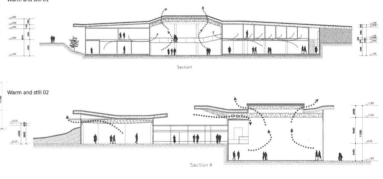

Warm and still 01

Section1

Warm and still 02

Section 4

The living roof does not connect with floor slab directly. There is a narrow space between the roof and the slab, being controlled by double glazing windows and fans. When this space is close, the green roof can be a good insulation shelf. When it opens, this space will enhance the ventilation in ETRI building. Prototyping hall have big volume, such huge space can influence the microclimate in this building. We try to use it in our environmental design.

On breezy days the movement of air across the narrow space increases the "stack" effect.

At night, opening windows in the narrow space can let cooler air in.

At night 01

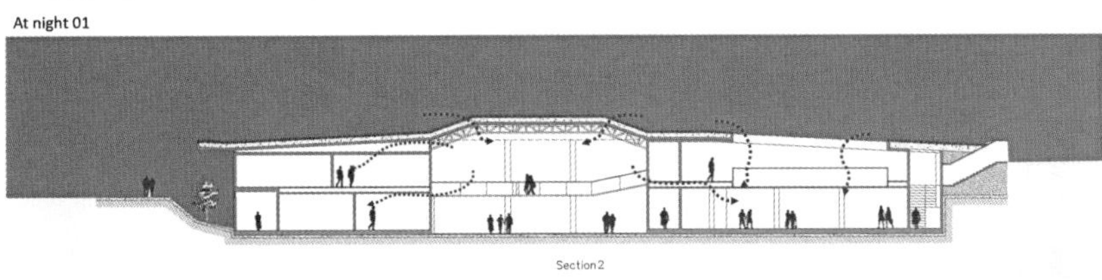

At night 02

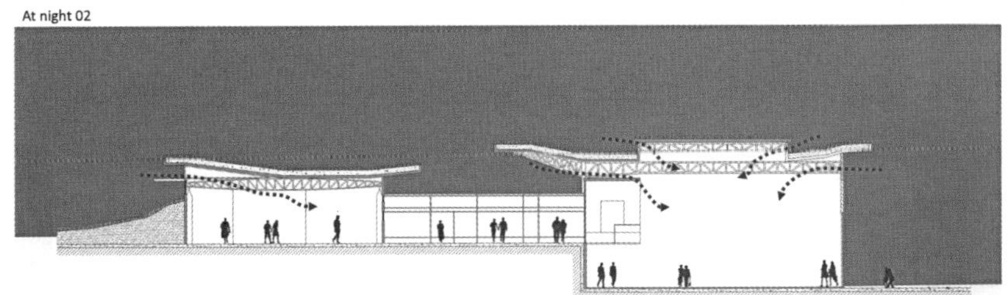

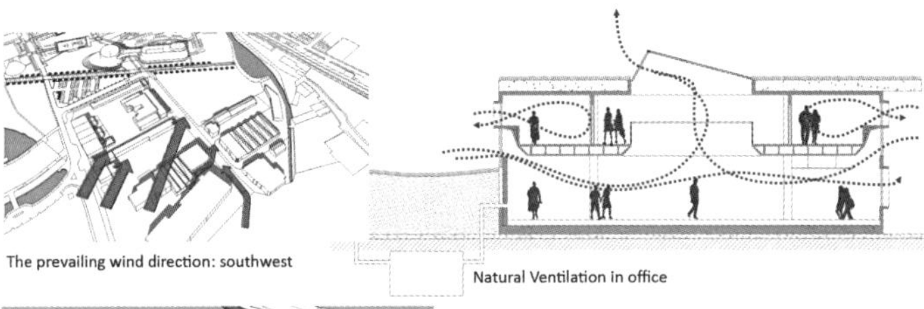

The prevailing wind direction: southwest

Natural Ventilation in office

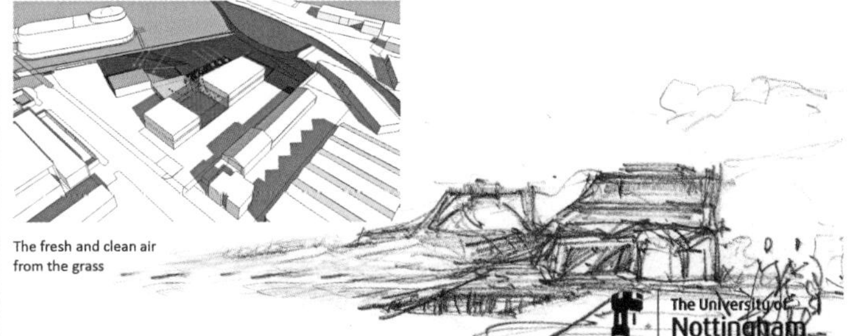

The fresh and clean air from the grass

059

DESIGNERS-设计人: Harriet Palmer, Catherine Legg, Rahul Suralkar

A Sixth Form Centre for a Comprehensive School, Derbyshire

The project developed in part from an analysis of the site and microclimate, which promotes natural ventilation, and encourages daylight usage without excessive solar heat gains. The orientation of roof surfaces was considered to adapt future application of solar thermal and solar photovoltaic panels. Two new routes into the centre are created; one for Public use (ie. Community) and the other for Private (ie. Students). It is intended to provide a series of desirable journeys through spaces for reflection and learning. The proposal exploits the topography, and re-uses part of the existing building to create a new enclosed sixth form centre to bridge the divide between the younger and older academic years. The location of the Centre also created the opportunity to provide a new public face for the School. The point of arrival is celebrated through the arrangement of the auditorium and cascading landscaped steps, which form a visually striking communal space for interaction and public performance.

综合中学第六阶（高三）年级中心，德比郡

本方案的发展主要围绕着对场地与微气候的分析进行，提倡自然通风，鼓励在防止接受过多太阳热能的前提下鼓励自然采光。屋顶的朝向也考虑到将来安装太阳能热水器与太阳能发电板的可能。在中心内设计了两条道路；一条主要为公众使用（例如当地社区）而另一条为私人使用（例如学生）。这样做的目的是为了在可供思考与学习的场所内提供一系列有趣的路径。方案考虑了场地的地形特点，通过对现有建筑的一部分进行重新利用而创造出一处新的第六阶中心，从而可以消除高低年级间的隔阂，使其联系起来。该中心的位置也同时为学校提供了一个新的门面。入口处由大讲堂与层层叠叠的景观设计加以强调，既十分引人注目，又可形成一处社区的互动与公共表演空间。

▼ - Plan 平面图

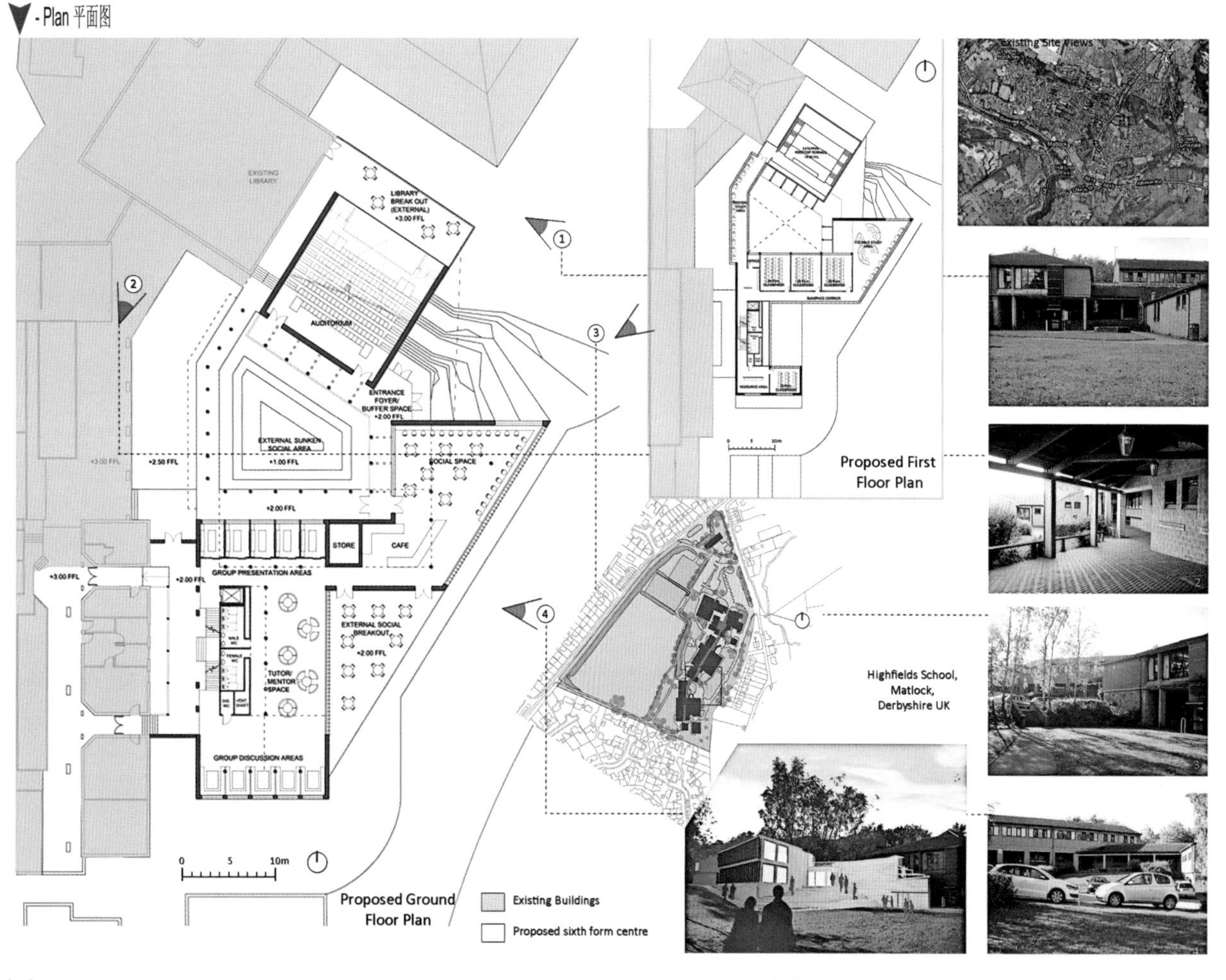

▼ - Perspective 透视图

▼ - Model 模型

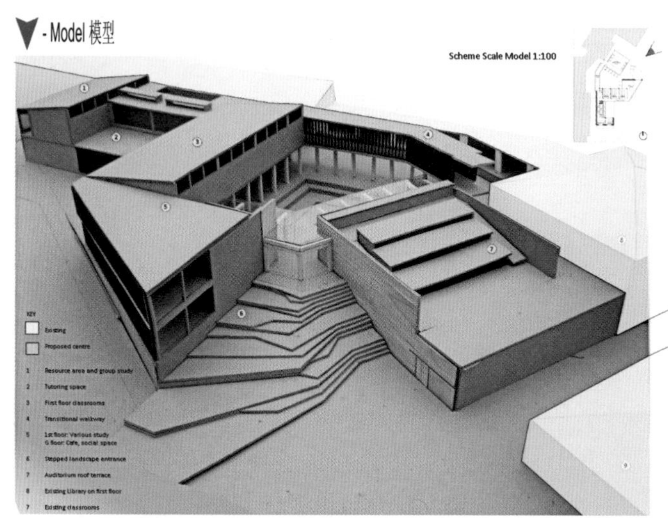

1 Front court landscaped area

2 Entrance to Auditorium

3 Internal view of Auditorium

Public Route LG to Ground Floor

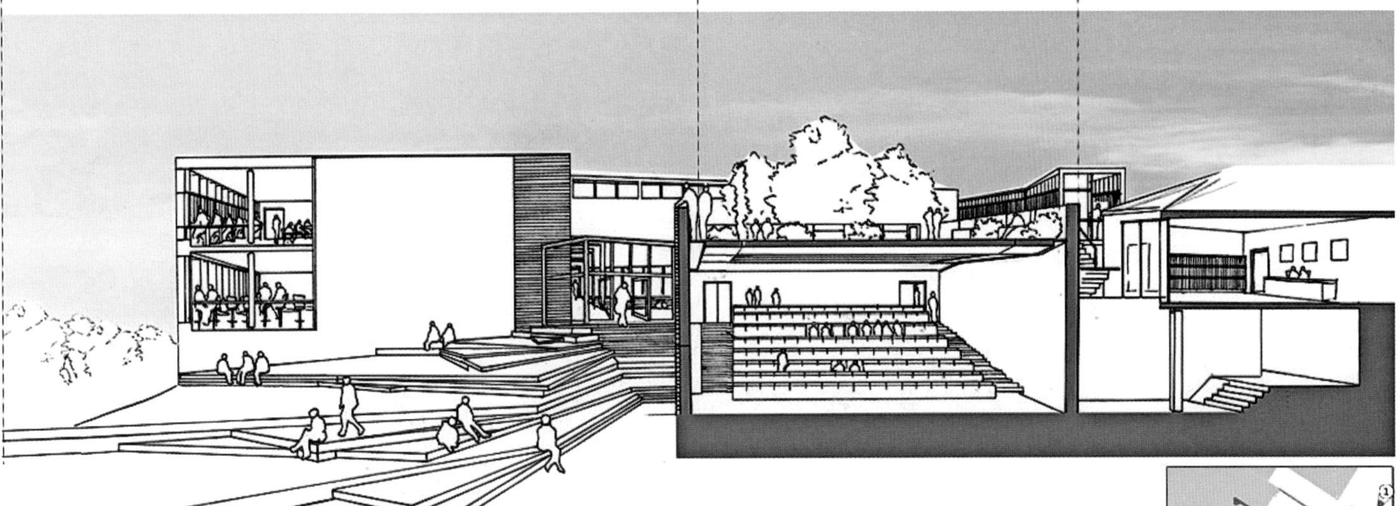

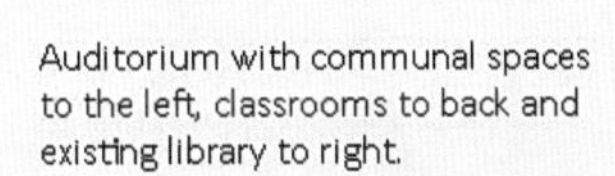

Short Section, Public Entrance

Auditorium with communal spaces to the left, classrooms to back and existing library to right.

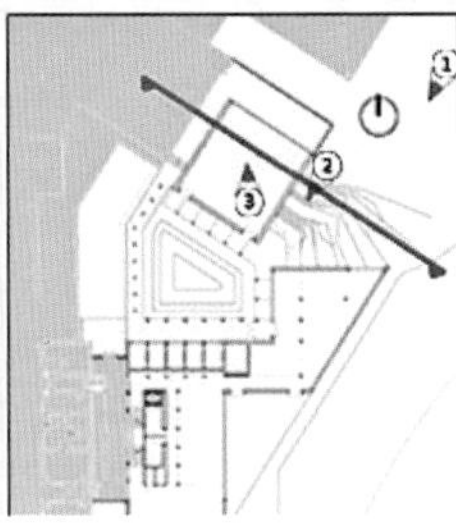

1 View from Auditorium Roof Terrace

2 External Courtyard

3 Transitional space between teaching space and roof terrace

First Floor Student Route

4 External Cafe / Tutoring Breakout Space

5 Student Entrance, Circulation Core

Long Section Student Areas

showing main circulation core with transitional space in front of courtyard to the left and resource room and group study to right.

Ground Floor Student Route

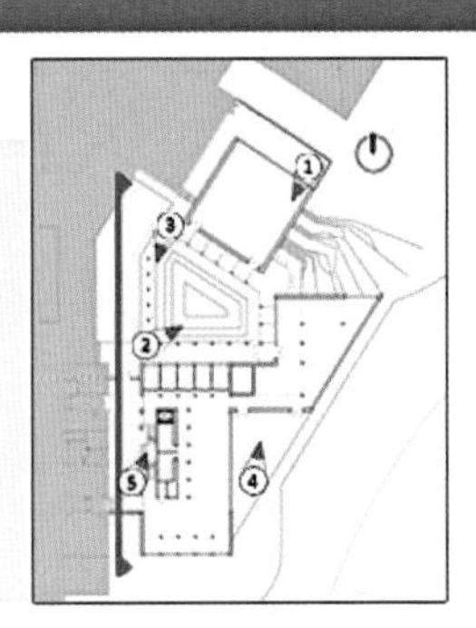

060

DESIGNERS-设计人: Fiona Tan, Stavroula Yiannakou, Phan Anh Nguyen, Yi-li Hseih

A Sixth Form Centre for a Comprehensive School, Derbyshire

The aim of this project is to design a high performance educational building which will achieve 'Zero Carbon' and provide inspiring and stimulating learning and teaching environment. The key design challenge is to create a new entrance to the existing school as a new focal point and to link up different parts of the existing building in a more coherent manner. The architectural intervention focuses on three points: expressing the continuity of the existing circulation spine which links different parts of the school; emphasizing the organic quality of a performing art institution by using organic geometries within a rigid form; using dynamic indoor and outdoor spaces to provide stimulating learning environment for the incubation of creativity in performing art. By adopting the integrated environmental design approach, the architectural ideas and environmental strategies developed were testing qualitatively and quantitatively. Detailed testing on selected spaces is presented to illustrate the integrated design process.

综合中学第六阶（高三）年级中心，德比郡

本项目旨在设计一种高性能教学楼，使其实现“零碳”效果并且提供具有启发性和激励作用的教学环境。关键的设计挑战包括创建适用于现有学校的新型入口使其成为该校的新聚焦点，以及使用更加协调一致的方式将现有建筑物的不同部分连接起来。这里，建筑物所要体现的干预性特点集中于三点：表现连接学校不同部位的已有中轴流线的连续性；通过使用严格的有机几何形式强调表演艺术机构的有机品质；使用动态的室内外空间提供对培育表演艺术创造力具有激励作用的学习环境。通过采取综合性的环境设计方法，成熟的建筑理念和环境策略得以定性和定量地测试。针对不同选定空间内详细测试的描述将（在这里）被用来展示综合设计过程。

- Plan 平面图

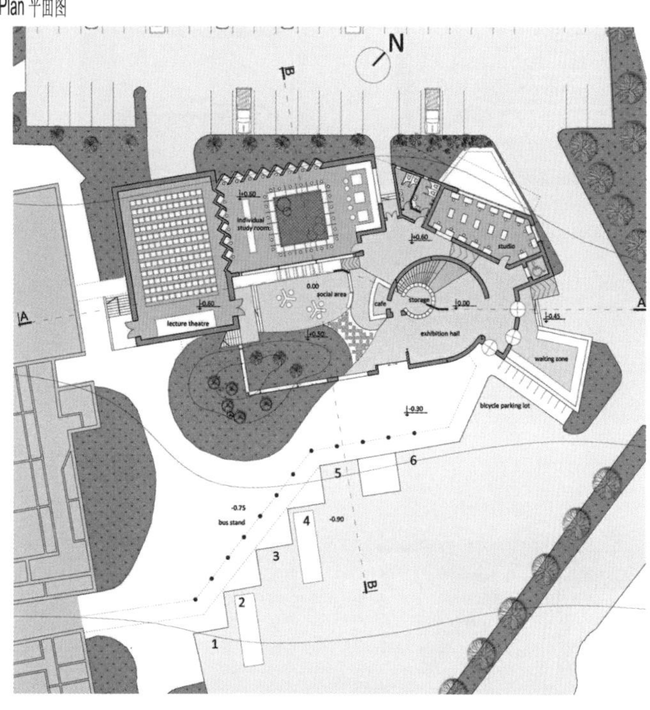

- Sketches 工作草图

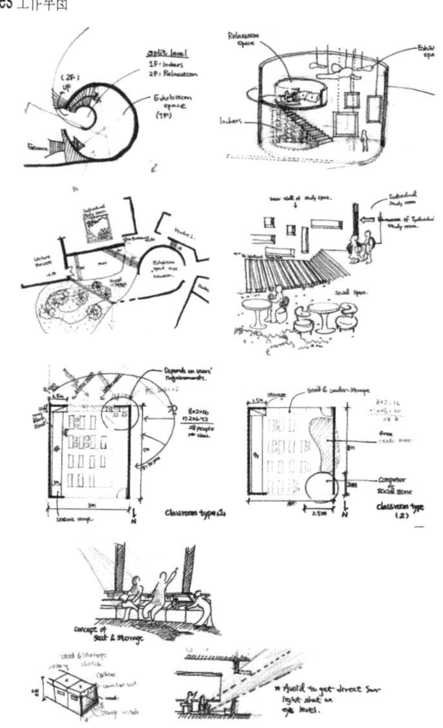

- Elevations and Models 立面图与实体模型

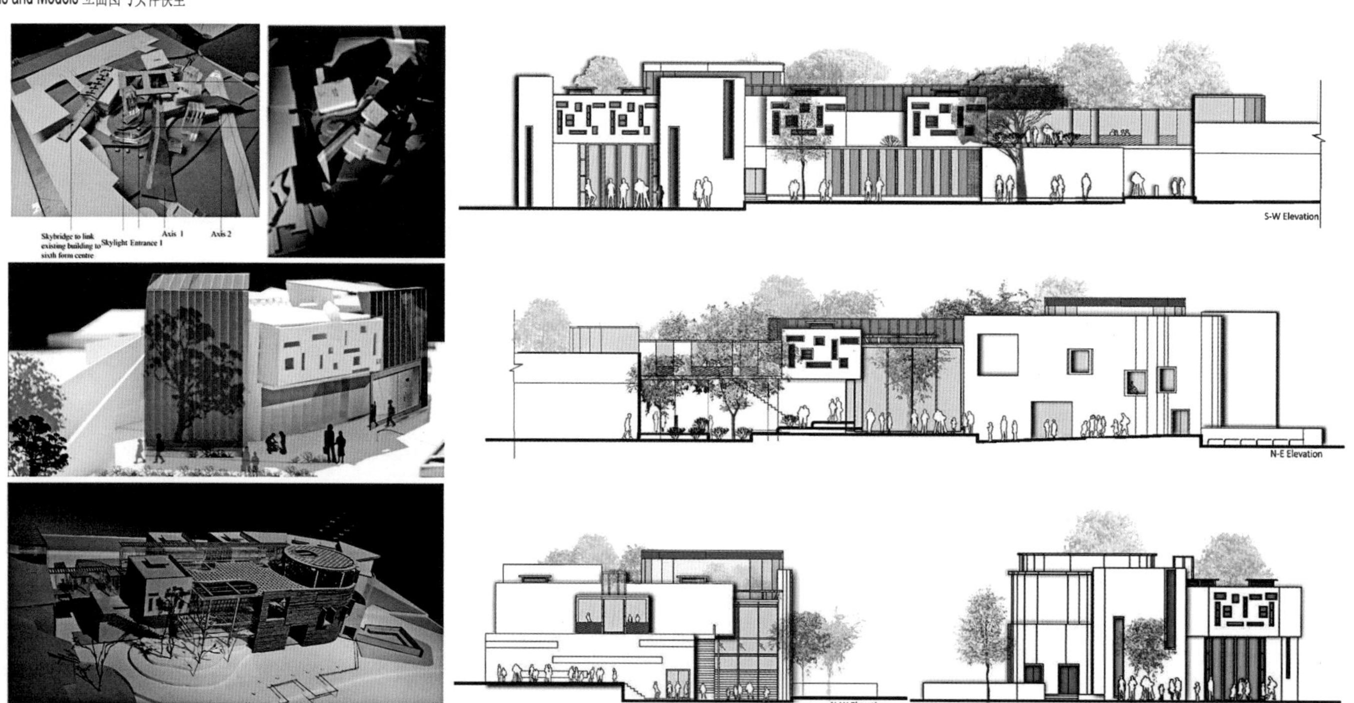

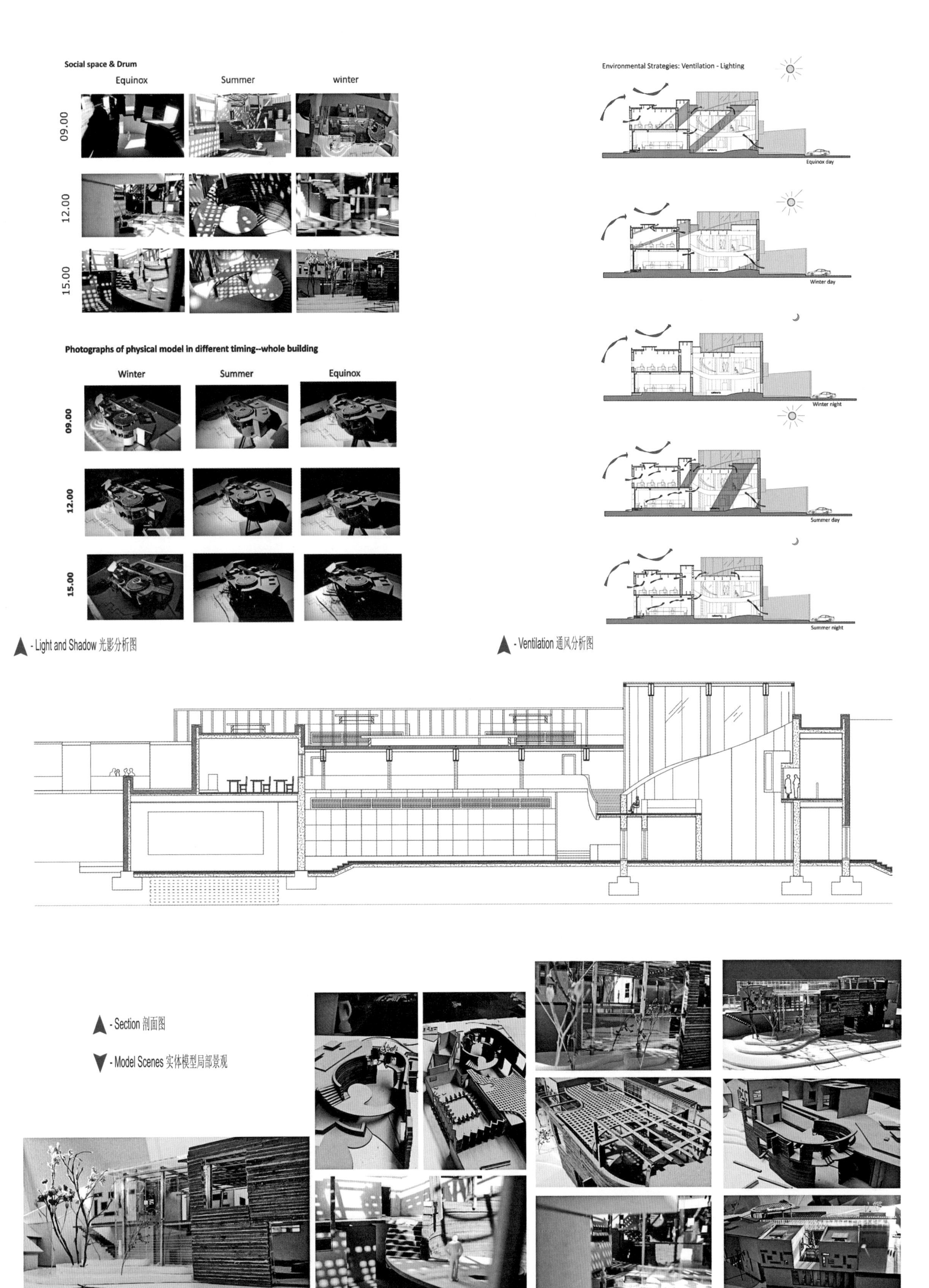

- Light and Shadow 光影分析图

- Ventilation 通风分析图

- Section 剖面图

- Model Scenes 实体模型局部景观

DESIGNERS-设计人: Phillip Oluwatoyin, Viren Mahida, Mohammed Patherya, Dishant Jariwala

A Sixth Form Centre for a Comprehensive School, Derbyshire

In the UK, all new school buildings will be designed to 'zero-carbon' standards from 2016. With this in mind, and in responding to the need for a new Sixth Form Centre for Highfields School, Matlock, we wanted to demonstrate that 'high performance' can be combined with design excellence. The new building provides learning and teaching facilities for about 300 sixth formers including individual study spaces and social areas. The school is proud of either its identity as a performing arts centre or the sixth formers set the right example to the younger students, through participation in community service programmes and organised charity events with the younger students. The project developed from an analysis of the site and microclimate, which revealed opportunities to protect the new centre from the prevailing winds, while promoting natural ventilation, and encouraging daylight penetration without excessive solar heat gains. These principles informed the design of classroom, study areas and the main lecture theatre. The orientation of roof surfaces was also considered in relation to the future application of renewable energy technologies such as solar thermal and solar photovoltaic panels.

综合中学第六阶（高三）年级中心，德比郡

在英国，到2016年所有新的学校建筑设计为都必须符合“零碳”标准。基于此，我们希望在马特洛克（德比郡首府）的海菲尔德中学第六阶年级中心设计项目中证明生态高效性完全可以与优秀的建筑设计融为一体。新建筑将为300名第六阶（高三）学生提供教学设施，这其中包括个人学习空间与公共空间。这所学校以其表演艺术的课外活动而远近闻名，除此之外，它的第六阶学生也经常参与社区服务计划，并常带领低年级学生一起组织慈善活动，从而为师弟师妹们树立的良好的榜样。对于这些表现，校方深感骄傲。基于对场地与微气候的分析，本方案希望利用当地盛行风来促进建筑自然通风，且同时保护建筑不受过多过大的风吹影响；此外，本建筑也将尽量采用自然采光，但同时不至于吸收过多的太阳热能。这些设计原则均在教室，学习空间与大课堂设计中得到了贯彻。另外房顶朝向也根据将来可能采用的可再生能源技术而给予了特别考虑，太阳能热水器与太阳能发电板均可安装其上。

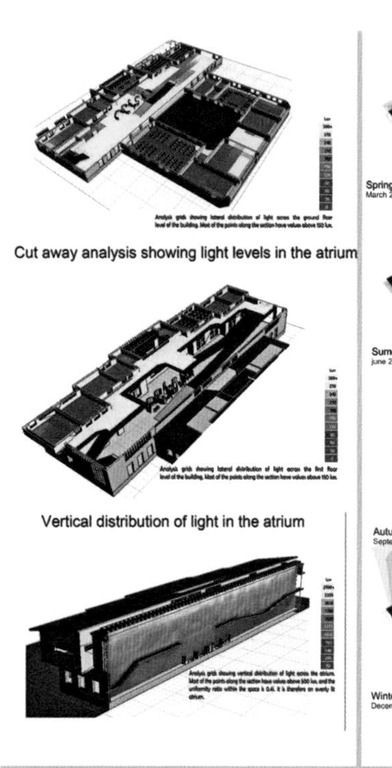

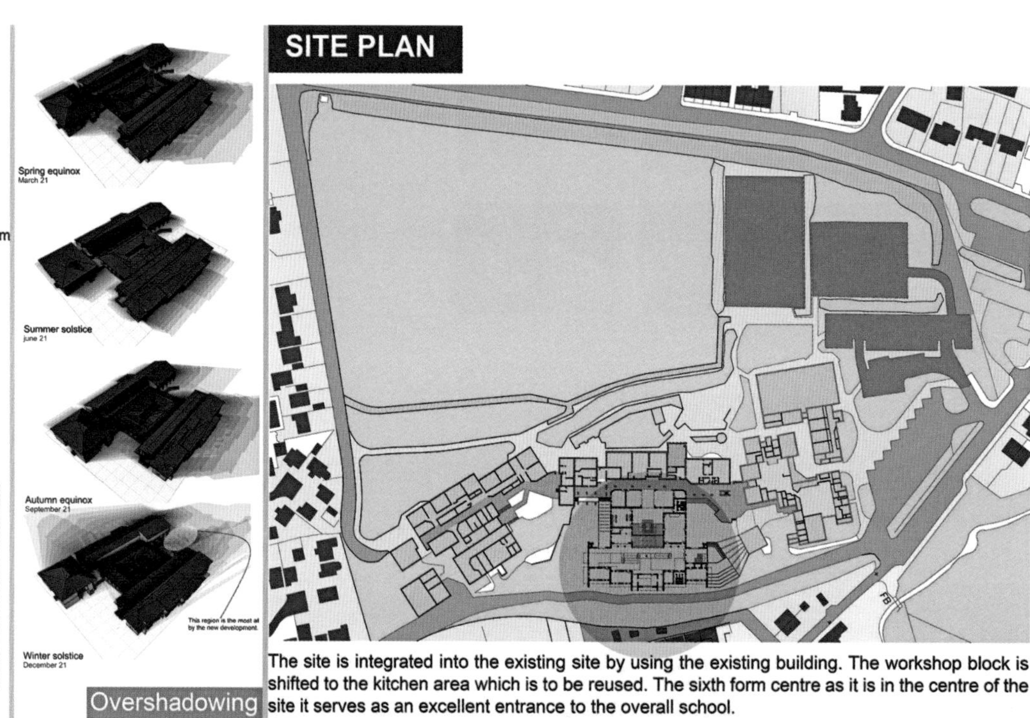
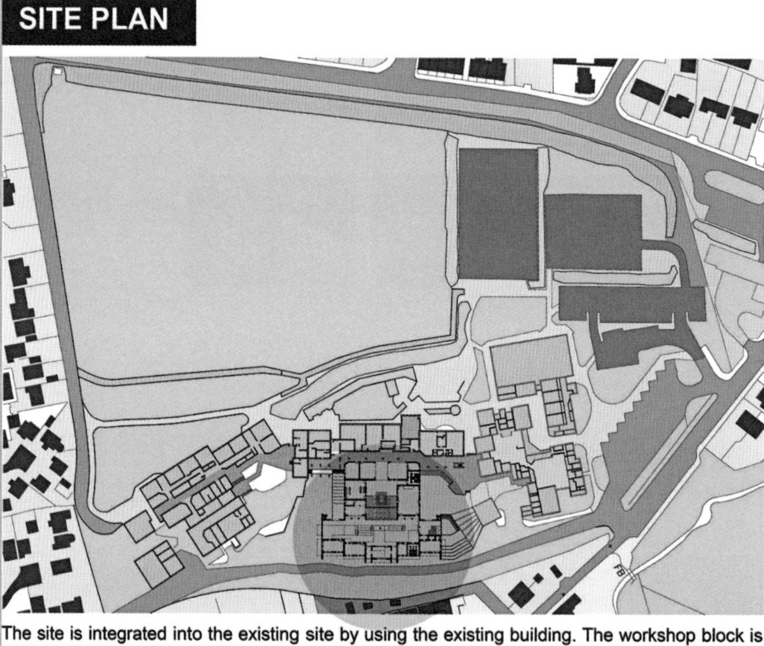

The site is integrated into the existing site by using the existing building. The workshop block is shifted to the kitchen area which is to be reused. The sixth form centre as it is in the centre of the site it serves as an excellent entrance to the overall school.

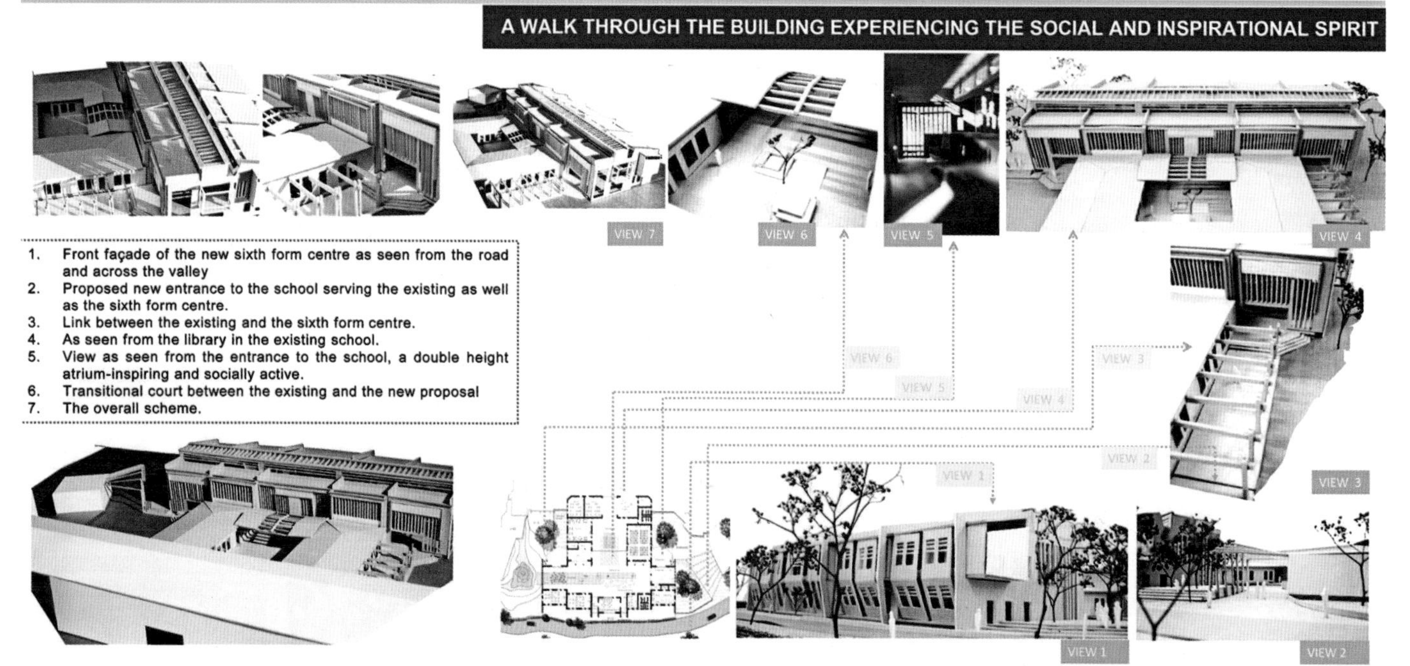

1. Front façade of the new sixth form centre as seen from the road and across the valley
2. Proposed new entrance to the school serving the existing as well as the sixth form centre.
3. Link between the existing and the sixth form centre.
4. As seen from the library in the existing school.
5. View as seen from the entrance to the school, a double height atrium-inspiring and socially active.
6. Transitional court between the existing and the new proposal
7. The overall scheme.

TUTORS - 指导教师: *Brian Ford, Benson Lau, Lucelia Rodrigues*

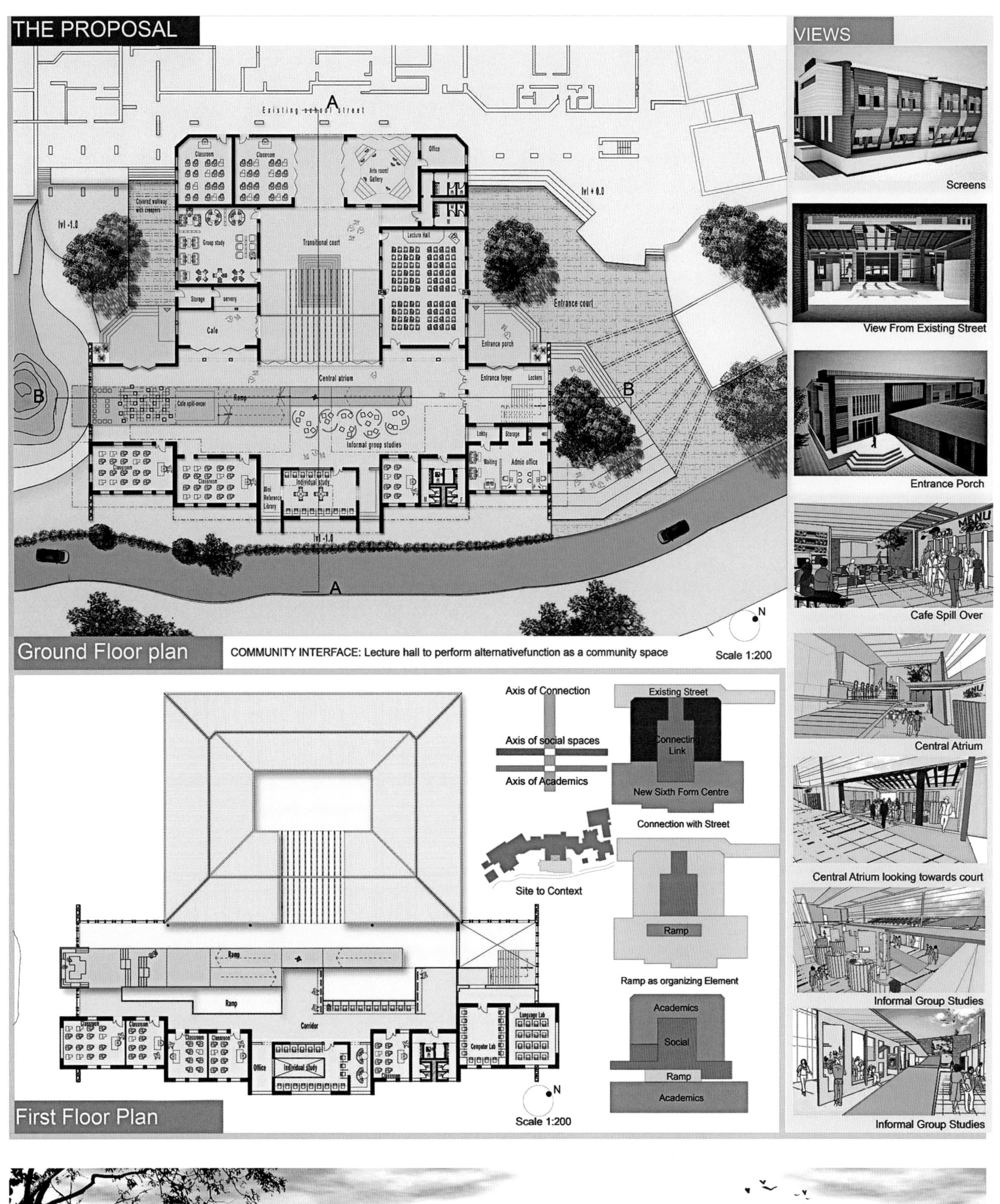
THE PROPOSAL
Existing school street
Classroom
Arts room/ Gallery
Office
lvl + 0.0
lvl -1.0
Covered walkway with creepers
Group study
Transitional court
Lecture Hall
Storage
servery
Cafe
Entrance court
Entrance porch
Central atrium
Entrance foyer
Lockers
Cafe spill-over
Ramp
Informal group studies
Lobby
Storage
Waiting
Admin office
Mini Reference Library
Individual study
A
B
N
Ground Floor plan
COMMUNITY INTERFACE: Lecture hall to perform alternativefunction as a community space
Scale 1:200
Axis of Connection
Axis of social spaces
Axis of Academics
Existing Street
Connecting Link
New Sixth Form Centre
Connection with Street
Site to Context
Ramp
Ramp as organizing Element
Academics
Social
Ramp
Academics
Corridor
Classroom
Office
Individual Study
Computer Lab
Language Lab
First Floor Plan
Scale 1:200
VIEWS
Screens
View From Existing Street
Entrance Porch
Cafe Spill Over
Central Atrium
Central Atrium looking towards court
Informal Group Studies
Informal Group Studies

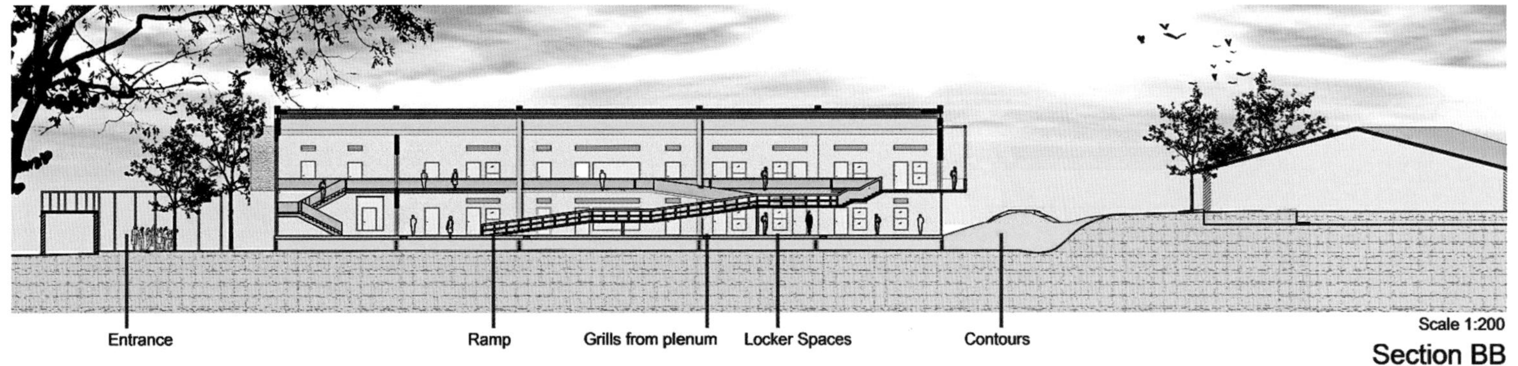
Entrance
Ramp
Grills from plenum
Locker Spaces
Contours
Scale 1:200

Section BB

—— ARCHITECTURE AND TECTONICS
—— 建筑与建构篇

Although the Architecture and Tectonics Research Group [ATRG] aims to the core of architecture and to the issues that are fundamental to the discipline of architecture, its research work spans from human ecology, philosophy, social science to engineering and materials science. Reflecting the polyvalent nature of architecture and the need for transdisciplinary design teams to deliver contemporary architecture, ATRG often works in collaboration with the other research groups and related disciplines from anthropology to engineering and science.

Tectonics can be described as the art and science of construction. Sverre Fehn eloquently states 'all architecture is dependent on construction. Construction seeks the earth; it falls upon it. The eye, light, and thought, that which spatially disturbs these words, [is] construction.' But the origins of Tectonics reside in the Greek word tekton, meaning carpenter and builder and the Sanskrit word taksan, the craft of carpentry. In defining tectonics, K. Frampton notes the poetic intent evoked by this term. G. Semper in the *Four Elements of Architecture* identifies the fundamental basis of architecture as: earthworks, hearth, framework/roof and light-weight enclosing fabric. The light-weight fames of superstructures are tectonic, and heavy-weight earth-based elements are steriotomic. This reflects the emphasis within ATRG on prefabrication, although research includes steriotomics as in Professor Michael Stacey's book *Concrete: a studio design guide*.

ATRG's research seeks to address the fundamental issues of contemporary architecture: re-linking architects and architecture students to a clear understanding of materials and the means of construction, combined with the need to demystify technology. Sverre Fehn observed of the work of Knut Knutsen who taught him architecture "as an architect you scribe your tales into the sand of the earth and the language is the language of materials. In the beginning you meet materials as a challenge. You try with all your power. You force it. But the significant architects develop a dialog with materials."

A fundamental well spring of ATRG research is making and how the decisions related to architecture develop as physical realization is approached. Making is exploring the form of reviving a craft tradition or state of the art through sustainable manufacturing processes. To facilitate this research the University of Nottingham has constructed a Prototyping Hall, with a related external south facing Prototyping Area. A recent example is the development of a low carbon high quality double façade, which minimises the need for central plant, with distributed and integrated services combined with a low carbon footprint, providing benefits to occupants and a lower energy requirement. Some prototypes address specific tectonics issues and many prototypes seek to reduce the demand side energy requirements of architecture and the built environment.

The members of ATRG contribute to the teaching of architecture throughout the years of the school from First Years to supervising PhD students. It is at Fifth Year and Sixth Year for Maters and Diploma [RIBA Part 2] that there is the clearest interface between research and teaching, with Digital Architecture Studio and Zero Carbon Architecture Research Studio [ZCARS] at fifth year and Making Architecture Research Studio [MARS] at Sixth Year.

The projects included in this chapter span from intense urban investigations and intimate caring environments, to new architectural typologies required to deliver sustainability. The quality of this work has been recognized in national awards including two winners of SPAB's Philip Webb Award, Jennifer Routledge in 2008 and Clara Byrne in 2012.

尽管建筑与建构研究组(ATRG)重点着手于建筑的核心内容与建筑学科的基本问题，其研究范围却横跨了人类生态学，哲学，社会科学到工程学与材料科学等。ATRG经常与其他研究小组，以及从人类学到工程学，科学等相关学科领域进行协同合作，从而不仅体现出建筑的多样性本质。而且强调了当代建筑设计中跨学科设计团队的必要性。

建构可以被描述为构造的艺术与科学。斯维勒•费恩曾极具说服力地描述道：“所有建筑均依赖于构造。构造追求与大地的接触；立于其上。眼睛，光线，与思想，能够在空间上影响它们的，唯有构造。”而建构一词则源于希腊单词tekton，意思是木匠和建筑工，以及梵语taksan，意思是木工手艺。关于建构的定义，K•弗兰普顿注意到了其反映出的诗意。G•散帕尔在《建筑四要素》中则确认了建筑的基础：土方工程，火塘，框架/屋顶，与轻质围护材料。这其中，那些体量巨大却相对质量较轻的结构框架体本身就是“建构”，而那些源于大地的沉重建筑元素则被称为“石构”。在ATRG，尽管在迈克尔•斯泰西教授的《混凝土：设计指南》一书中，涉及了对石构的研究，但是我们的主要精力还是集中在建构之上，这一差别主要体现在ATRG对预制技术的重视上。

ATRG的研究主要着眼于当代建筑的基本问题：重新将建筑师/建筑学学生与对材料和构造手法的明确认识联系起来，并根据需要展示技术。斯维勒•费恩就在研习他的建筑学老师——科纳特•库特森的作品时受到了启发：“作为建筑师，你的故事应该书写在大地的沙子上，而所用的语言则是材料的语言。刚开始时，你会把材料看作是一种挑战。你会竭尽所能，甚至强行使用它。但是一个伟大的建筑师则可以与材料直接对话。”

ATRG研究的一个根本源泉是制造，以及如何做出相应的决定，使建筑方案可以逐步发展为具体现实。制造的本源即是探索如何在可持续生产的过程中复苏传统手工艺或艺术的地位。为了支持这些研究，诺丁汉大学特地建造了一座建筑原型生产车间，而与之相连的南向室外延伸则是建筑原型建造场。最近的一个实例是低碳高质双层立面项目，它可以分散与整合各种服务，降低建筑碳足迹，进而使对中央环境控制设备的需求降到最低，为住户提供好处并减少对能源的消耗。一些建筑原型是处理具体的建构问题，而大多数原型则是用来测试如何减少建筑和建造环境对能量的消耗。

从学院的第一学年到博士生课程，ATRG的成员在建筑学教育上均有贡献。在硕士课程与英国皇家建筑师协会第二阶段职业建筑师培养计划——即五年级与六年级，研究和教学会结合得更加紧密。其中，数字化建筑设计组和零碳建筑设计研究组[ZCARS]是五年级课程，而建筑制造研究组(MARS)是六年级课程。

本章中的项目涵盖了从对城市的强化调研与对环境的热切关注，到可表达可持续发展思想的新型建筑类型。这些工作成果也获得了国家级奖项的认可，其中包括古建筑保护协会所颁发的菲利普•韦伯奖的两名获奖者——2008年的詹妮弗•罗德里奇和2012年的克拉拉•伯恩。

—— Prof. Michael Stacey
迈克尔 · 斯泰西 教授

The Head of the Architecture and Tectonics Research Group
建筑与建构研究组主任

062

DESIGNER-设计人: Eleanor Farrant

Tectonic Salvage

St Andrews Church sits precariously on a fast encroaching coastline. Through the incorporation of an archaeologist working above and below ground, the history of the Covehithe Church is explored and displayed using salvaged iconographical sculpture from within St Andrews. Architectural salvage, including that of surrounding churches is brought to the site and sold on, with the occasional use of an auctioneer, whilst the archaeologist records the former journey of the items through time. The design process involved an early use of tectonic ideas at strategic and detailed scales. This method directly informed the end of year exhibition which combined 1:1 tectonic fragments alongside spatial studies. Technology, space and atmosphere, locked together as developmental strands, subvert the notion of instrumental technology leading to the idea of material practice as a humanistic pursuit. The objective of the process was to reinstigate a dialogue between culture and technology.

建构性拯救

圣安德鲁教堂坐落于正被急剧侵蚀的海岸线上，状况十分危险。与一位在地上与地下工作的考古学家合作，我们凭借存储在圣安德鲁教堂内的废弃肖像了解并展示了科夫海斯教堂的历史。旧建筑的很多遗存，包括来自周边教堂的物品，都被带到此处出售，有时候还需要一位拍卖师，而这位考古学家则记录下了这些物品历经沧桑的旅程。该设计过程包括在战略层面与细节层面上对构筑观念的早期使用。这一方法直接影响了学年末设计展的布置——将一个1:1的构筑体片段模型与空间分析相结合考虑。技术、空间和氛围，绞锁在一起如同发展的脉络，颠覆了对将物质实践看作人文追求目标的工具依赖化技术的信仰。而这一方案的目标则在于重新鼓励文化与技术之间的对话。

▼ - Archive Section

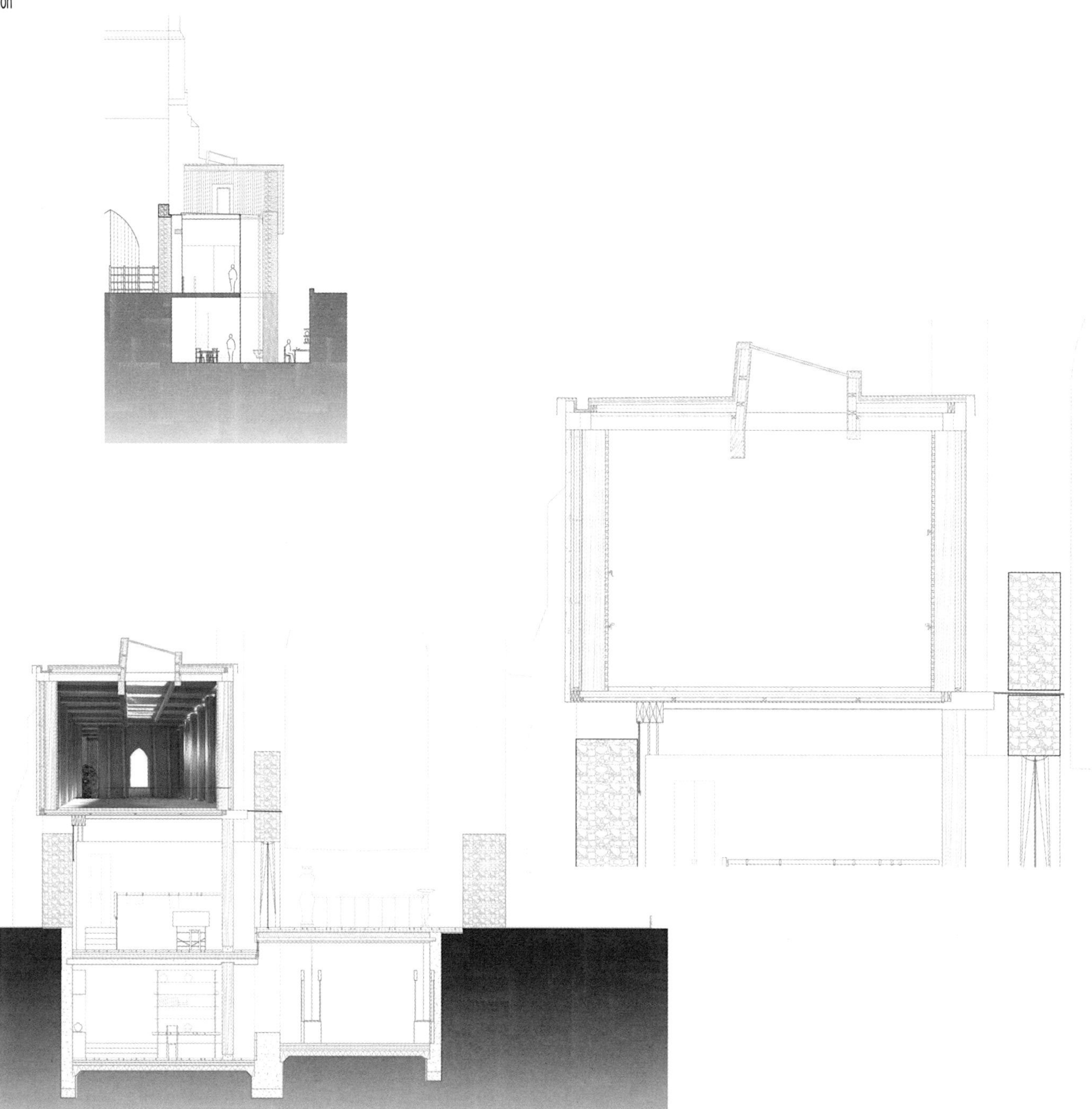

▲ - Section Development 剖面图

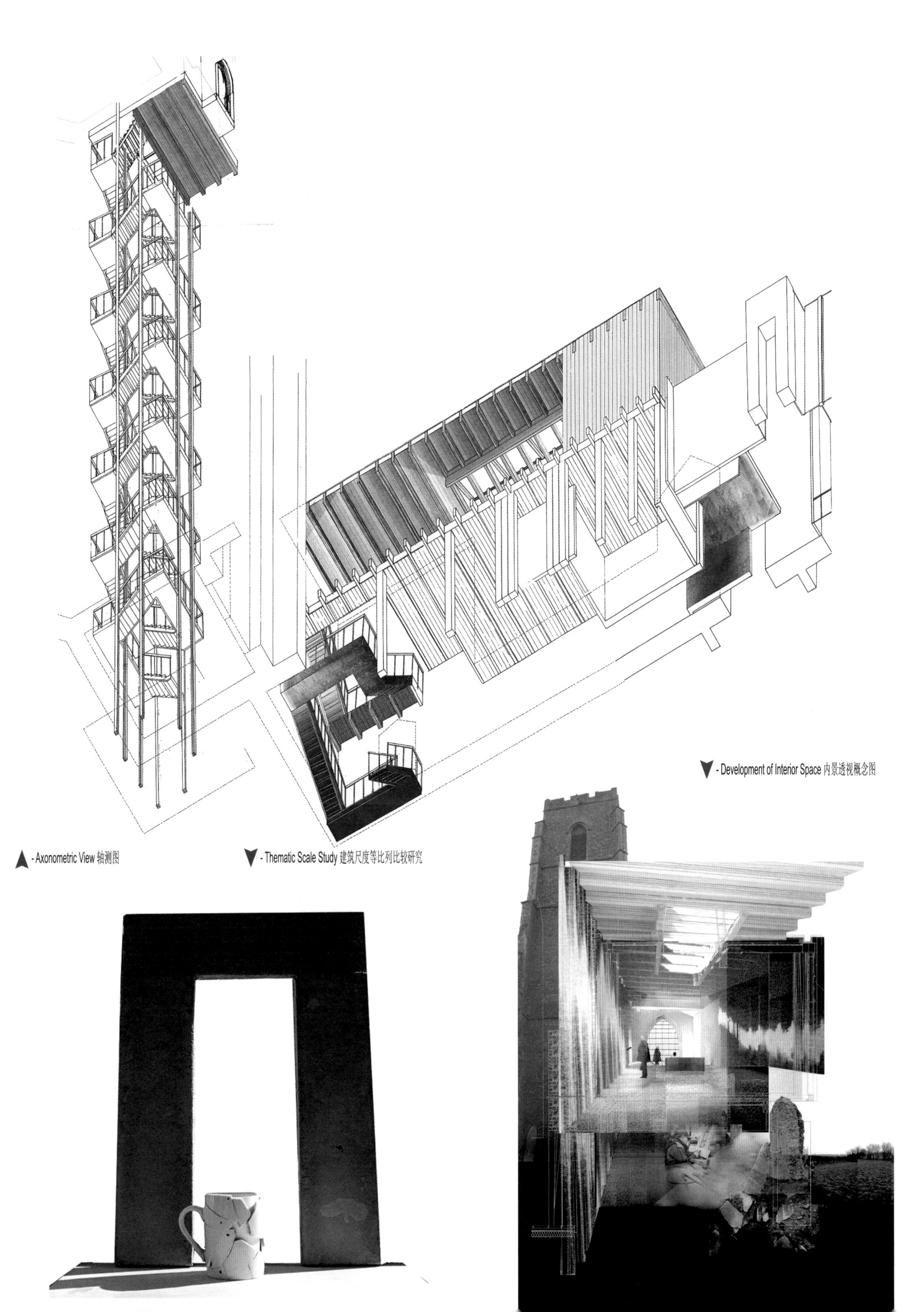

▼ - Development of Interior Space 内景透视概念图

▲ - Axonometric View 轴测图

▼ - Thematic Scale Study 建筑尺度等比列比较研究

063

DESIGNER-设计人: Victoria Fabron

Stone Repairation

The stone restoration centre is a programme that mediates between the laboratory process and the practical production process. Two coexisting programmes allow a stone carver to work on replacement stones with the conservator. The different 'languages' of scientist and tradesman enter into a rich discourse, through the programme that forges lateral connections between them. A laboratory of archaeological and scientific documentation contains equipment for analysing material qualities of stone, converting them into digital, written, drawn and photographic data. This is done in the attempt to preserve not its physical being, which is "earthly", but its non-physical, "ideal/divine" being. The stonemason's knowledge of stone is incomparable with any kind of equipment, laboratory testing and observation. He has experienced it, learnt it, in terms of its resistance against himself, his methods, skills and tools.

石材整修

在石料修复中心里，实验室中的研究工作与实践生产过程将被中和为一体。这两种过程被并置在一起，从而使雕塑家与保护专家可以一起工作，替换建筑上的石材，此外，科学家与商人的不同“语言”也将在此融汇形成丰富的积淀。通过这一计划，所有这些不同领域将会被锻造成一体，实现一种横向的联系。这里有一间考古与科研记录实验室，室内装备着用来研究石料材质质量的设备，并可将这些信息数码化，转化为文字记录，绘图与照片。实验室并不试图保护石材的物理特性，因为这仅仅意味着它的尘世意义，相反这里很注重对象的非物理特性，因为那包含有理想与神圣的寓意。石匠对石头的认识是任何设备，实验测试与观察所不能比拟的，因为坚固的顽石不断抗拒着雕刻方法、技艺、工具以及石匠本人，而正是在这种交手过程中，石匠获得了对石料最深入的体验与了解。

▼ - Site Diagram 场地分析图

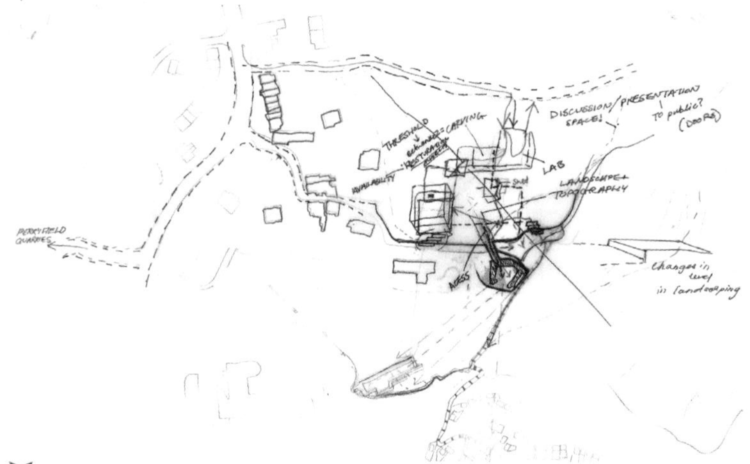

▼ - Strategic Model 建筑概念模型

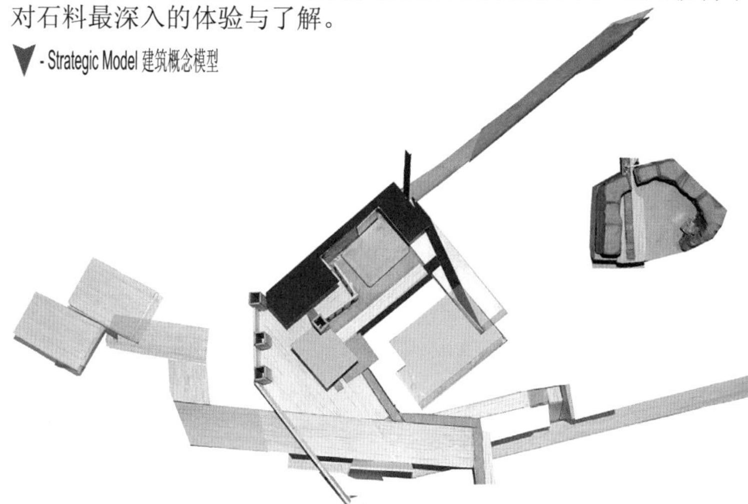

▼ - Site Relief Model 场地立体模型

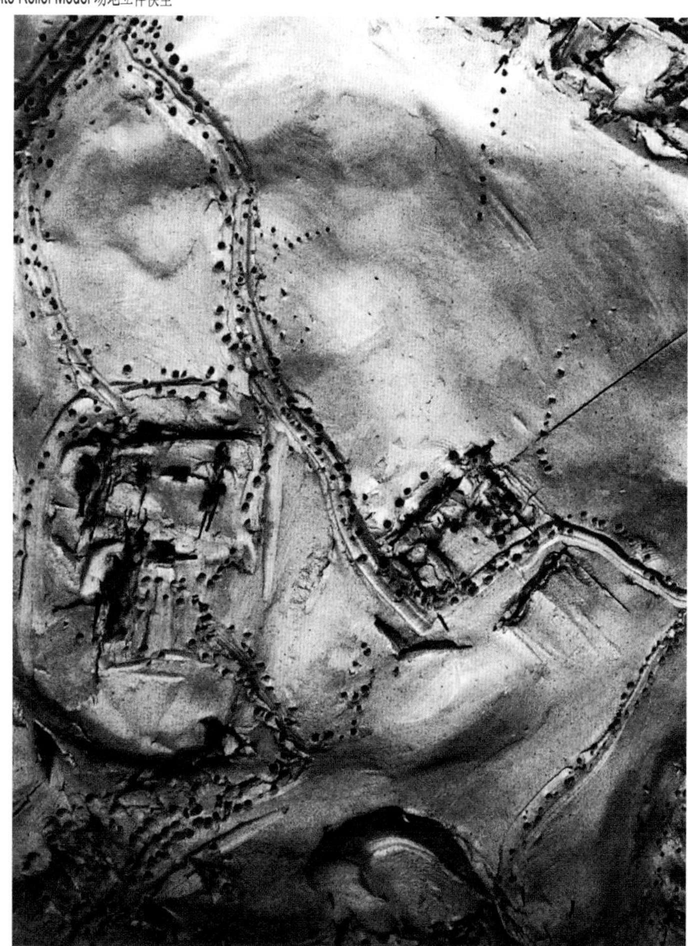

▼ - Material Sudy 材料研究

▲ - Conceptual Section 概念剖面图

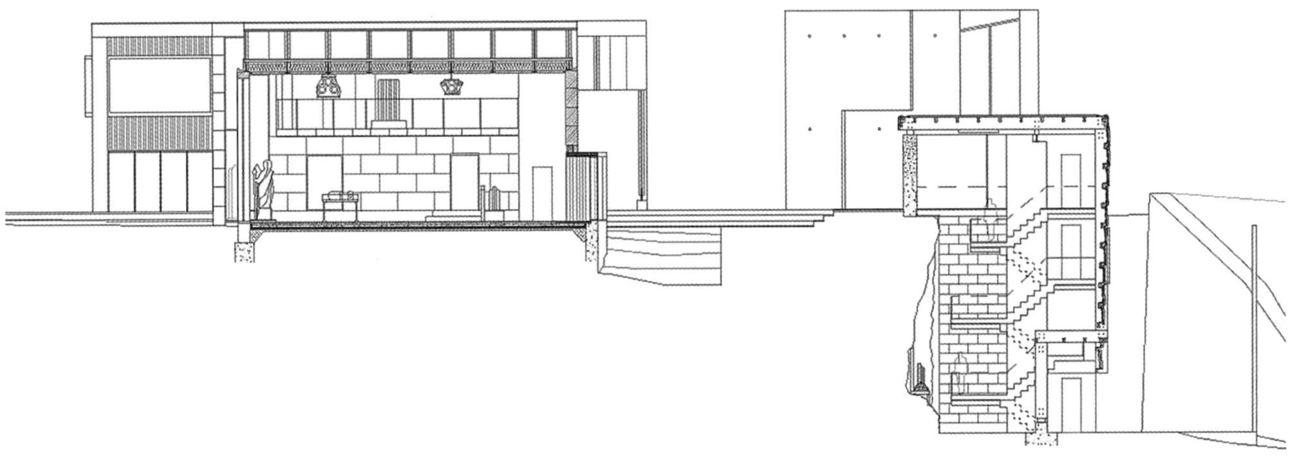

▲ - Topographic Section 技术性剖面图

▼ - Workshop Courtyard Perspective 工作室中庭透视图

DESIGNER-设计人: James Boon

Public Records Office

In the Doomesday Book Portland was the first entry for the Dorset area, sighting its importance as a major political power. Nowadays this power has been stripped from it. There is a great resentment on the island that it has no self-government having lost its Town Halls and legal institutions in the 1980s to a nearby larger town. A new Public Records Office is proposed where people can register births, deaths and marriages, thus restoring control of their own records and reinstating a spatial focus to the island. The Public Records office would be a permanent space aimed at locals who go there for a reason, this creates an interesting contrast to the transient space of tourism. Although both elements have different qualities, their programs would both overlap with a reading room where older records would be accessible to visitors of the hostel, offering people who are passing through the island for a few days a chance to catch a glimpse of Portland's rich history and culture.

公共注册办公室

在末日审判书中，波特兰是进入多赛特地区的首要通衢，从而使之成为一方重要的政治力量。今天，这种力量已经剥离消失。然而，自从1980年代起，这里的市政厅与法律部门就被并入了临近的大城市，从而使岛上充斥着一种由于没有当地政府机构而造成的不满。因此，一所新的公众注册办公室被纳入设计。在这里，人们可以注册出生、死亡和婚姻，从而既恢复了当地人对自己档案的控制，又复苏了这座小岛的空间焦点。这座公众注册办公室对于当地人而言将是一处永久性机构。大家为了各种目的前来办事，而旅游者也会把它当作一处短暂停留的景点来参观，从而制造出了一种有趣的对比。尽管这两种空间元素有着不同的要求，但在一间阅读室中二者却能得以中和。在那里，古老的档案可以同被住在附近旅馆中的旅游者查阅，从而为这些仅在岛上逗留几天的匆匆过客创造了一睹波特兰悠久历史与文化的机会。

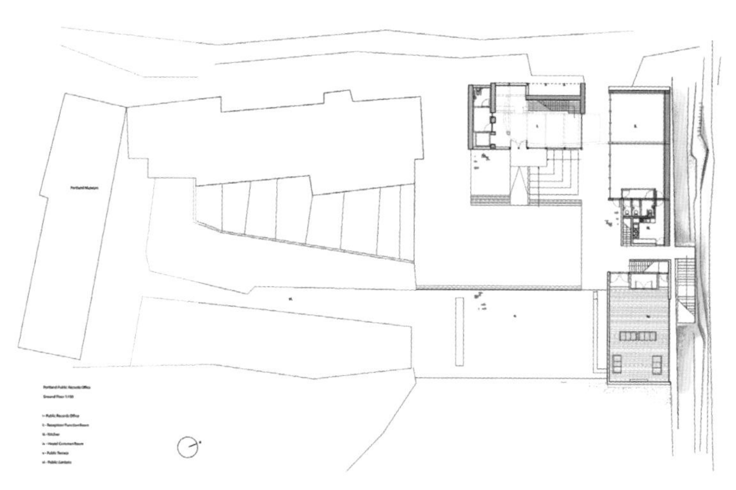

▲ - Ground Floor Plan 底层平面图

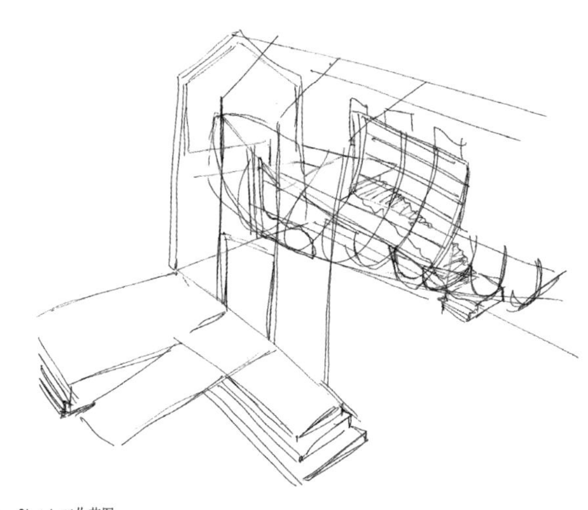

▲ - Sketch 工作草图

▲ - Developmental Model 工作模型

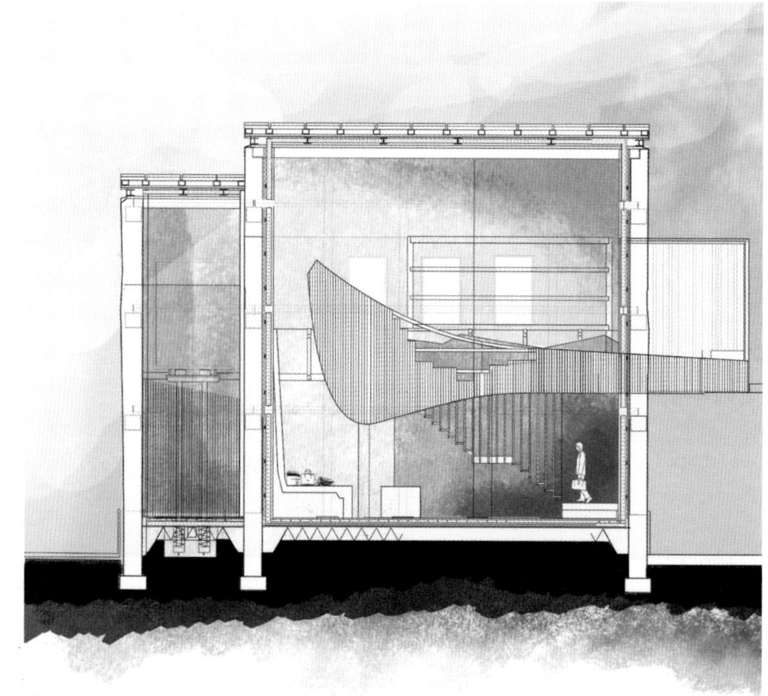

▲ - Tectonic Idea and Resolution 构造研究

▲ - Studio Practice 设计组工作场景

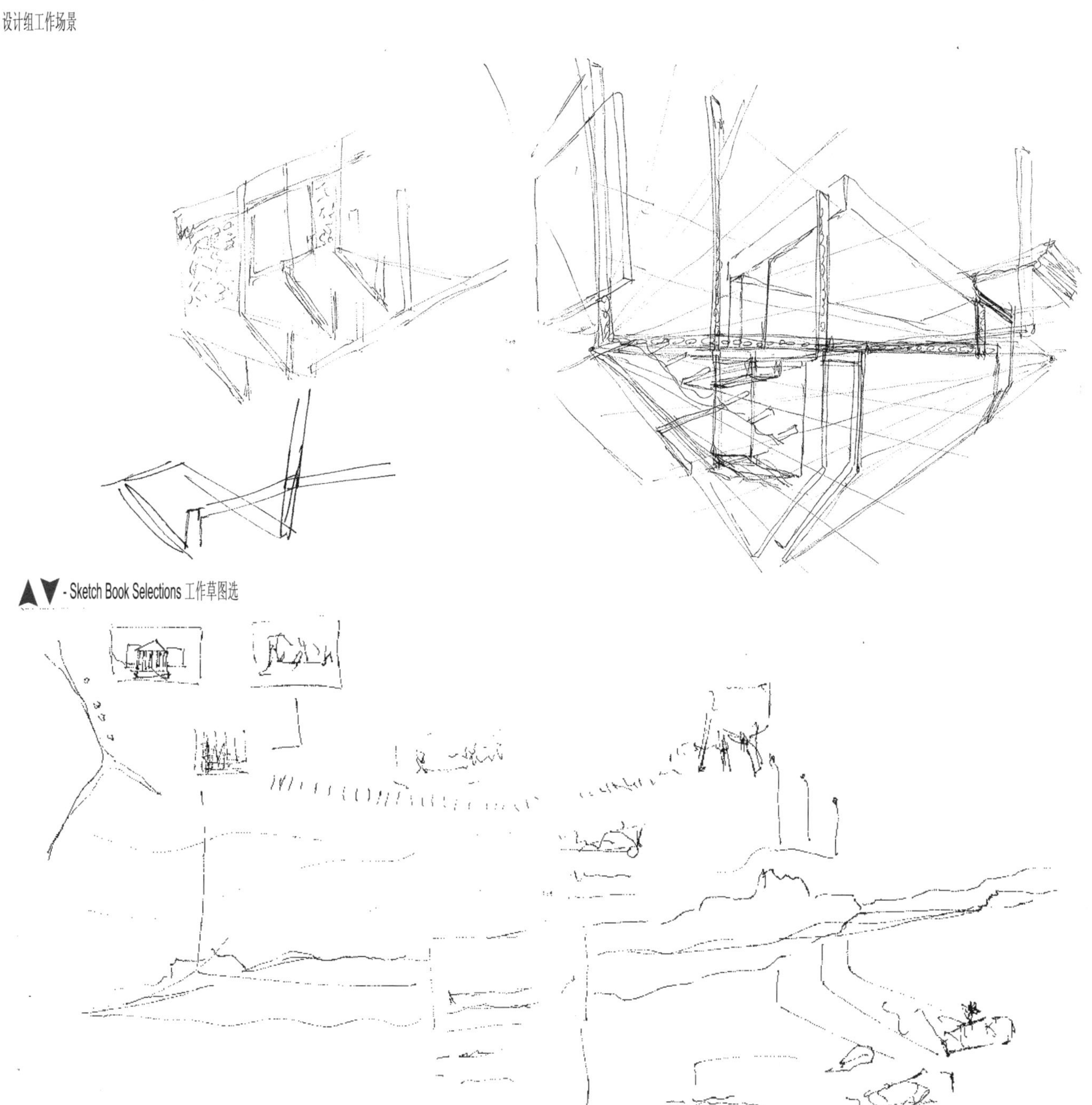

▲▼ - Sketch Book Selections 工作草图选

DESIGNER-设计人: Samuel Clarke

Jurassic Coast Geology Laboratory

The programme developed from an interest in memory, preservation and time, into an archaeological research institute that overlaps with geology. The core of the programme revolves around private research facilities and laboratories, conference rooms and areas where a range of scientists can study and carry out conservation/ preservation work. The space houses a team of experts monitoring the entire stretch of the Jurassic coast and as such provides a central networking place for discoveries and documentation of the changing coastline. This results in a vast archive space which can be publicly accessed unlike much of the world heritage analysis. The new condition encourages people to use the site at different time scales, from the rambler to the resident and researcher. This integration and cross-over of two user groups is what laterally enriches the programming and creates a conversation between the institute and the outside world.

侏罗纪海岸生态实验室

基于对记忆、保护与时间的兴趣，本项目旨在发展一处功能上与地质学领域相互重叠的考古研究中心。方案的核心部分围绕着一些私人研究设施与实验室，会议室和适合科学家们开展研究保护工作的空间布置。这里可为一队监控着整条侏罗纪海岸线的专家提供工作场所，并使他们可以借助一处中心网络工作室来发现与记录岸线的变化。这些科研活动必然要求有一处巨大的档案空间，但是与大多数世界遗产的科学分析成果不同，这里的档案向公众开放。研究中心所提供的这些新条件也鼓励所有的人们——不管是漫步者，当地居民，还是研究者——根据使用时间的长短来尽量利用资源。这种对两类不同使用者群体的整合与交融从侧面丰富了对建筑的使用计划，也在研究中心与外部世界之间创造了一种“保护”。

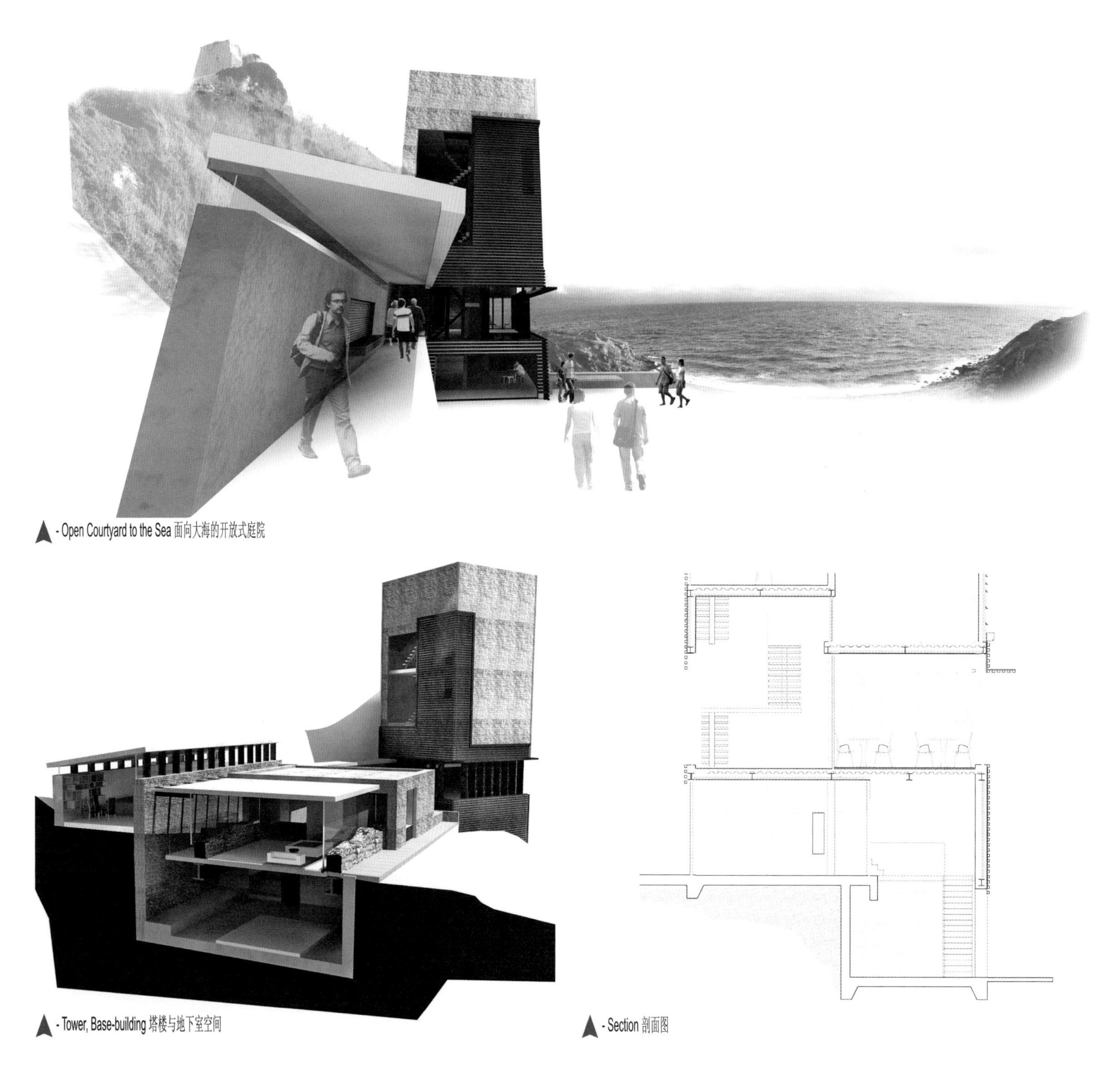

▲ - Open Courtyard to the Sea 面向大海的开放式庭院

▲ - Tower, Base-building 塔楼与地下室空间

▲ - Section 剖面图

- Interior Concept 内景概念

- Courtyard and Cafe 庭院与咖啡屋

066

DESIGNER-设计人: Paul Ornsby

Parametric Train Station, Nottingham

Our lives are increasingly affected by the networking of our world. The Internet, the media and mobile communication are collapsing on sense of scale and distance. However, in contrast to this our cities are often made up of a succession of isolated elements within static boundaries. The architecture of future needs to increase connectivity within the city through the creation of fluidic boundaries, to 'weave' a new urban environment at a local and global scale. Issues of connectivity within the city and the use of a 'textile tectonic' for parametric architecture have been explored through the design of a tram terminus incorporating an informal performance space at Nottingham Train Station. The building is a weaving of transitory events and a point at which different scales of network within the city are linked, allowing new and unexpected connections to be forged.

参数化火车站，诺丁汉

存在于世界中的网络已对我们的生活产生了越来越多的影响。例如，互联网已经是交流在规模与距离上将媒体与移动通讯远远抛在了后面。但是，与之截然相反，我们的城市却常常由一系列被限制在既定边界内的孤立元素所构成。于是通过创造一种流动性边界，未来的建筑需要加强在城市内的相互联系，从而在地区与世界维度上“编织”出一种新的城市环境。通过将一处位于诺丁汉火车站内的非正式表演空间重新设计成一座站台，本方案希望对创建城市中联系的观点和将“编织建构”概念在参数化建筑上应用的观点进行探索。建筑本身其实就可以被看作既是一系列交织在一起的瞬时性活动，又是城市网络在各种不同尺度下交汇的一个点。在这里，崭新的与意想不到的联系将彼此熔合。

- External Perspective 外景透视图

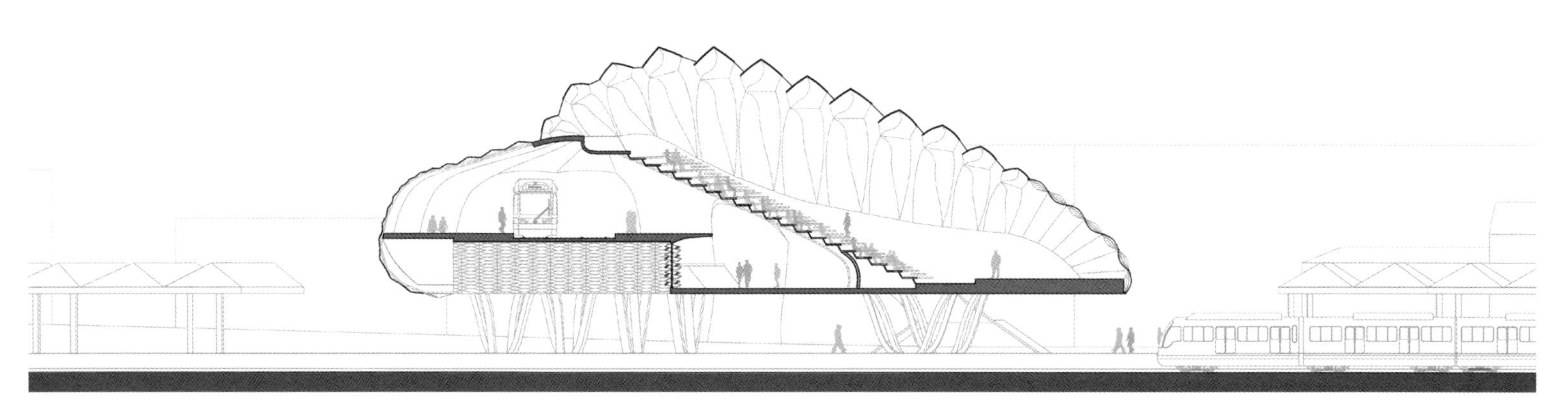

- Section 剖面图

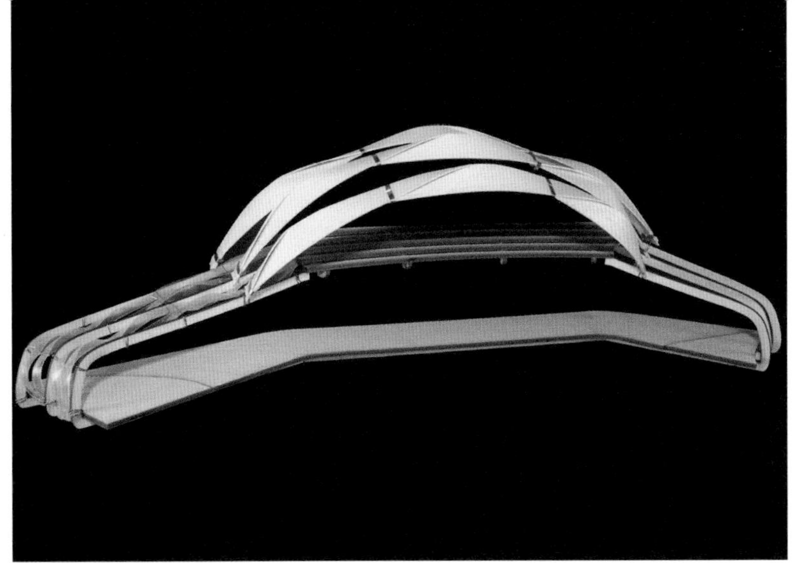

▲ - Physical Models 实体模型

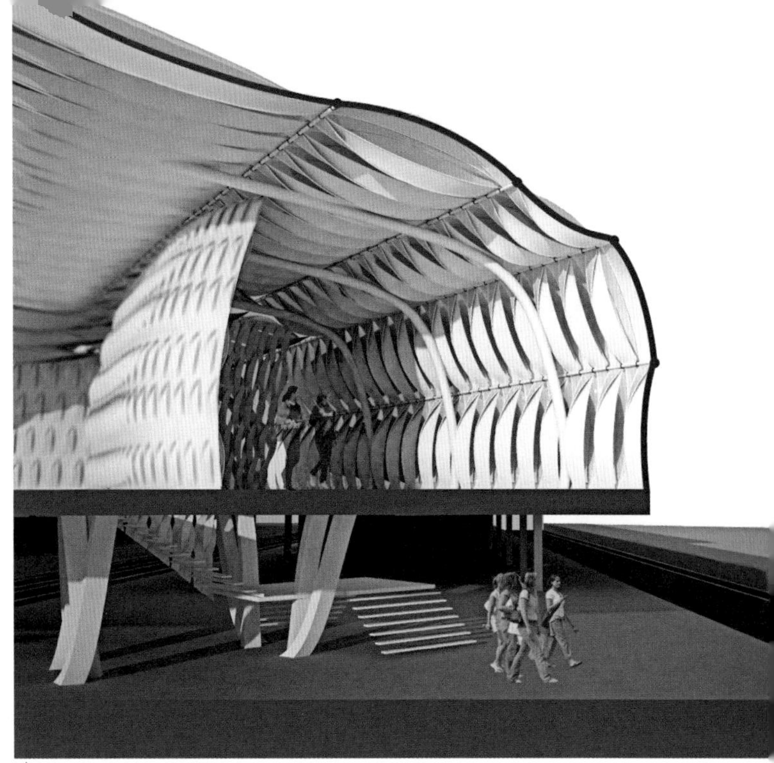

▲ - Perspective Section 剖透视图

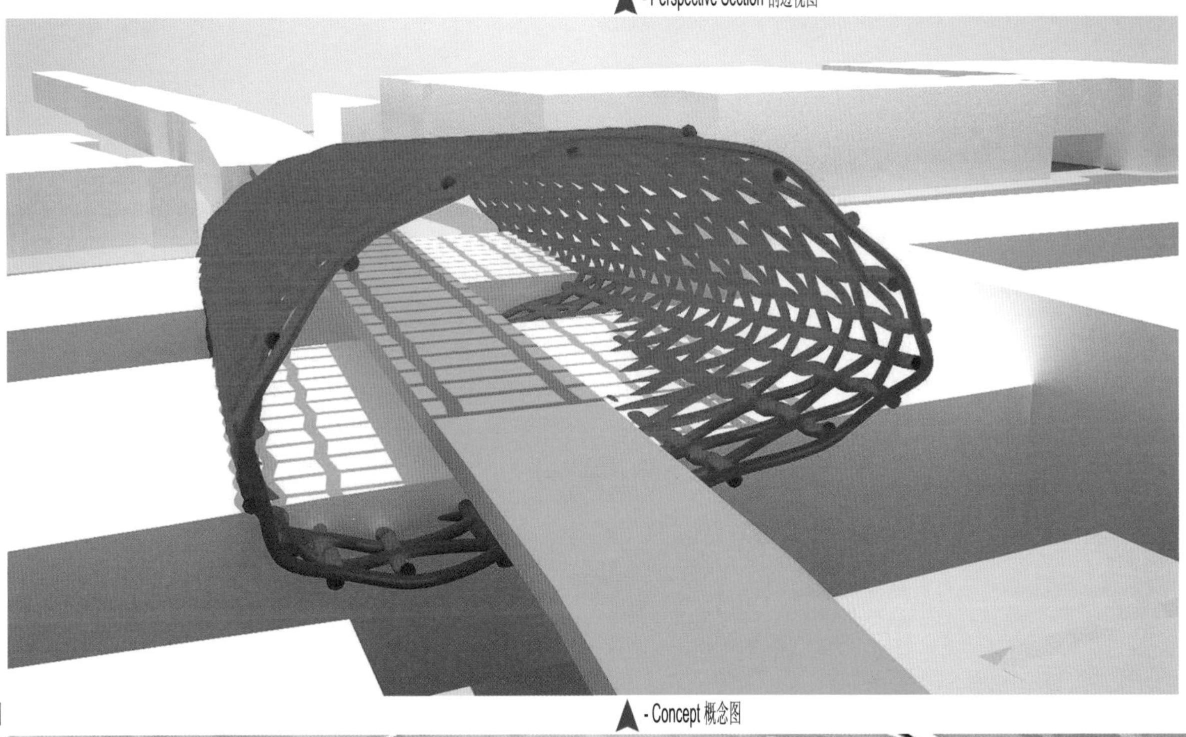

▼ - Audiotorium 观众看台透视图

▲ - Concept 概念图

067

DESIGNER-设计人: Marc Mathias-Williams

The Buoyant Bridge

The bridge that currently crosses the channel connecting the River Trent to Iremongers Pond, Wilford, is low and floods regularly during the winter; breaking in half a footpath that loops around the pond. This project seeks to create a connection that acts both as a reliable passage across the channel, as well as a barometer in the landscape, to the changing water levels. A sign to an environment in flux. The design employs a system of buoys, sitting on a lock that filters the water entering and exiting the pond. When water levels in the lock are low, the load of those crossing the bridge is carried by the cables connecting the buoys – acting like a rope-bridge. As the water levels rise the load begins to be taken by the buoys themselves as they rest on the water. This system is supported by an inflatable/pneumatic lock that also rises and falls with the water levels. When the lock reaches capacity platforms are deployed to provide anglers and walkers with a new viewpoint of the pond and the surrounding landscape.

浮桥

在石料修复中心里，实验室中的研究工作与实践生产过程将被中和为一体。这两种过程被并置在一起，从而使雕塑家与保护专家可以一起工作，替换建筑上的石材，此外，科学家与商人的不同“语言”也将在此融汇形成丰富的积淀。通过这一计划，所有这些不同领域将会被锻造成一体，实现一种横向的联系。这里有一间考古与科研记录实验室，室内装备着用来研究石料材质质量的设备，并可将这些信息数码化，转化为文字记录，绘图与照片。实验室并不试图保护石材的物理特性，因为这仅仅意味着它的尘世意义，相反这里很注重对象的非物理特性，因为那包含有理想与神圣的寓意。石匠对石头的认识是任何设备，实验测试与观察所不能比拟的，因为坚固的顽石不断抗拒着雕刻方法，技艺，工具以及石匠本人，而正是在这种交手过程中，石匠获得了对石料最深入的体验与了解。

- Elevations at Low and High Tide 高潮位与低潮位立面图

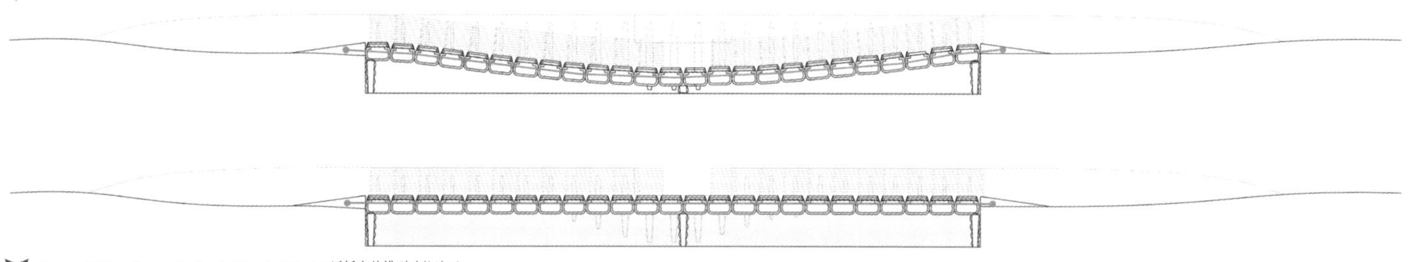

- Buoyant Bridge Demonstration by Physical Model 浮桥实体模型功能演示

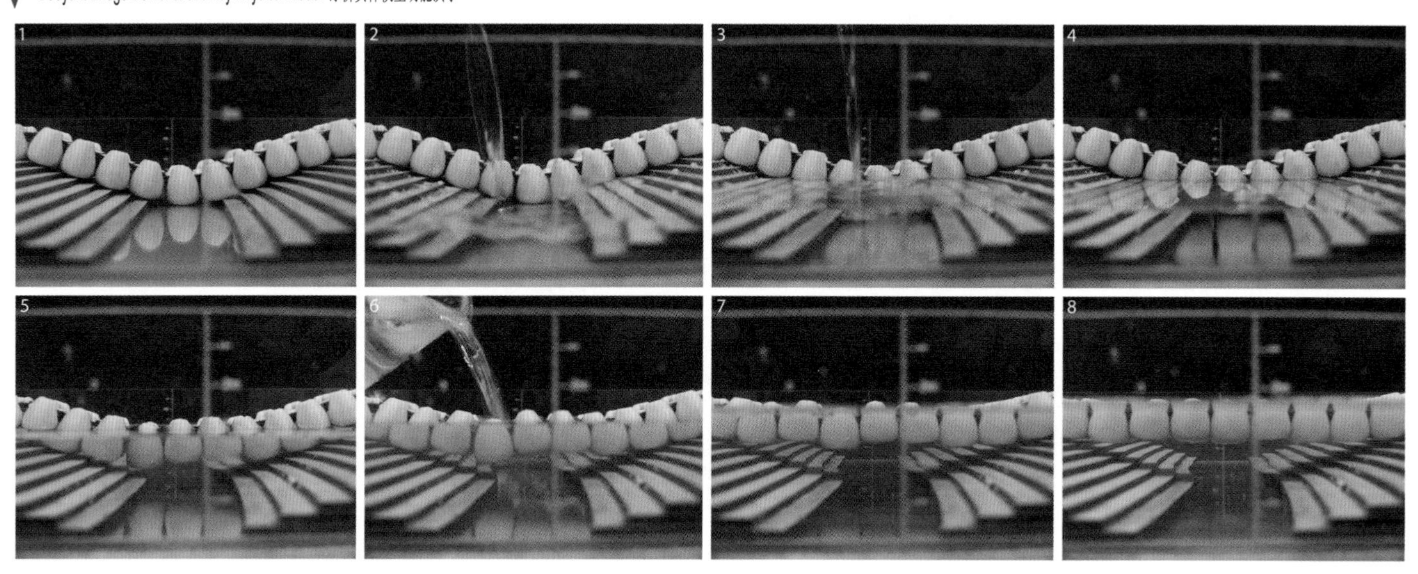

- Perspective

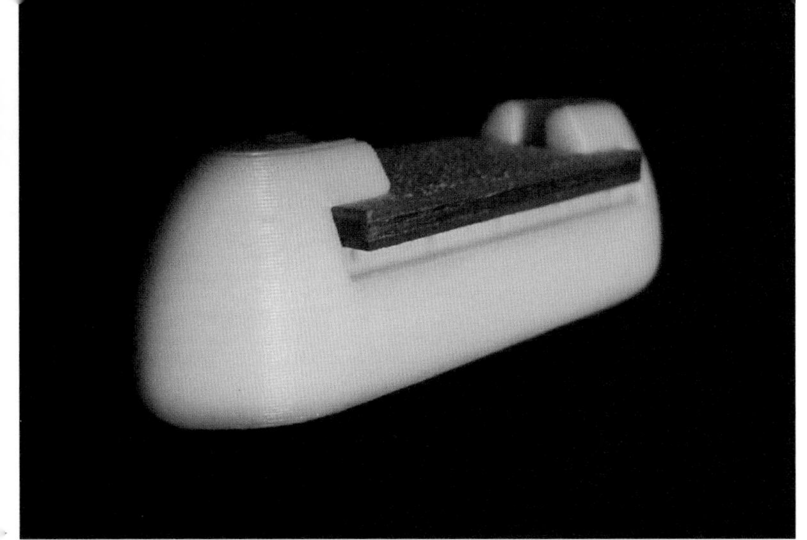

- Rapid Prototype Details 由三维塑形机建立的桥体单元模型

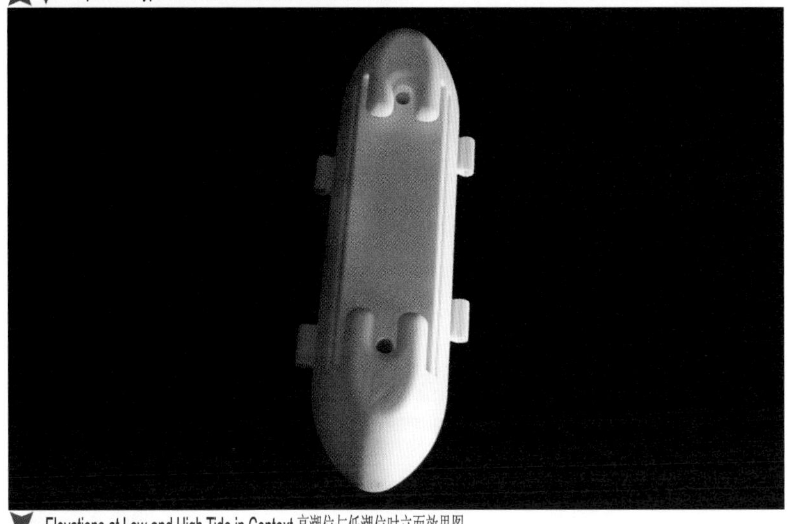

- Elevations at Low and High Tide in Context 高潮位与低潮位时立面效果图

- Perspective 透视图

DESIGNER-设计人: Georgia Christodoulou

Christmas Stalls at Old Market Square, Nottingham

My project is about redesigning the stalls of Nottingham Christmas Wonderland. The aim of this project is to design a unique modular kiosk with a complicated façade that will create interesting light patterns. The kiosks will be repeatedly used throughout the year by the City Council. The kiosks will be rented to the different markets that take place in the old market square instead of the usual marquees that they currently have for rent. The modular kiosks that will be frequently used during weekends will be stored on site in the form of a street furniture. This will avoid the transportation of all components from a storage room to the site and will speed up construction and deconstruction process. Since the kiosks are primarily designed for Christmas Wonderland the presence of light patterns is important. On the one hand, light patterns are spread outwards when the stalls are used at nighttime during Christmas. On the other hand, light patterns are spread inwards when the kiosks are used at daytime for different events. The light effects from the façade pattern were tested through Rhino and physical models.

诺丁汉老市场广场的圣诞节摊位

我的设计针对诺丁汉圣诞游乐场进行改造，其目的在于设计出一个独特的标准化售货亭，其复杂的立面上能打出丰富的灯光图案效果，供市政府全年重复使用。售货亭面向旧集市广场上的各种集市活动进行出租，以代替目前供租赁的小木棚。售货亭的不同模块形式通常在周末时使用，平时则作为街道装饰就地存放；这样可避免组件在储藏点和场地间的运输，也节省了搭建及拆除的时间。由于售货亭主要为圣诞游乐场而设计，灯光效果尤为重要。一方面，圣诞期间摊位会在夜晚向投射出灯光图案；另一方面，在白天，当售货亭进行各种活动时阳光会将图案投射在其内部。立面的灯光效果已通过“犀牛”软件及实际模型的检验。

▲ - Elevation 立面图

▲ - Details 建筑细节

▼ - Site Scan 城市实景扫描合成效果图

▲ - Night View 夜景效果图

▼ - Combinations 体块组合

069

DESIGNER-设计人: Kaplan Pirgon

Hexagonal + Expandable

The transitory event building is designed to expand and contract to facilitate for the wide variety of events that take place on the Old Market Square, Nottingham. The rotational and expandable joints on the hexagonal frame allow for these changes in form. A multi layered facade system comprised of neoprene elastic skin and rigid smart plastic provide the flexibility in the required areas, whilst also controlling the lighting levels and offering views of the existing sites main features. The internal stages are located at either end of the building, and through parametrically defined cut-outs in the skin, impressive backdrops of the town hall and fountain are created. Combined with adaptable furniture the building can be completely transformed with minimal time and effort. The default form of the building has been informed by the sites geometry, circulation of the public and the spatial requirements of the GameCity event.

六角形+可扩展性

这个可伸展的临时建筑面向诺丁汉老市场广场的各种活动。六角形结构的节点能够旋转并扩展，以便在形式上适应各种需求。由氯丁橡胶弹性薄膜以及硬性智能聚合材料而成的多层立面系统提供了充足灵活的空间，它同时控制着照明强度，并突出了场地的景观特点。内部舞台位于建筑的两端，而经参数化设计而形成的孔洞则会成为市政厅和喷泉的一处引人注目的背景。建筑内布置了可变式家具，能以最少时间和人力完成形式转换；而基地几何形态、公共流线以及“GameCity”节的空间要求则限定了建筑的基本形式。

▼- Frame 网架三维模型

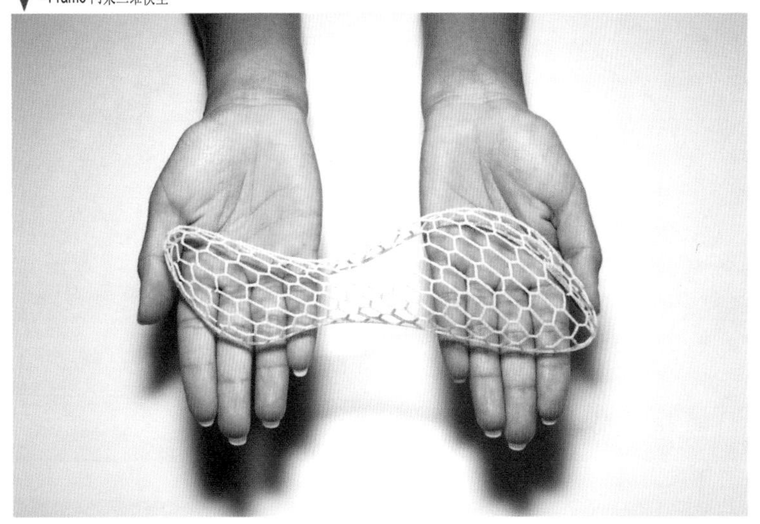

▼- Skin Geometry Definition 表皮几何形态研究

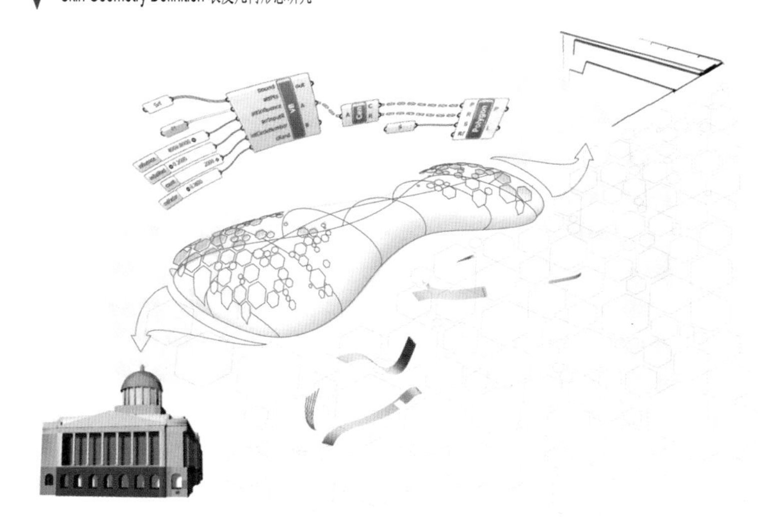

▼- Gamecity 2012

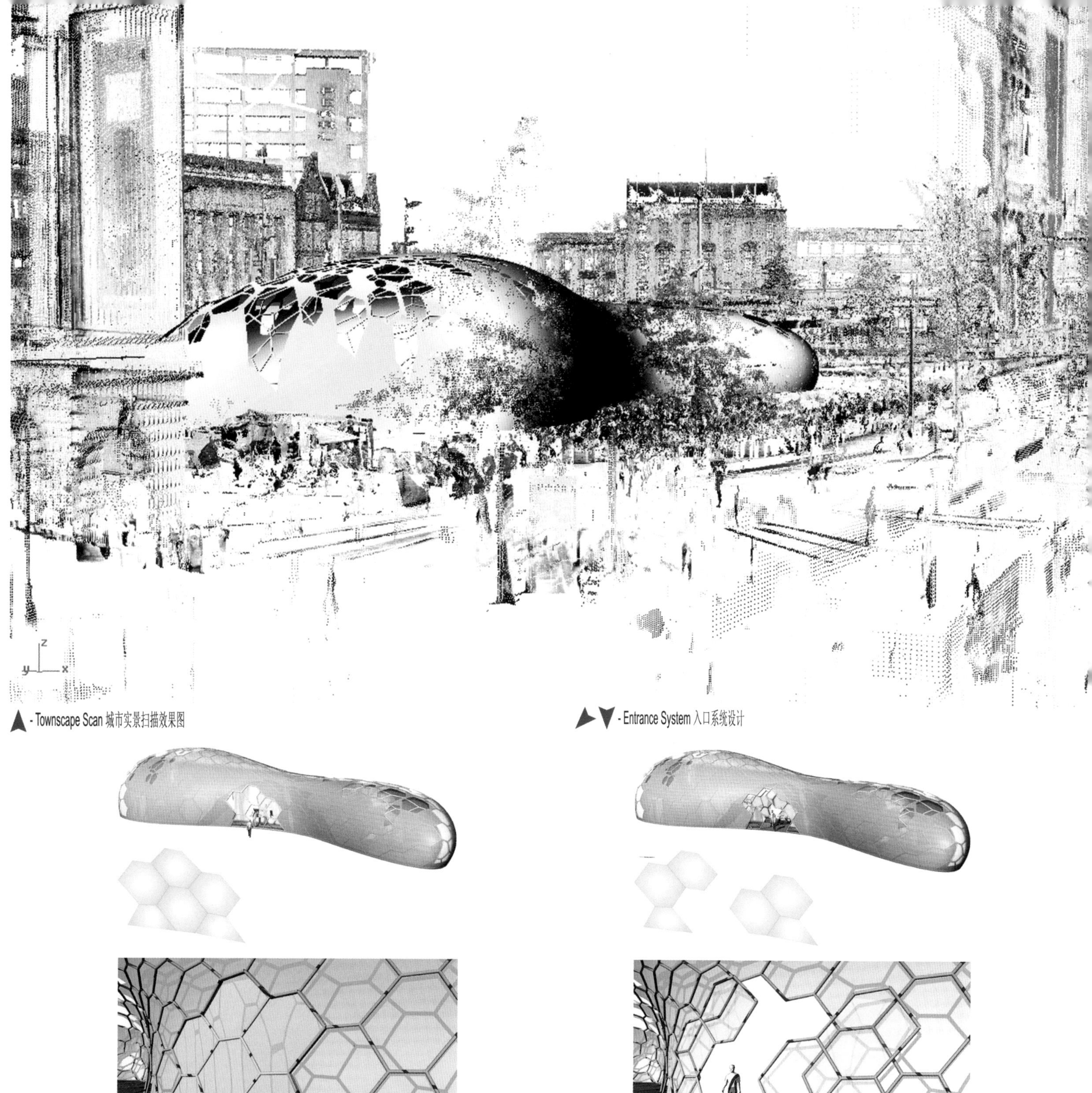

▲ - Townscape Scan 城市实景扫描效果图

▶▼ - Entrance System 入口系统设计

▼ - Section 剖面图

DESIGNER-设计人: Mia A Tedjosaputro

Fashion Pavilion at Old Market Square, Nottingham

The generated individual brief is a (mobile) Fashion Pavilion which exhibits students' fashion design projects. The main objective is to expose students' projects to the public. This is a golden opportunity both for fresh and exciting young students and artists to showcase their talent, and for prospective employers and those in the fashion industry to explore this vibrant and energetic new work. Old Market Square- Nottingham is the primary site, with a high degree of flexibility for the pavilion to be re-used in other cities. Design inspiration came from ruffles by looking at their spatial quality, and their ubiquity as a substantial fashion element since the15th century. A triangular tessellation system appears as the selected form of built feature as well as the primary structural system, and the façade components (smart joints) allow freedom of alteration from one form to another desired form. Additionally, wrapped in a stretchable PVC membrane, the façade components create the impression of movement. The proposed interior arrangement is constructed in such a way that the pavilion is not an internal-oriented exhibition only, but also external-oriented, a feature which may help to attract the public.

诺丁汉老市场广场时尚亭

设计旨在创造一个展示学生时装设计作品的流动展厅。由于其主要目的在于向公众展示学生的作品，这对于初出茅庐但热情无限的年轻学生和艺术家来说是展示其才华的绝佳机会，服装产业的雇主也能对此充满生机与活力的项目加以利用。诺丁汉老市场广场是本作品的主要所在地，但其高度灵活性使得该建筑也可在不同城市内重复使用。褶皱作为服装元素从15世纪起便被广泛使用，从而成为设计关注空间品质的灵感来源。三角形的镶嵌式系统成为建筑形体及结构的特征，立面的智能连接构件方便了依需要而变化的形式。另外，立面构件因外部包裹着可伸展的PVC膜而动感十足。室内布置依展示厅功能而定，不仅满足内部展示，而且对外有利于吸引公众。

- Plan 平面图

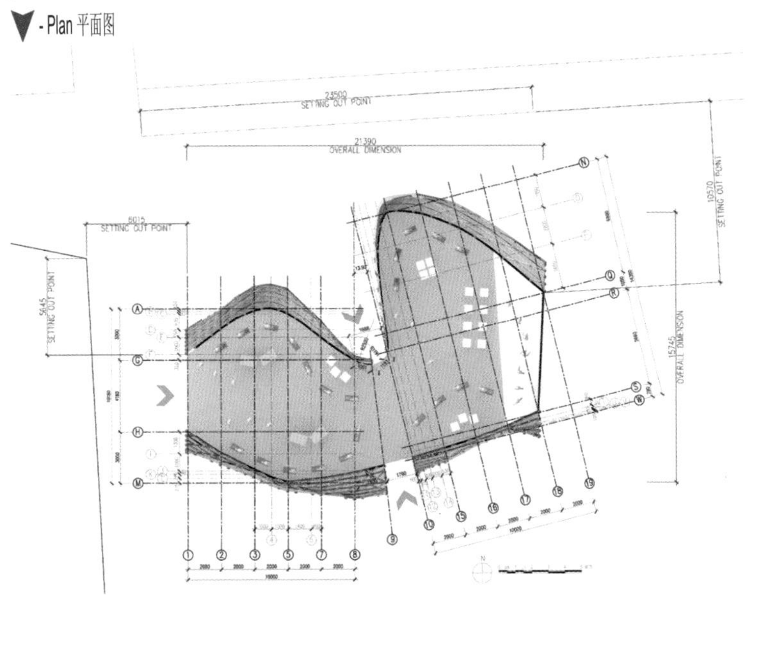

- Master Plan 总平面图

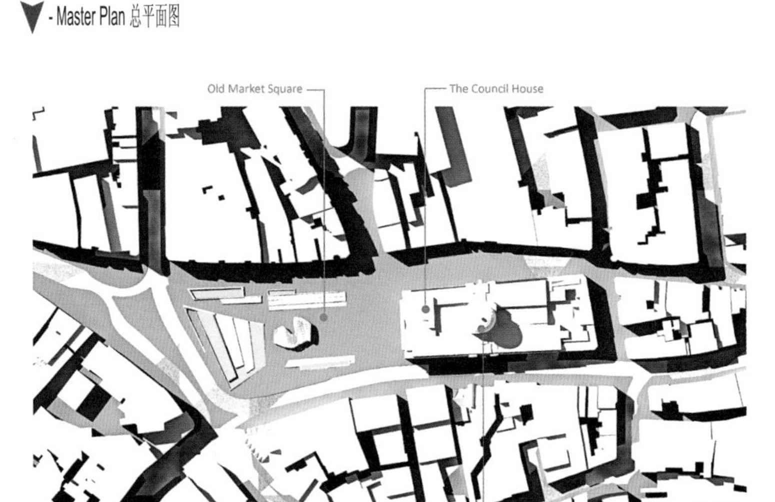

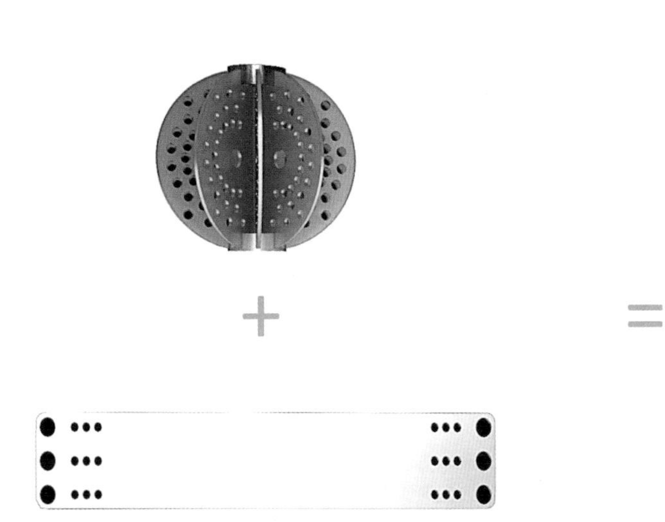

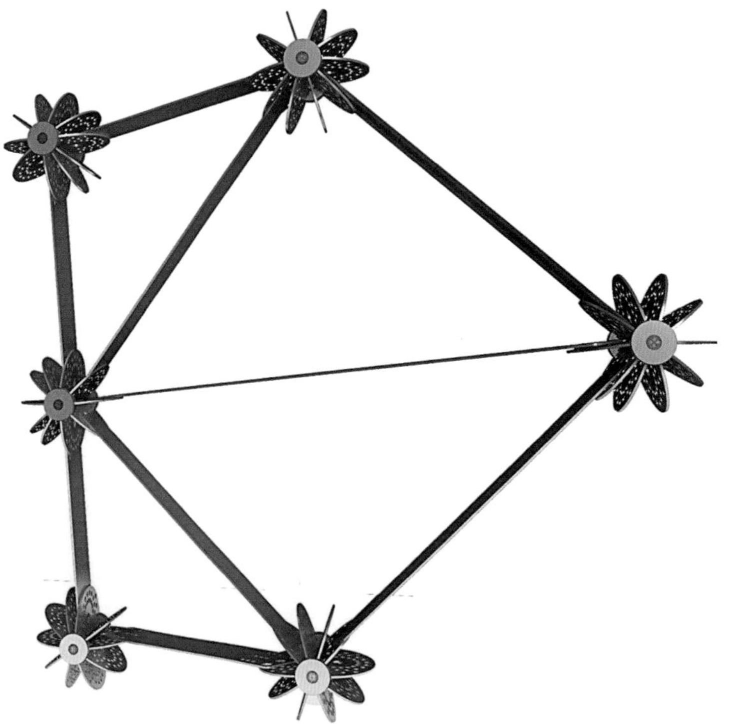

- Facade Component 立面构造

- Exterior Perspectives 外景透视图

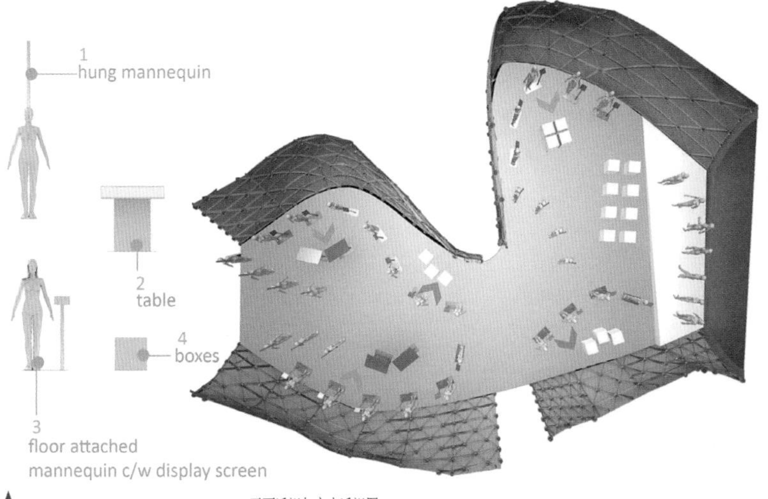

- Interior Plan and Perspectives 平面透视与室内透视图

- Sections 剖面图

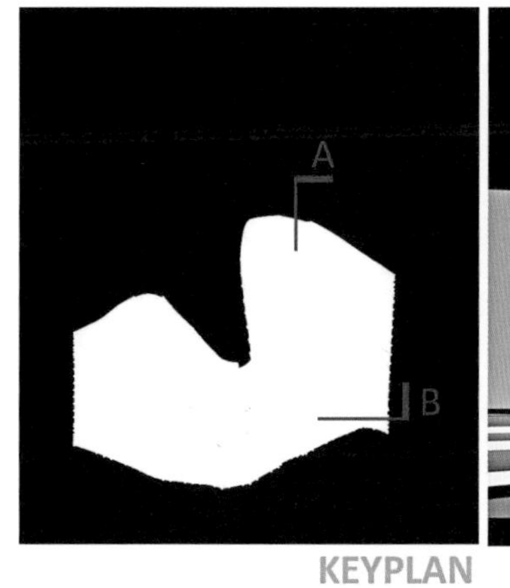

KEYPLAN

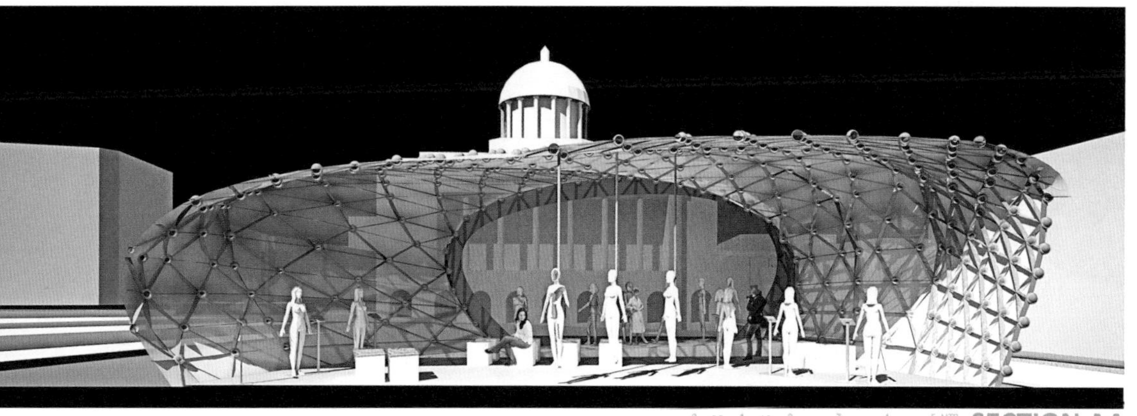

SECTION AA

0 0.5 1 1.5 2 3 4 5 MTS

SECTION BB

071

DESIGNER-设计人: Sheldon Brown

Learning Through Materials

A new Institute of Masonry set in a post-industrial landscape north of Manchester is the core of this proposal. The Institute of Masonry combines scientific research into stone and related materials with exploration of the design techniques appropriate to such materials and direct hands workshop experience in using the materials. The institute facilitates a symbiosis of 'traditional' techniques developed in tandem with digitally controlled mechanical processes. Direct work-experience as an apprentice stonemason with the National Trust at Hardwick Hall formed part of the research phase. This task was undertaken to engage directly with the material. The design is conceived in layers beginning with the hewn rock face that forms the rear edge of the spaces, the next wall is formed from traditionally constructed ashlar stone incorporating reliving arches. The outer edge of the building, which includes the CAD CAM facilities, is formed of smooth monolithic concrete with an intrinsically low U-value.

领悟材质

这个方案的核心是在曼彻斯特北部的后工业区域中建立一所新的砖石研究所。通过科学研究石材和相关材料，这个砖石研究所可以开发出新的工程技术，并指导人们如何使用这些材料。传统工艺与数字化控制机械生产过程在此被有机地结合在了一起。此外，与国家信托基金在哈德威克大厅所组织的石匠学徒班合作也是整个研究的一部分。该方案致力于能够与材质直接发生联系。方案的设计在数个层面上得以发展，空间的后部边缘采用了劈裂的岩石表面，而另一面墙则采用传统的方石与附有浮雕的拱门构成。建筑的外缘由光滑的整块混凝土构成，其中包括CAD CAM设备，同时其固有的热惰性系数也非常低。

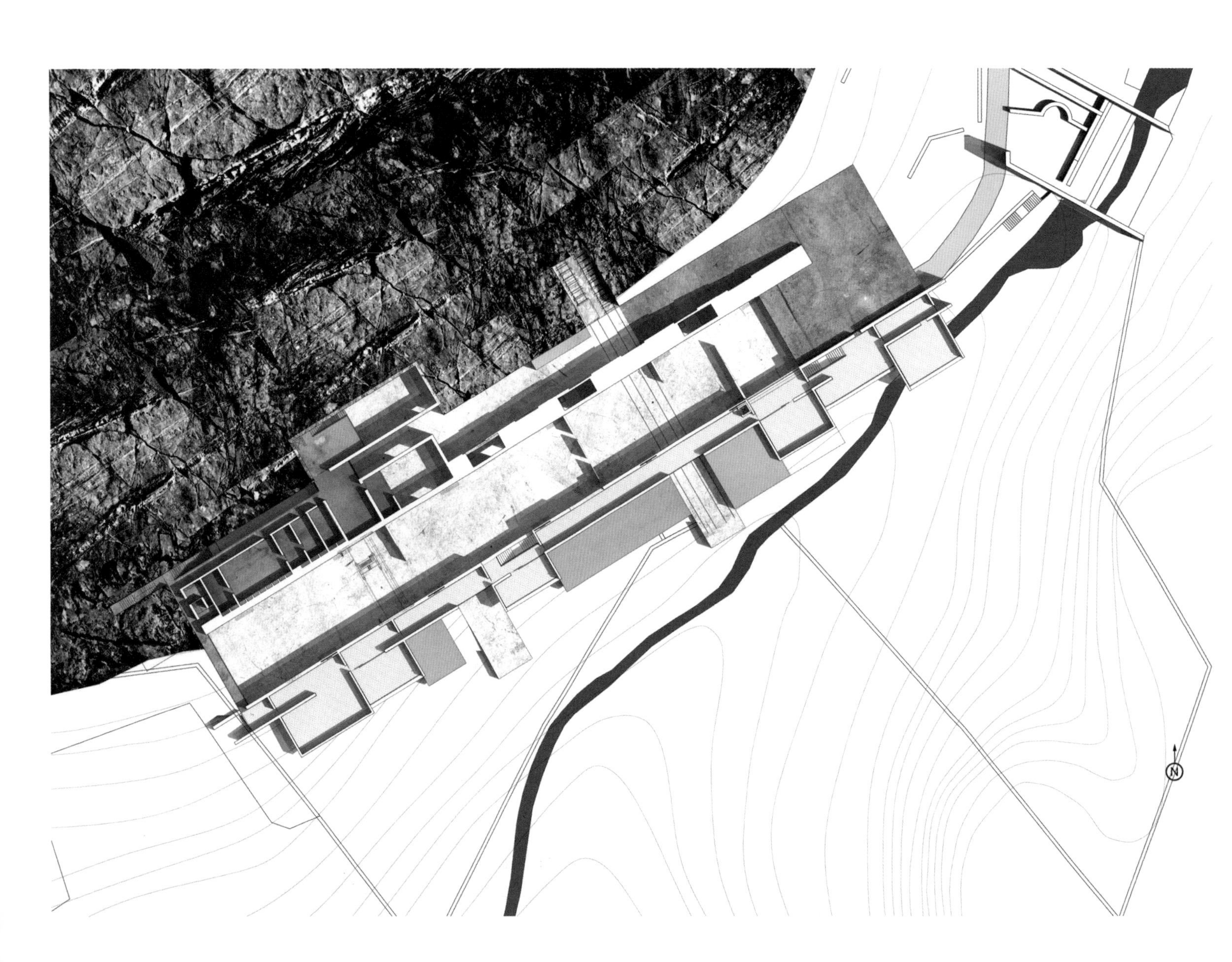

▲ - Lithic Base Datum (Plan) 石之书（平面图）

- Rock, Stone, Concrete and Extrusion-Sections 岩石，石材，混凝土与逐层剖面图

DESIGNER-设计人: Lizzie Webster PRIZE-获奖: Lord Mayor of Nottingham Urban Design Award 2007

Lost Subterranean City Space

Why are we building in the present day with little attention to the past? Layers to the urban landscape are regularly sealed over, often trapping a layer of history below us. As we walk through the streets of Nottingham, hidden beneath our feet is a collection of over four hundred man-made caves. By reintegrating the caves back into society a significant historical layer can be explored, creating a rich context to the city. Nottingham sits on a large sandstone cliff that cut across the city now masked by a shopping centre which denies the potential of the caves. The proposal is to tear the large shopping centre away from the cliff, opening-up a canyon between the sandstone cliff and an archaeological centre. The architecture will highlight the beautiful tactility of these dark forgotten spaces.

失去的地下世界

为什么我们在设计当代建筑时很少去关注那逝去的历史呢？层层的城市的景象总是不断地被掩盖，厚重的历史慢慢积淀在了我们脚下。漫步在诺丁汉街道上，你会惊奇地发现，脚下竟埋藏着四百多处人造洞穴。通过将这些具有重要历史价值的洞窟重新融入现实生活中，我们将能够对这一重要的历史层面进行探索，同时也将为社会添加一道独特的文脉。诺丁汉坐落在一座横穿整个市区的砂岩峭壁上，然而一座巨大的商业中心完全抹杀了地下洞穴的潜在价值。于是该方案计划把商业中心从砂岩壁上移除，并在砂岩壁和考古中心之间开出一条峡谷，从而使新的建筑得以着重体现这个失落的地下世界所独有的厚重质感。

▼- Canyon Level 峡谷层平面图

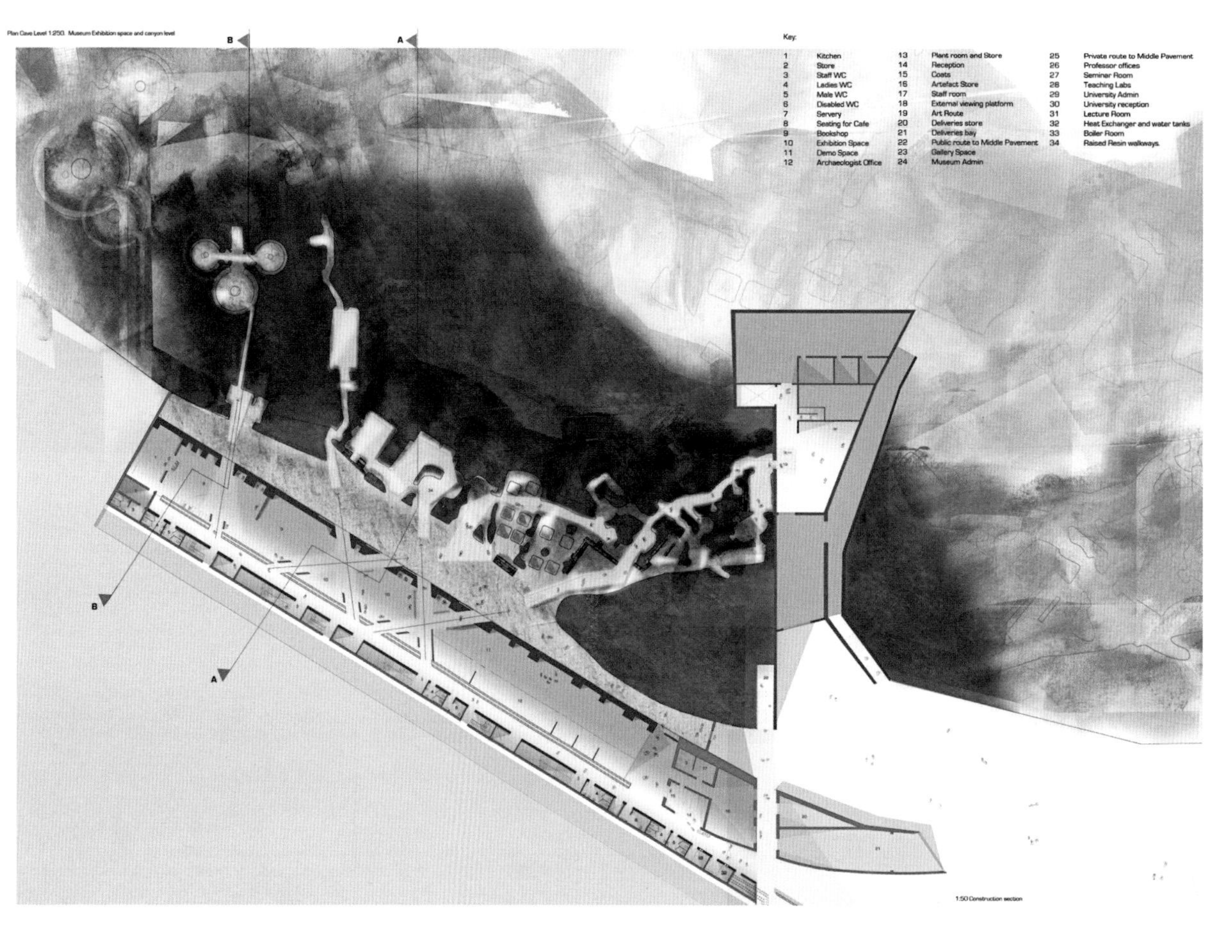

▼- Canyon Section 峡谷层剖面图

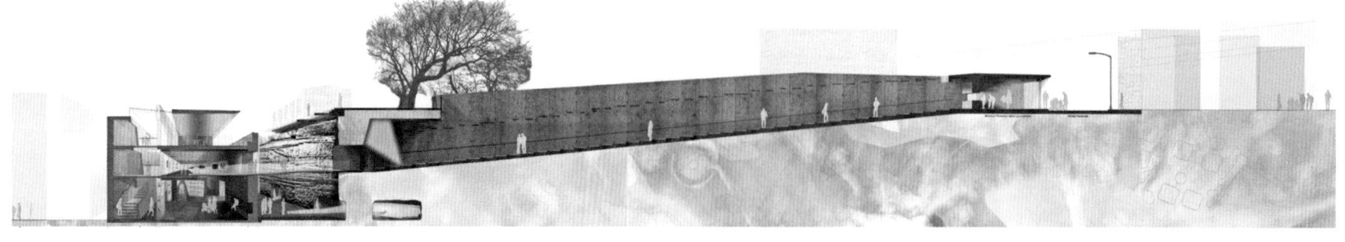

▼- Subterranean Access 地下层入口

Above ground reception cave. Light projections will welcome visitors. Excavated recesses will provide waiting spaces for visitors
Once visitors have experienced the above ground reception space, they will proceed down an open air staircase accompanied by panels of information about the 45 cave uses.
Bridge link over
Cliff
Museum
Glass bridge link deliveries visitors from the entrance bridge over into the circulation wall
Visitors descend through the circulation wall into the cave level exhibition space
Visitors pass into the canyon from the exhibition space created by the building and the sandstone cliff
Exhibits lie in the polycarbonate wall. Information is etched into the concrete display panels.
Cast resin walkways lit from below guide visitors around the caves
Archaeologists travel up through the circulation wall to the archaeology department at the top of the building

073

DESIGNER-设计人: Adam Chambers

Solar Hydrogen Centre

The pending energy crisis begs society to rethink its energy habits. Western society is now reliant on importing fuels to feed its energy needs. Even with energy-saving strategies, society's fundamental functionality, communications and economies would not survive without electricity. With profit making companies controlling our national grid's production we must seek a more localised solution, by harnessing renewable energy tailored to each individual community depending on location and resources available. Architecture can play a key role in making the way energy is produced and consumed, 'visible' within society. This project investigates the importance of the integration of renewable energy into Nottingham Meadows community, focusing primarily on the storage of solar energy through hydrogen. The solar hydrogen plant will function as a test bed research facility for the solar hydrogen production. It will produce energy for the community as well provides an area where the community can interact both with the process and each other.

太阳能制氢中心

即将到来的能源危机迫使我们的社会重新思考利用能源的方式。西方社会对化石燃料有着相当高的依赖度。即使制定了能源节约战略，整个社会的一些基本功能，如通信和经济等运行仍然离不了电力的支持。然而，控制着整个国家电网的那些公司都是些利欲熏心的集团，因此我们必须寻求一个更加地方化的解决方式，根据每个独立社区的位置与资源需求，来合理地利用可再生能源。这时，建筑就可以在能源的生产和消费中扮演至关重要的角色，在社区中起到表率作用。该项目探讨了在诺丁汉，梅多斯社区中实现可再生能源综合利用的重要性，并着重考虑了通过氢来存储太阳能的方法。这种太阳能制氢装置可以作为研究太阳能氢产品的测试平台。它将可为这个小区提供能源，同时也能为社区提供一处场所，使居民能够与相关生产过程互动，且促进他们的相互交流。

▲ - Technological Appearance 建筑构建外观

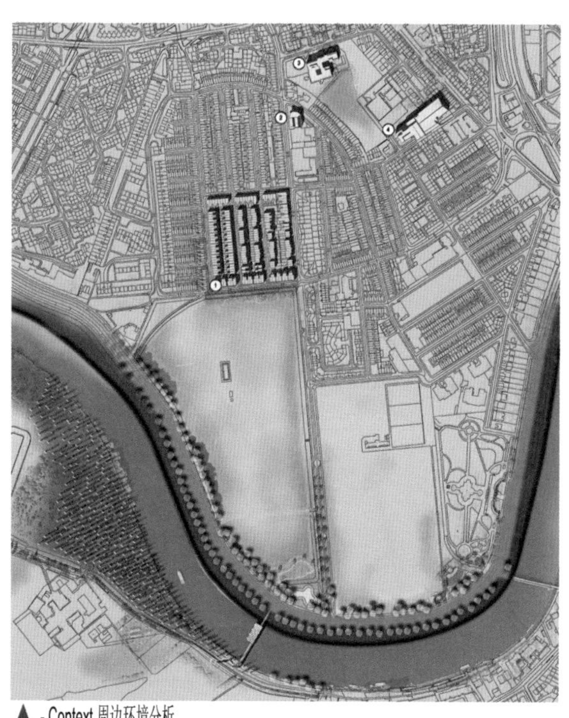

▲ - Context 周边环境分析

▼ - Perspective

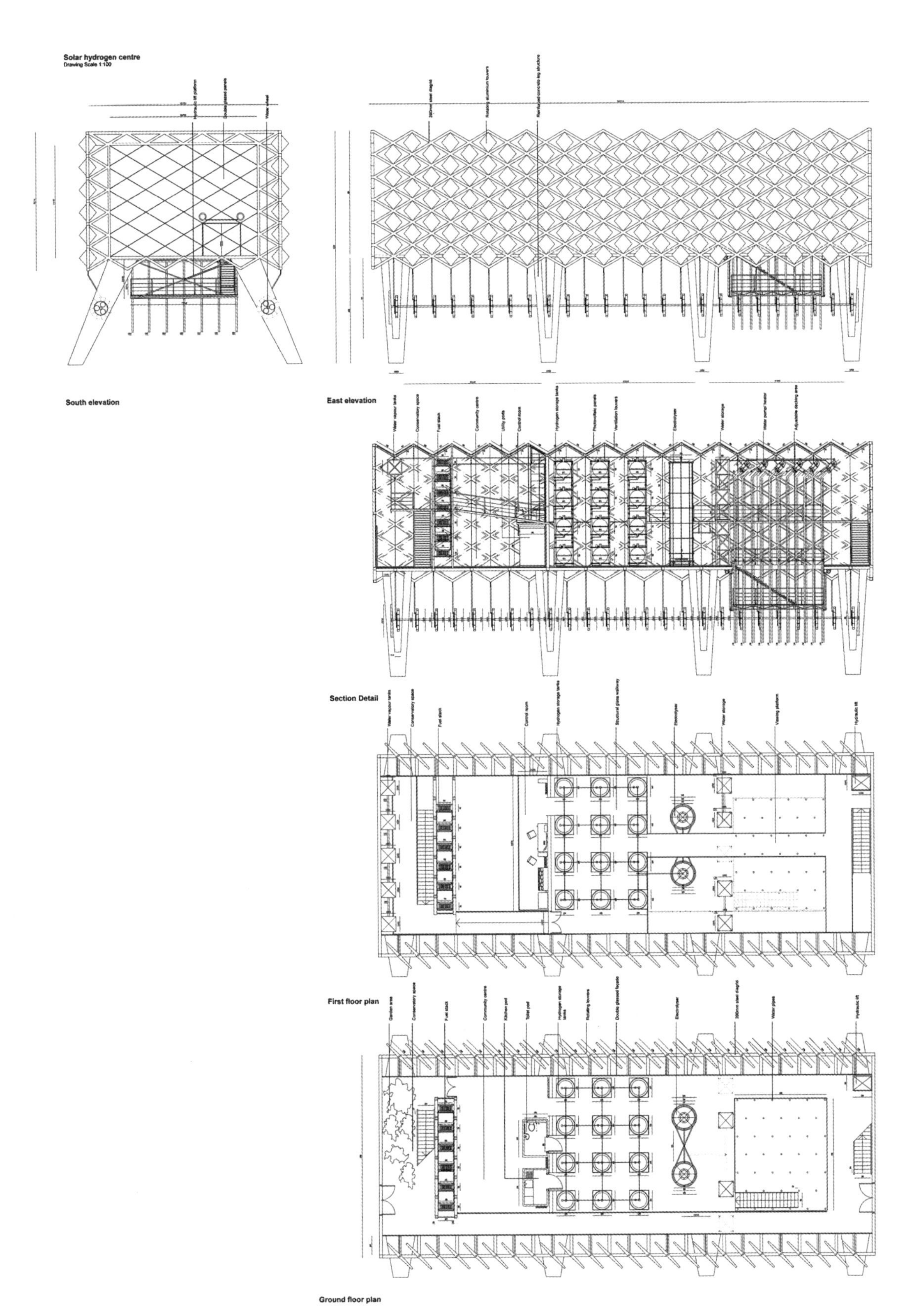

▲ - Detailed Plans and Section 平面与剖面详图

▼ - Equipmentality 建筑配套设备

DESIGNER-设计人: Alex Dale-Jones

The Water Squares of Liverpool

This project examines the potential to exploit the developing activity of the Tidal Stream Energy Industry by introducing a regional 'hub' or 'observatory' for commercial testing of tidal stream devices beyond the prototype stage in conjunction with associated research by an established local oceanography facility. The site of interest is The Liverpool Docklands with the test area the Mersey Estuary and Liverpool Bay; an area with an excellence in tidal research, a city with an outstanding history of marine and tidal monitoring, possibilities for additional grid capacity with public and private sector investment opportunities following the recent acquisition of the Mersey Docks and Harbour Company by Peel Holdings Limited. A large area of derelict river frontage is available to the north of the city centre with an existing historic wet dock framework in place capable of providing the necessary facilities for such an intervention.

利物浦水之广场

该项目旨在挖掘潮汐能源产业的发展潜力。它包括一个区域性的枢纽平台或观察站，可在原型机的基础上，对潮汐能装置做进一步的商业测试，此外，它还与当地现有的一所海洋学研究所进行合作研究工作。适合的场地位于利物浦的港口区内，而测试区则为默西河口和利物浦湾的水域。这里是进行潮汐研究的极佳区域，而利物浦城又拥有杰出远洋历史和潮汐监测经验。由于最近皮尔控股有限公司收购了默西码头和港口，今后这里将有可能吸引更多的公共和私人投资，从而增加国家电网的容量。在市中心的北面，沿河有一大片可利用的干涸区域，其中还有一个历史悠久的湿船坞框架，可以为整个嵌入项目提供必要的设备支持。

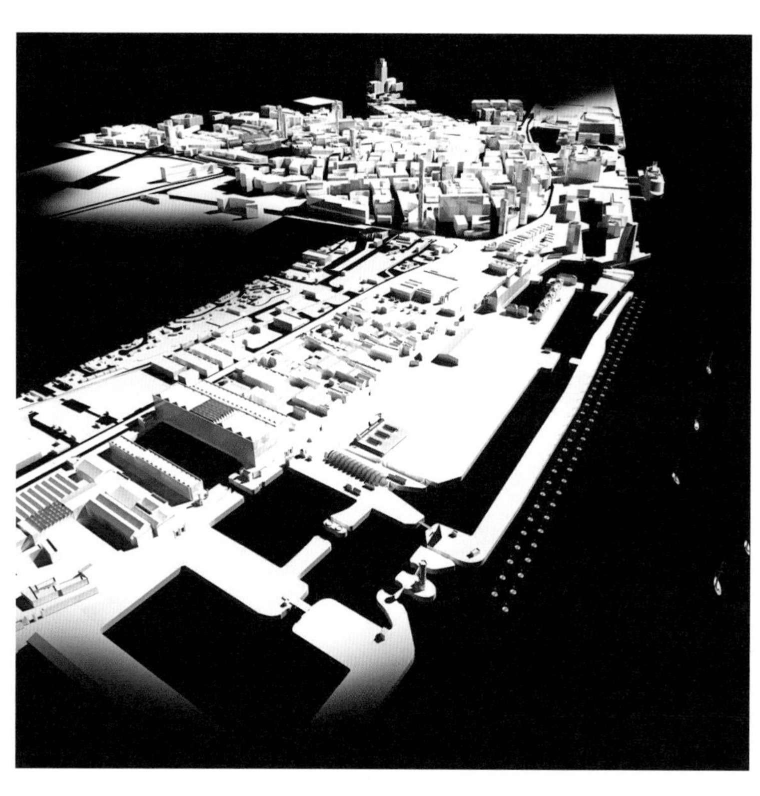

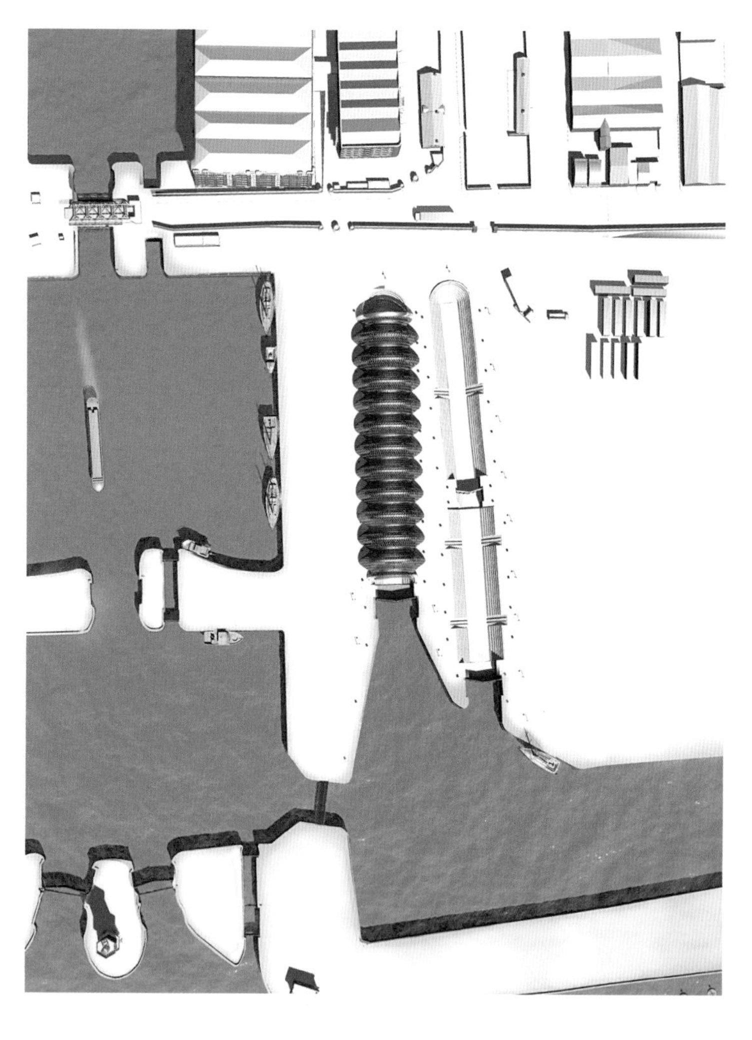

▲ - Context 周边环境分析

► - Dock Location 港口位置

▼ - Topography 地形与高程分析

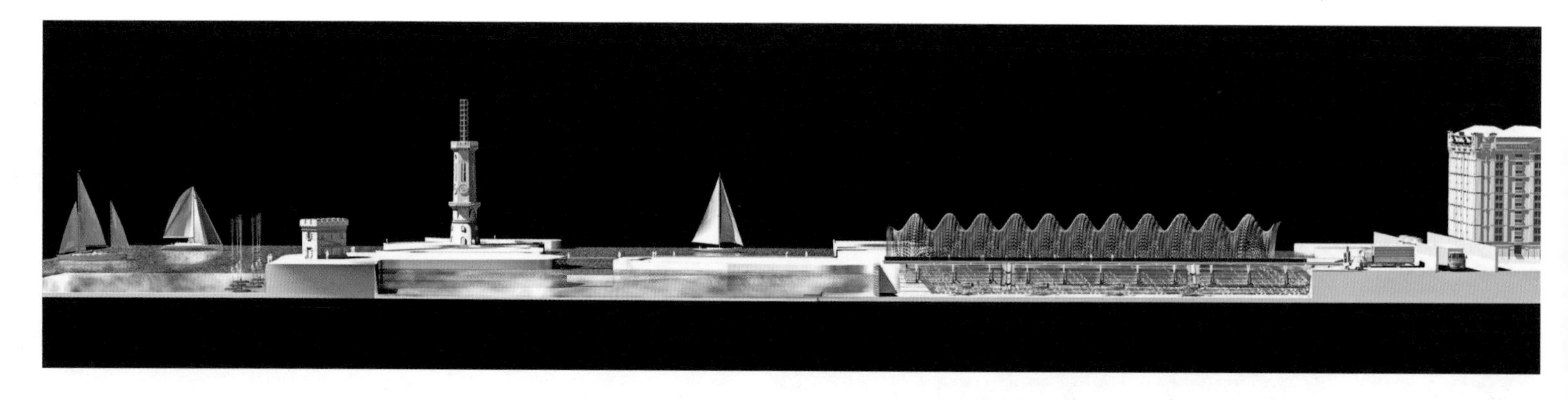

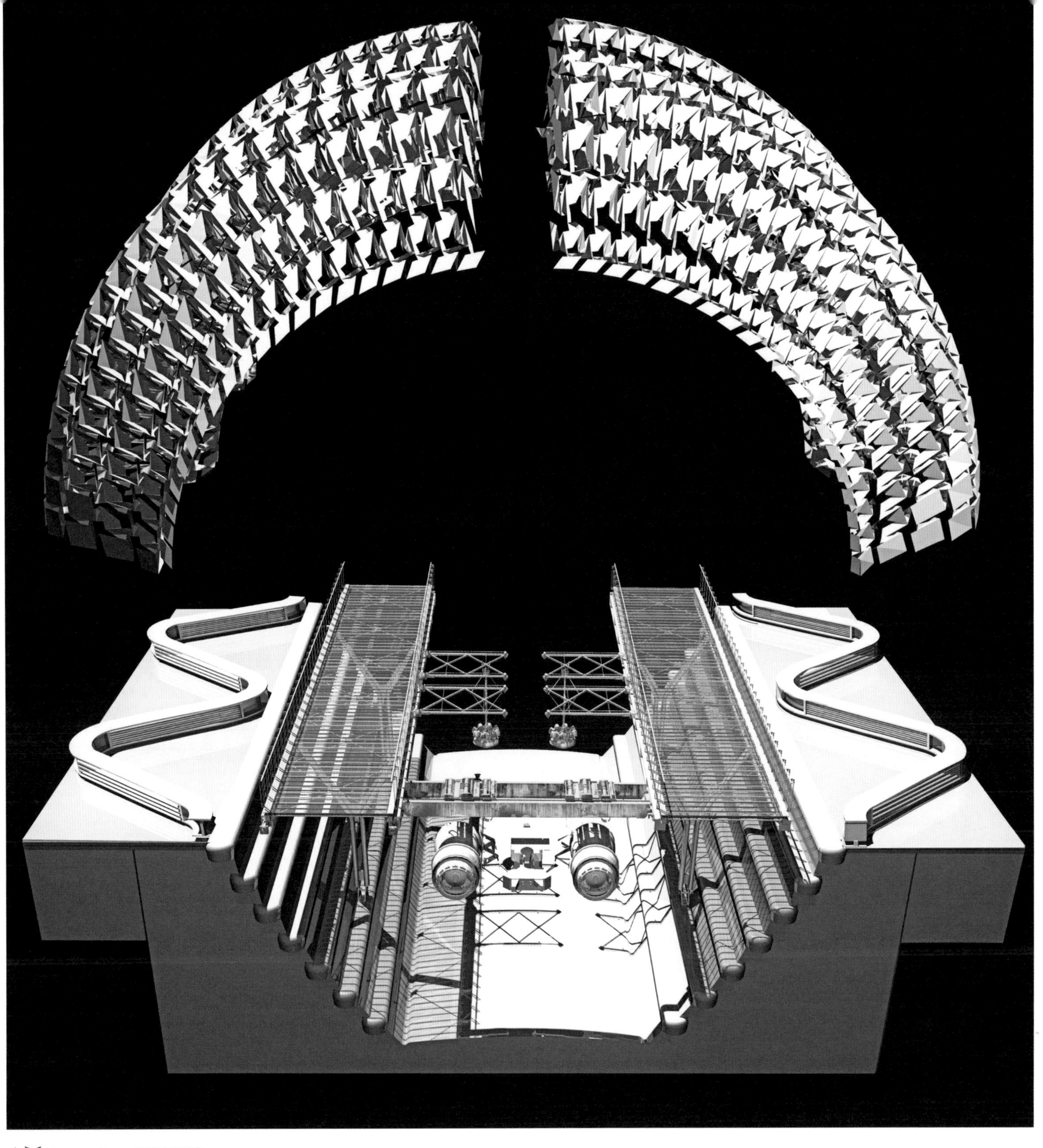

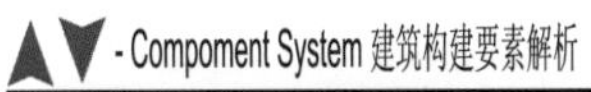

- Compoment System 建筑构建要素解析

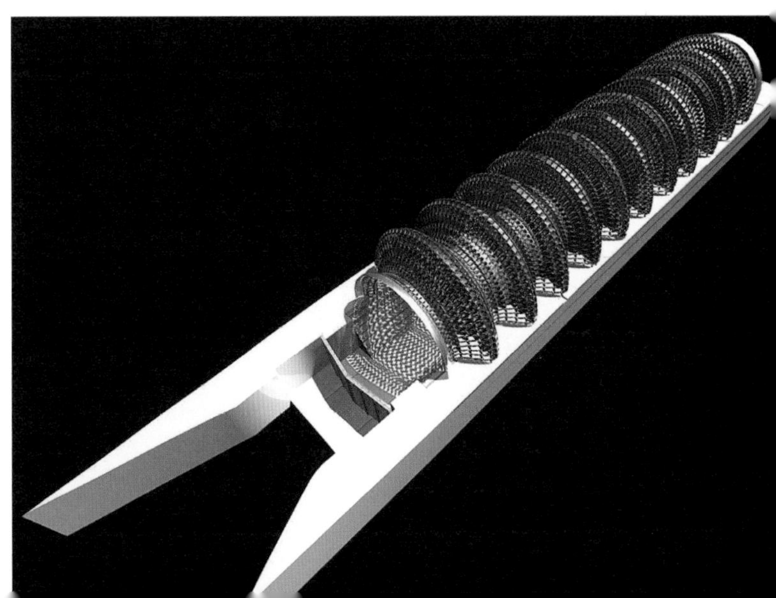

DESIGNER-设计人: Jennifer Routledge PRIZE-获奖: Philip Webb Award 2008

Newhaven - Passé Nouveau

The crumbling marine workshops of Newhaven are the focus of a prize-winning scheme judged by the Society for the Protection of Ancient buildings (SPAB) who awarded it their prestigious Philip Webb Award. The design proposes to repair and extend two listed workshop buildings on the railway quay to form a new Maritime and Transport museum, reviving links to Newhaven's shipping past. The converted marine workshops and new museum building would show Newhaven's history as a port and particularly the importance of marine and railway engineering in its past. A new pedestrian link from Newhaven Station to the town centre features a high tech movable footbridge over the harbour. A new artificial dock between the new and original quayside buildings houses the restored PS Ryde paddle steamer, which is currently in very poor condition on The Isle of Wight.

纽黑文-逝去的辉煌

古建筑保护协会（SPAB）将著名的菲利普•韦伯奖颁发给了这个方案，以此来表彰它对纽黑文荒废的海事车间所提出的创造性建议。这个方案计划对两个位于铁路码头的厂房（历史保护建筑）进行修复和扩建，使之成为新的航海与运输博物馆，以此来展现纽黑文历史上辉煌的航运业。改造后的海事车间与新博物馆将主要展示纽黑文的港口史，及其海运与铁路工程在历史上的重要性。一条新的人行步道将纽黑文火车站与市中心连接起来，而其中跨越港口的可移动高科技人行天桥更是格外引人注目。在岸边的新旧建筑之间还搭建起了新船坞，用来存放莱德号明轮蒸汽客船，以此来为这条目前正停靠在怀特岛，无人照料，破败不堪的客船提供一个新家。

▲ - Rear Entrance 后部入口

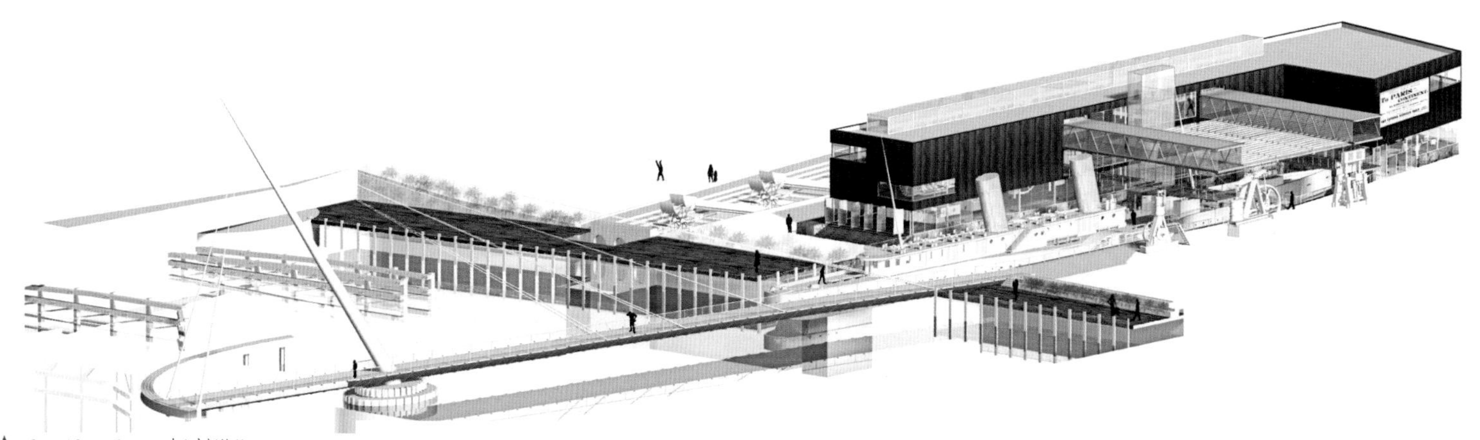

▲ - Central Space Strategy 中心空间处理

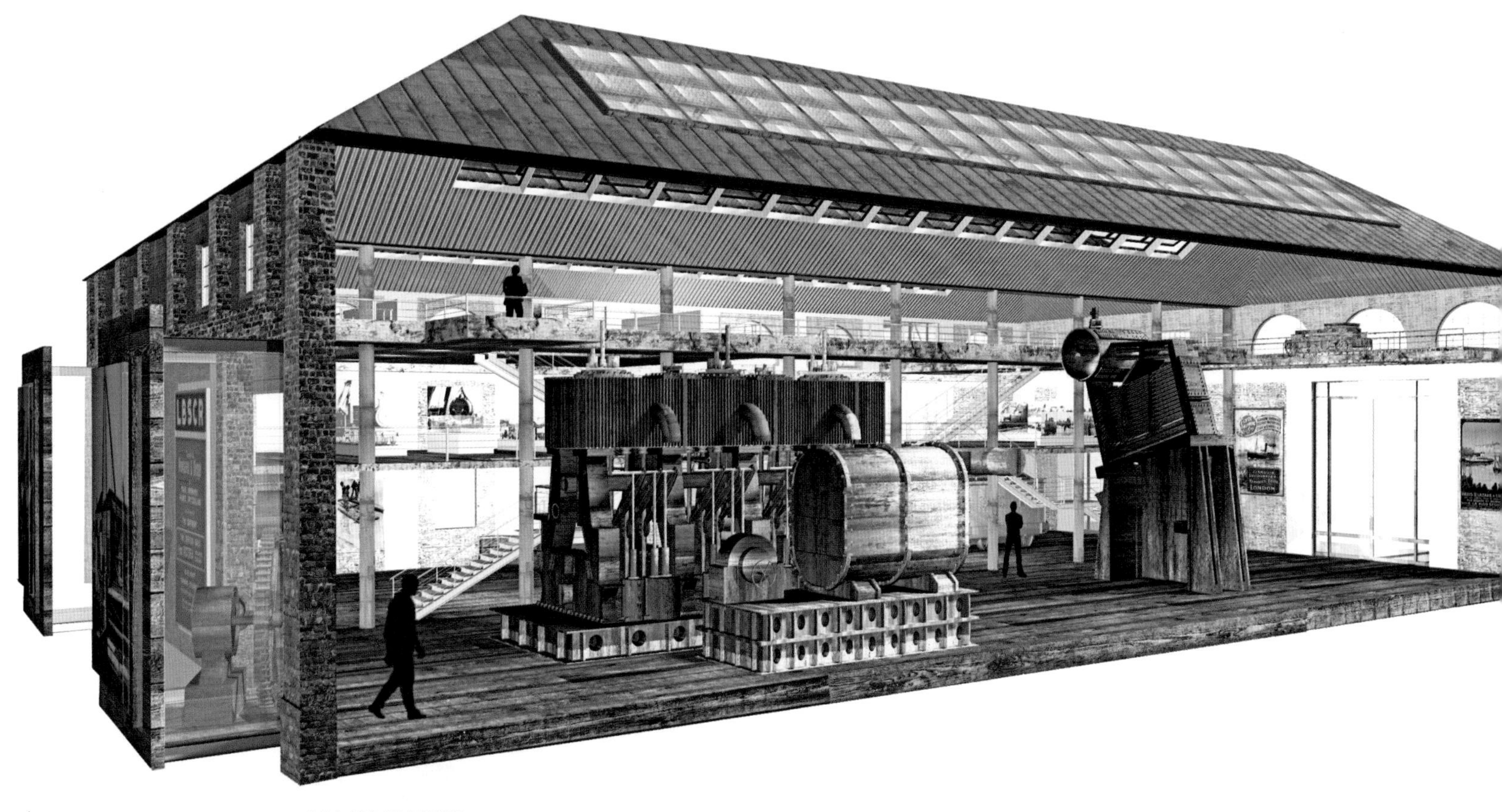

- Transformation of Existing Fabric into Museum 将现有建筑机理转变为博物馆

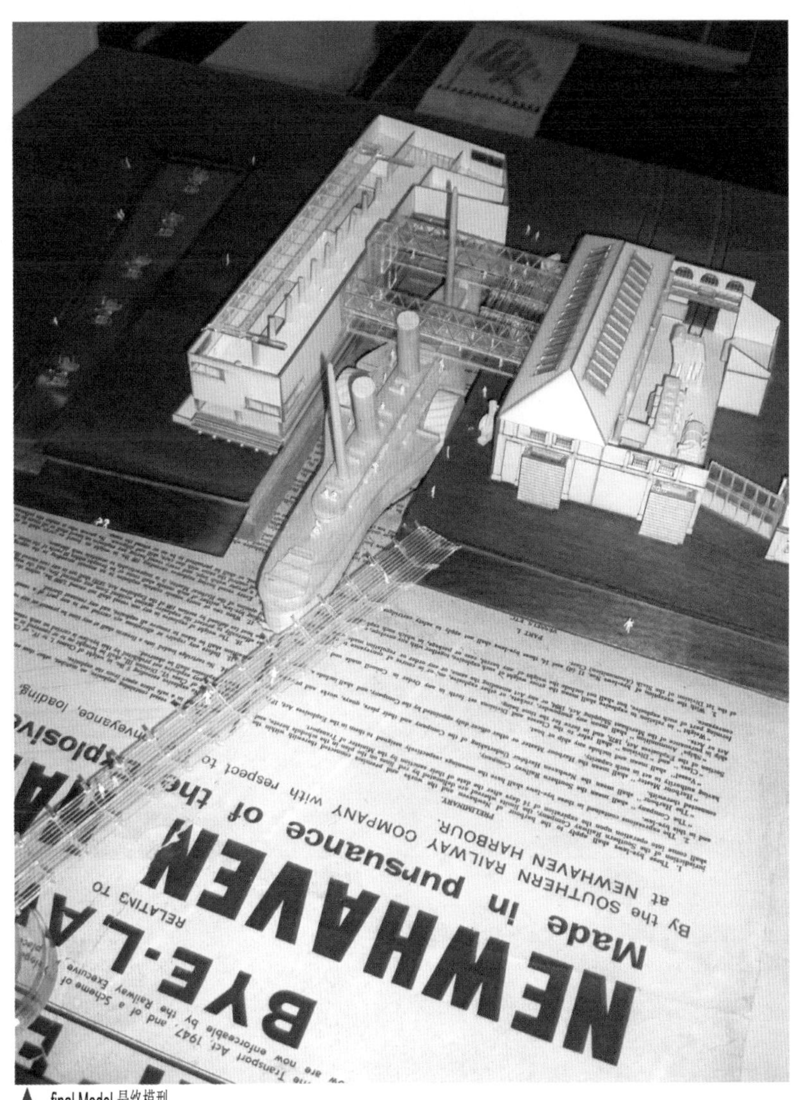

- final Model 最终模型

- 3-D Print of Museum Exhibit 博物馆展品三维塑形

DESIGNER-设计人: Steve Townsend PRIZE-获奖: Commended RIBA Silver Medal 2009

Digital Intimacy

The 'Kinaesthetic Interaction Space' is conceived as an interactive architectonic intervention aimed at children with autism, providing sensory stimulation to aid interaction with other children through shared kinaesthetic experience. The focus of the thesis is on the development of dynamic material systems that could enable new forms of interactive environment. Architecture is conceptualised as an embodied interface and physical space has been fused with digital media in order to stimulate the inhabitation of inhabitants. K.I.S. is intended to facilitate playful explorations and fluid dialogues between people. The user learns to interact with their environment through an intuitive process, engaging the physical presence of inhabitants and forming spatial narratives. The system is flexible, transformable and fully demountable, meaning that the same kit of parts can be assembled to adapt to a range of spatial requirements. The project was developed through a full-scale prototype that was constructed to enable the experience of the qualities of the surface, both visual and tactile, and the observation of its use, including people's responses.

数码拉近你我心

这一“动觉互动空间”被设想为一处可以提供互动机会的建构嵌入体。它旨在为患有孤僻症的儿童提供感官刺激，进而通过共享运动感觉体验来协助他们与其他孩子产生互动。本研究主要集中探讨可以帮助实现新式互动环境的动态材料体系发展。在这里，建筑被概念化为一种具象界面，而为了让用者享其用，其物理空间则与数码媒体融为一体。动觉互动空间意在人们之间促进充满乐趣的探索与流畅连贯的对话。使用者在此将学会如何凭借直觉与周围的环境互动，进而将使用者的物理性存在融入其中，形成富有意义的空间。灵活性，可变性与完全可拆卸性是这一系统的特点，而这也正意味着同样的一整套设备可以通过不同的组合来适应一系列不同的空间要求。整个方案均围绕着一套全尺寸原型模型的建立而发展，从而使我们可以在视觉与触觉领域切身体验空间界面的质量，并同时从第三方的角度观察它的使用，包括了解人们对其反应如何。

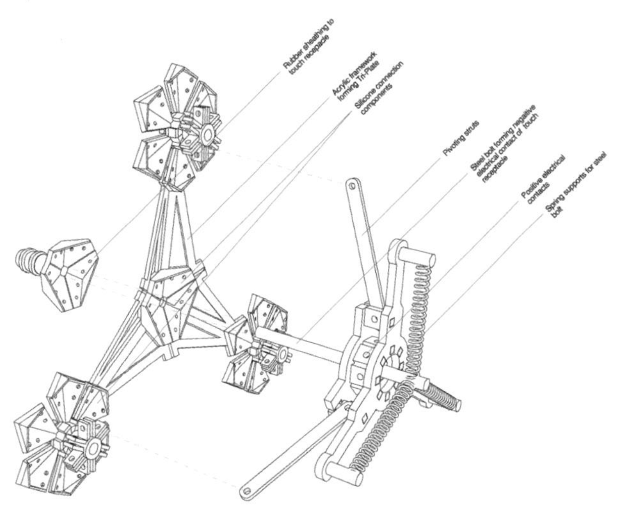

▲ - Tri-Plates Interaction 三向板构造详图

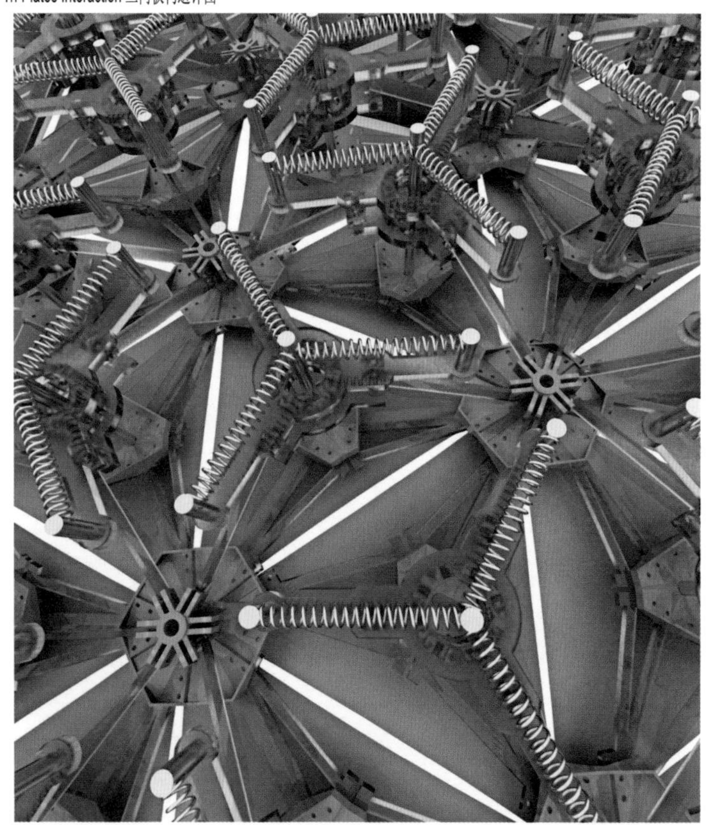

▲ - Sprung Interaction Mechanism 弹性互动装置

► - Kinaesthetic Simulation 动觉模拟装置

◄ - Rear Vew of the Sprung Interaction Mechanism 弹性互动装置背面细部

DESIGNER-设计人: Dominic Wilson　PRIZE-获奖: Shortlisted RIBA Silver Medal，2009

The Greenwich Archives

Starting with an interest in Baroque ecclesiastical architecture, this investigation encountered men 'decentred' from their geocentric homes by the post-Copernican revolution – a shift made manifest in the proliferation of perspective distortions in the Baroque arts. Simultaneously, the unifying capacity of the Baroque (the gesamkunstwerk) – especially seen in the light of the dualism of subsequent enlightenment thinking – fuelled an interest in architecture as a mode of reconciliation. Sited at Greenwich, this proposal attempts to reinstate such a mode, echoing, yet attempting to break down such dualism through their respective embodiments in an archive and a chapel. The archive, containing the written word, embodies reason, whilst the chapel, containing the word of God, embodies faith. In each case the project is concerned with making visible of the word. The proposal develops through a blurring of the 'horizon of visibility' – the dividing line between earth and the 'heavens', representing the limits of human understanding – acting to bring together earth and sky.

格林尼治档案馆

基于对巴洛克折中主义建筑的兴趣，本方案试图去研究那些受到后哥白尼时代革命学说——一种导致了扭曲透视学在巴洛克艺术中盛行的运动——的影响，从其几何中心域偏移出来的人们。而同时，巴洛克那一统万物的包容性（亦称作gesamkunstwerk）在随后的启蒙主义思想中特别体现为二元论，并在建筑领域内促进了一种调和模式的产生。该方案坐落于格林尼治区，试图重新诠释这种模式。在一座档案馆和一所小礼拜堂中，通过它们的透视表现，方案回应着，也同时试图去解析这种二元论。这座档案馆，保存着书写的文字，体现了理性，而小礼拜堂，拥有上帝的文字，体现了信仰。在每所建筑中，本项目均注重对文字的视觉表达。它通过一种模糊的视觉地平线来发展——那是凡间与天堂间的界线；它代表了人类认知的局限性——将天地合二为一。

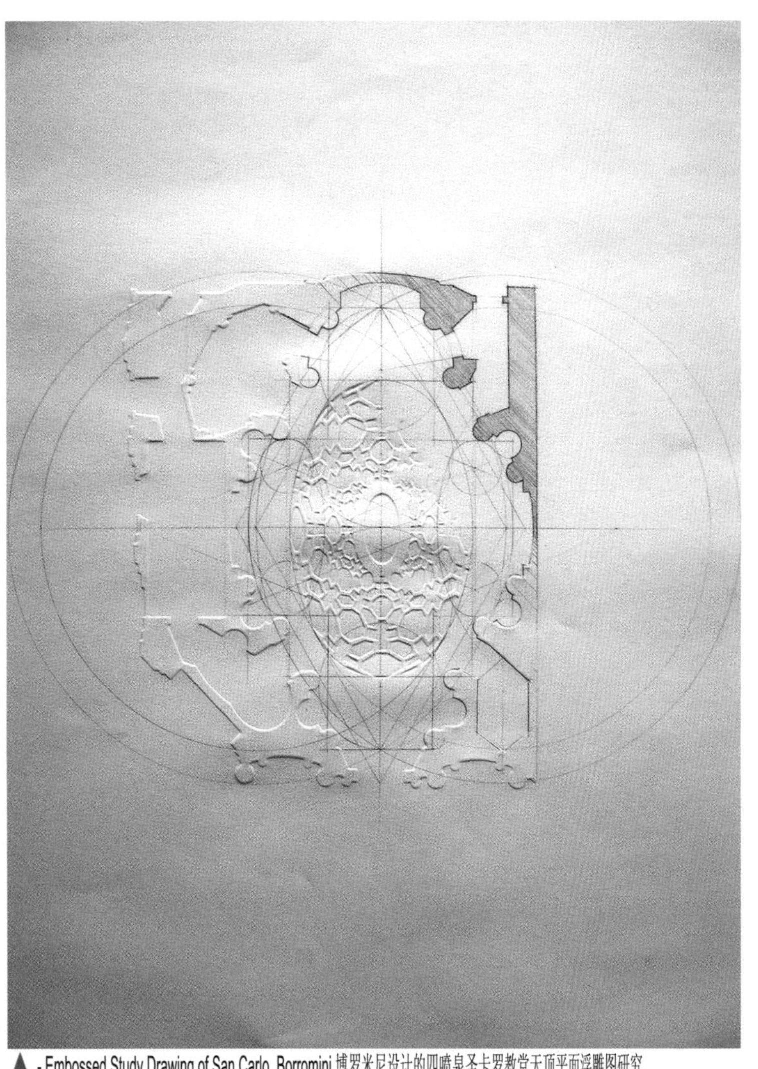

- Embossed Study Drawing of San Carlo, Borromini 博罗米尼设计的四喷泉圣卡罗教堂天顶平面浮雕图研究

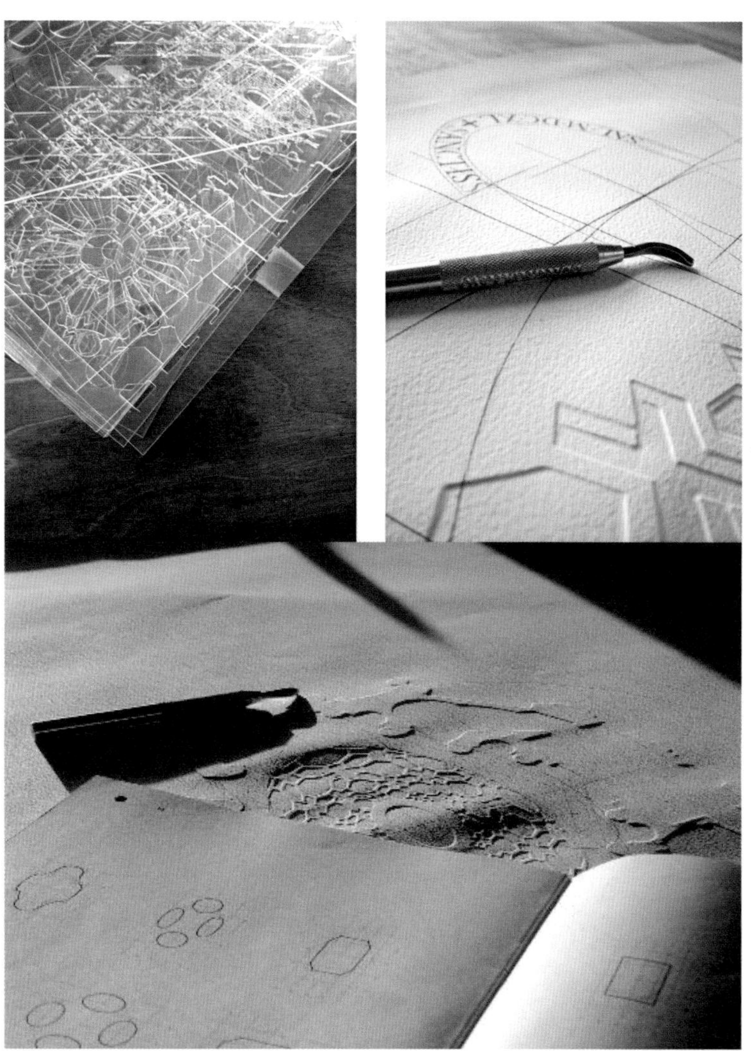

- Stencils, Tool and Process

- View from Bookstacks to Elliptical Wall 从书架望椭圆弧墙

- View of Gallery Bay towards Elliptical Wall 面向椭圆弧墙的展廊内景

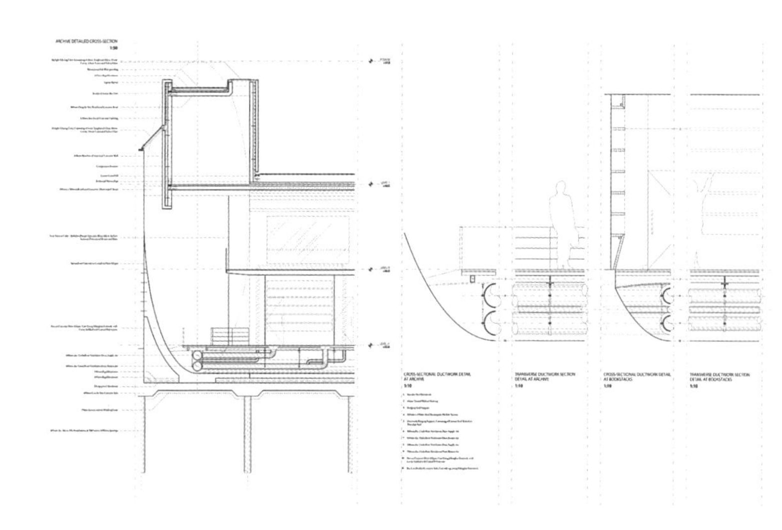

- Chapel 小礼拜堂

- Archive Detail 档案室细部详图

- Shadow Path Analysis 光影分析

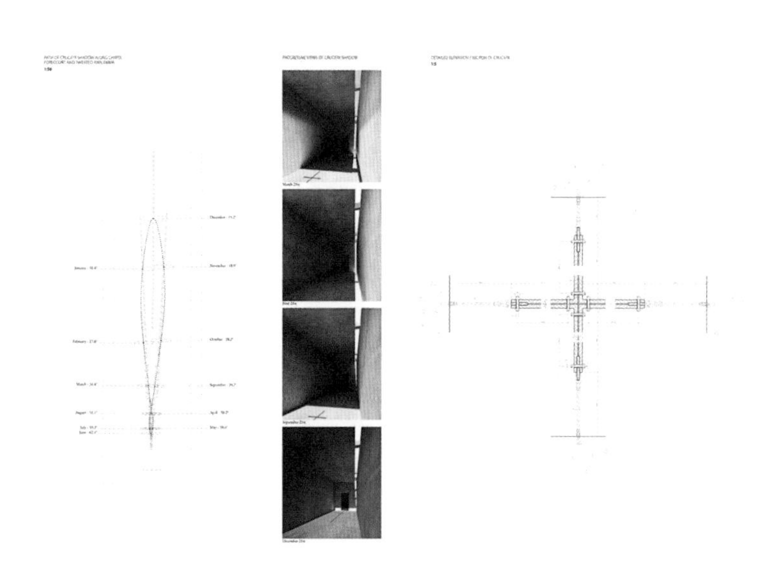

DESIGNER-设计人: Clara Byrne　PRIZE-获奖: Philip Webb Award 2012

Traces at 76 Dean Street

Although architecture has only a limited potential to directly challenge one's identity, this project acknowledges the physical and psychological connectivity we have with our surroundings, encouraging interactions to be created and nourished as a means through which identity can be informed. In a ruin, the masquerade of use disappears into a recondite object, yet dense with rich histories; it is a place of mysterious spirit and the intangible power of emotion. Traces of wear enable us to read the nature of past uses, and play a crucial role in our ability to interpret and appreciate our environment. Wear humanises architecture, feeding life into its inanimate form. No.76 Dean Street, Soho, a listed Georgian house, stands in a state of flux after a fire nearly destroyed it in 2009. This ruin is reinterpreted to create a safe environment for recovering drug addicts to challenge and enhance their identity.

追踪迪恩街76号

尽管建筑在直接改变一个人的个性方面只有很小的潜力，但是这个方案却呈现了存在于我们与周围事物之间的生理和心理上的联系，鼓励我们去创造和丰富与周围事物的互动，进而来塑造自我认知度。在一座废墟中，功能的浮华假象逐渐消隐在深奥的物质中，并随着悠久的历史愈发浓郁；这座废墟，拥有着神秘的灵气，无形的感染力。时光在建筑上消磨出的印记使得我们能够解读这里曾经拥有的生活，且能够为我们体会与理解周围的环境提供重要的帮助。慢慢地使用且一点点消磨一栋充满人性的建筑，就可以为那毫无生气的形式注入生命的活力。这座废墟，就是位于Soho区的迪恩街76号。2009年的一场毁灭性的大火，使得这栋乔治亚风格的历史保护建筑摇摇欲坠。但该方案意在为这里营造出一处安全的氛围，为吸毒者的戒毒康复提供服务，并同时加强整个区域的自身认同感。

▼ - Existing Ruin and Interior 现存的建筑废墟与内部状况

▼ - Circulation and Plans 内部交通设计与平面图

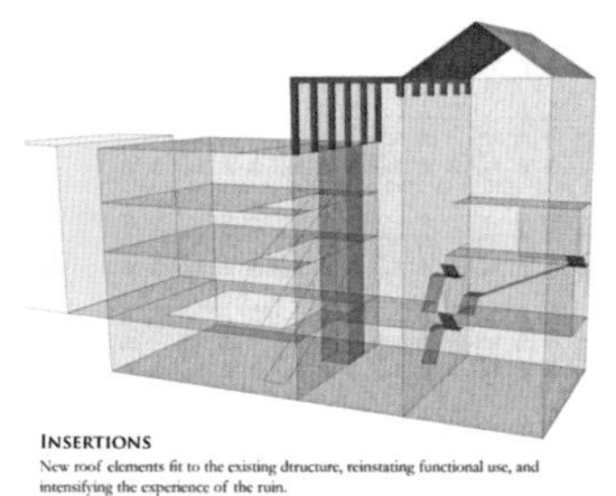

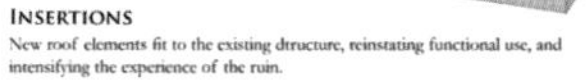

INSERTIONS
New roof elements fit to the existing dtructure, reinstating functional use, and intensifying the experience of the ruin.

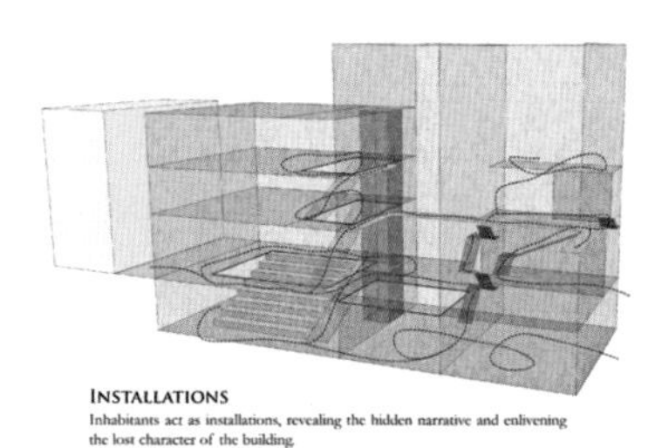

INSTALLATIONS
Inhabitants act as installations, revealing the hidden narrative and enlivening the lost character of the building.

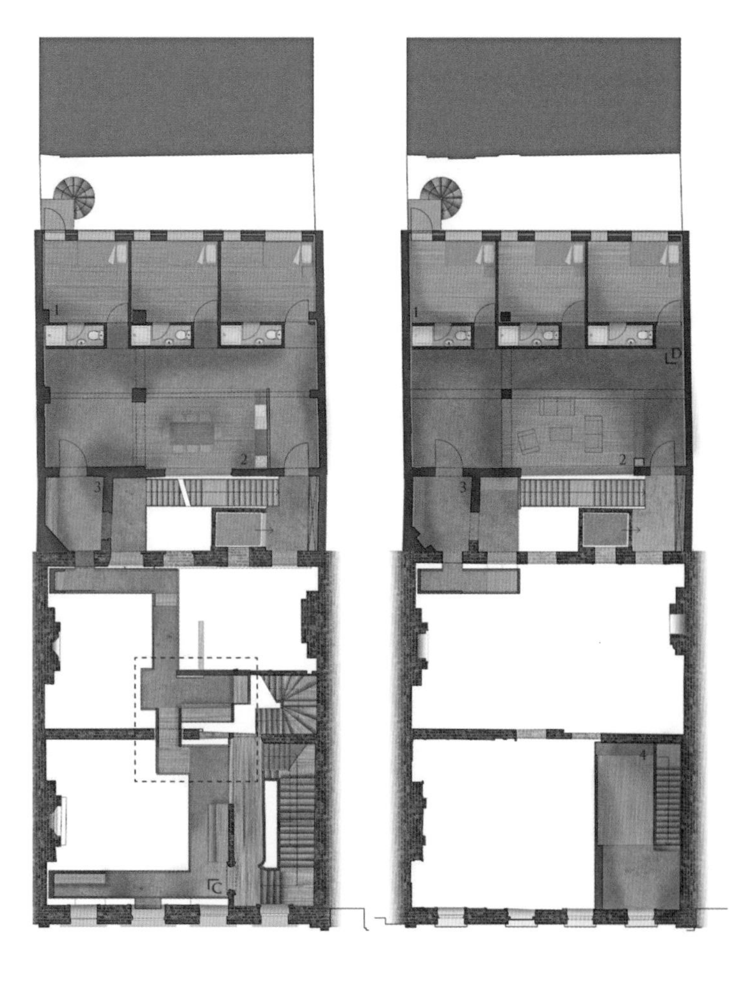

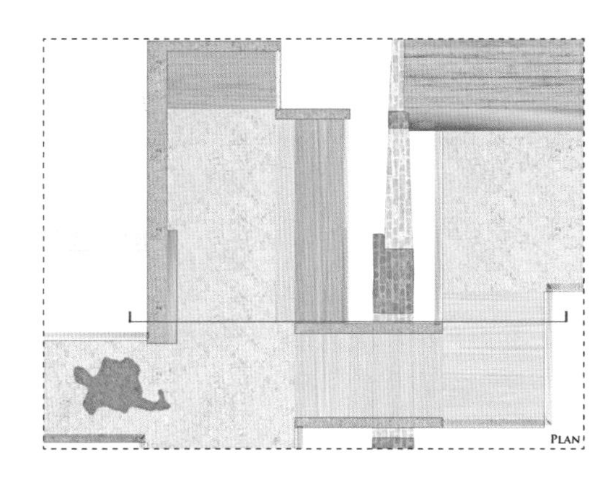

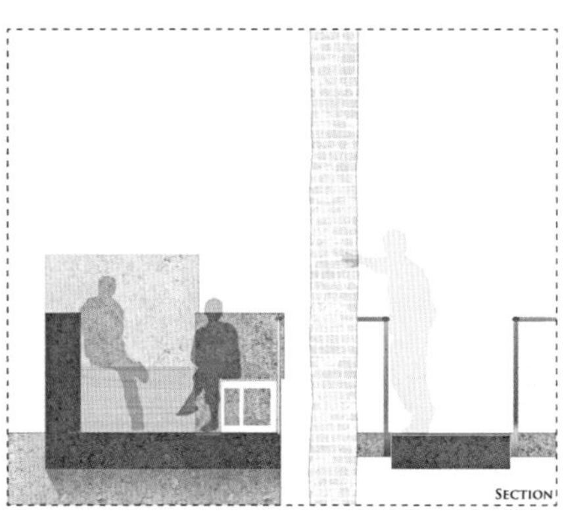

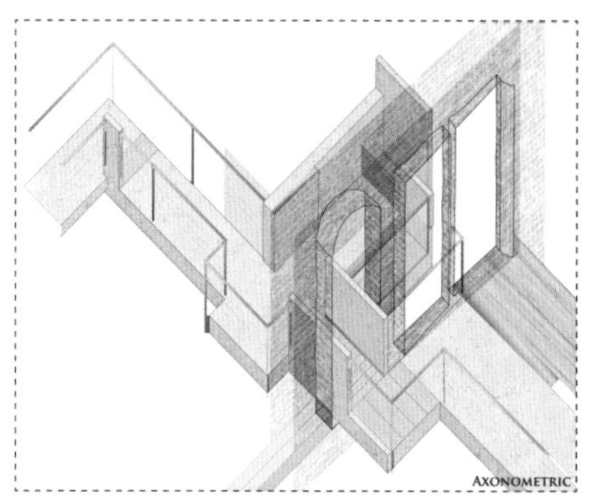

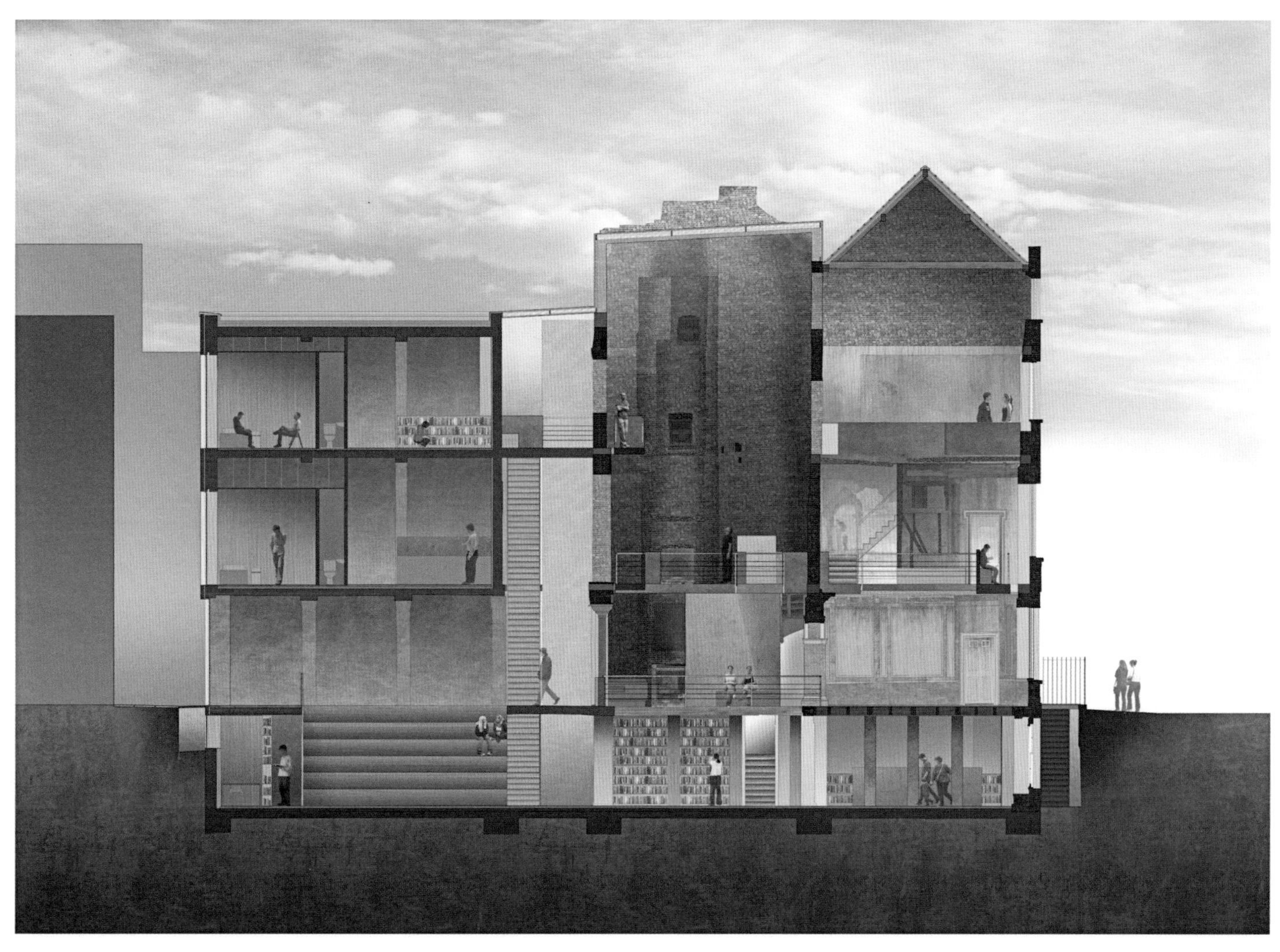

- Technological Insertion 嵌入体建构图

079

DESIGNERS-设计人: Jonathan Davey, Matthew Kidner　PRIZE-获奖: Nationwide Sustainable Housing Awards，2010

Green Street Housing

This is a high-density courtyard housing scheme for a site in the Meadows area of Nottingham. Our initial reaction to the site was the great potential offered by its proximity to valuable public green space and the possibility of restoring to the neighbourhood of the intimacy and conviviality which characterised the area prior to its redevelopment in the 1960s and 1970s. By re-opening the route through the site to an adjacent park, the park is extended into the site and a series of interconnected public spaces are formed, alongside more private walled gardens, reminiscent of the Victorian kitchen garden. The architectural strategy also draws extensively on the typology of the courtyard which allows for high-density living without a high-rise urban form, and the use of level changes at ground level to create thresholds between public and private spaces.

格林街住宅计划

这是一个位于诺丁汉梅多斯区的高密度住宅计划。我们最初的灵感源于那非凡的绿色公共空间所拥有的巨大潜力，以及重塑亲密而欢乐的邻里关系的可能性。而这些都曾经是该社区的显著特色，但却在20世纪60年代到70年代的现代化更新过程中遭到了破坏。通过开辟从社区到附近公园的道路，使得公园的景观可以延伸到社区内，并形成一系列相互联系的公共空间，除此之外，路旁还有矮墙围起来的私人花园，以及怀旧的维多利亚式菜园等等。在设计手法上，该方案强调了庭院类型的重要性，在避免建造都市类型高层的情况下，实现了一种更具亲和力的高密度模式。最后，利用地面的高差变化，在公共空间与私人空间之间实现了门槛效应，使之既相互联系又有所分离。

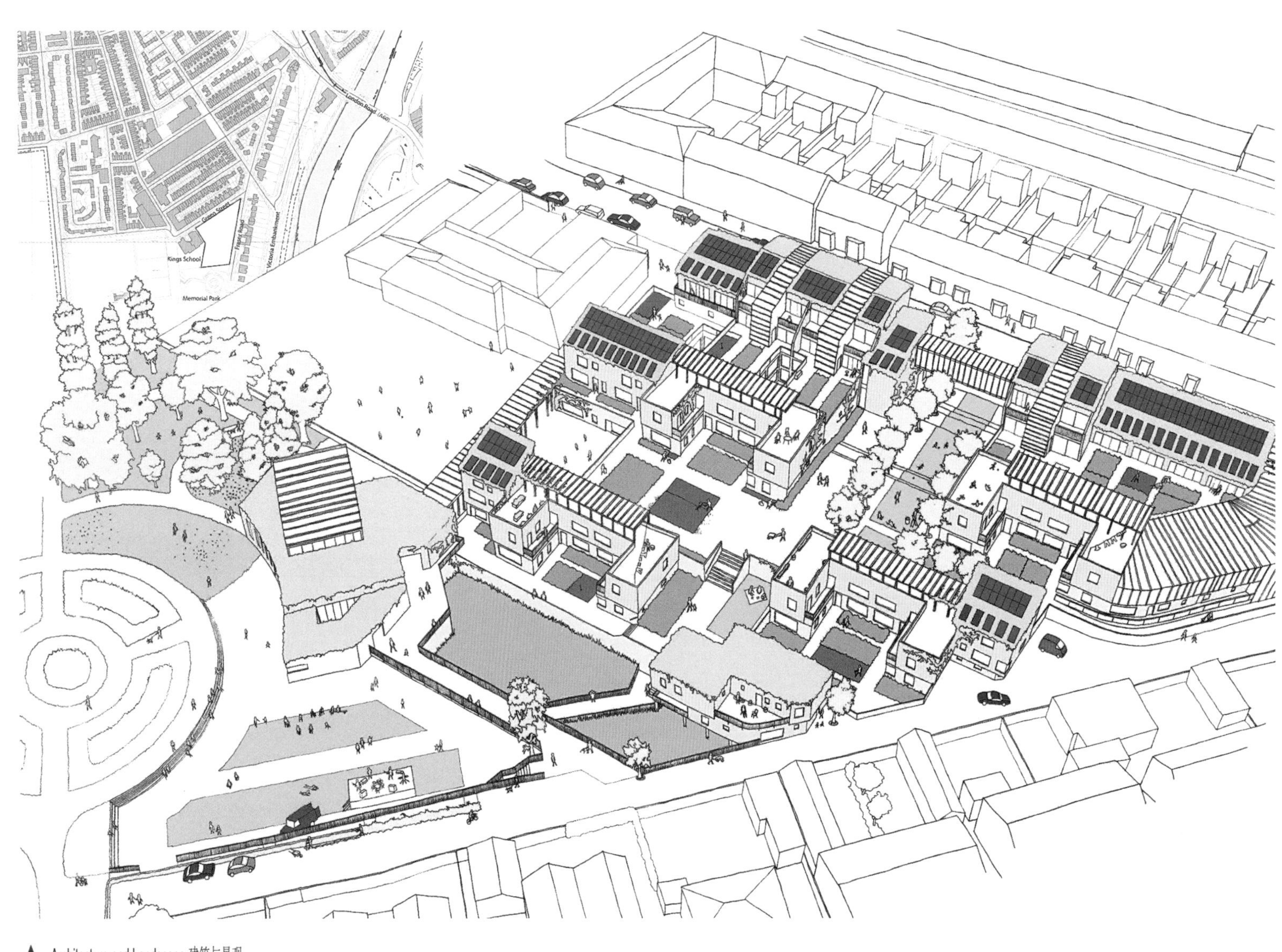

▲ - Architecture and Landscape 建筑与景观

▼ - Block Section 体块剖面图

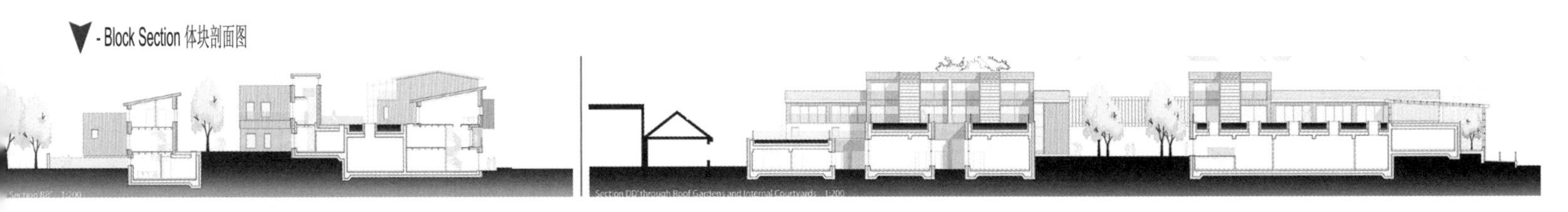

ZERO CARBON
HOUSING
Green Street | The Meadows | Nottingham
Level 2 Site Plan 1:200
- Level 2 Plan 二层平面图
- Level 1 Plan 一层平面图
ZERO CARBON
HOUSING
Green Street | The Meadows | Nottingham
Level 1 Site Plan 1:200
24.7m

DESIGNERS-设计人: James Freeman, Simon M.L Cheung, Matt Gilbody

Zero Carbon Housing Project in Meadows, Nottingham

The response to the site was an idea that the proposal should consider the context and not be isolated. There are two main architectural responses to the site. Firstly, a 'snake like' form that encourages the route of the social spine but at the same time responds to the urban grain of the site. Secondly, the new row houses connect the existing terrace housing and aims to create a series of courtyards on site. The 'social spine' is the primary access through the site and connects the local area to the park. Car access has been limited on site; no vehicles have access through the site, excluding maintenance and emergency services. Actually, the report has proved that zero carbon homes can be designed in tight urban sites, without compromising the architecture and cost of living with respect to lifetime use.

零碳住宅设计，诺丁汉梅多斯区

设计主导思想——“回应场地要求”要求本方案一定要考虑当地文脉，使之溶于周围环境之中。所设计的建筑在这里对场地有两个主要的回应。第一，蛇形的建筑形式鼓励了“社区脊柱”的形成，并同时与此处的城市机理相呼应。第二，新设计的成排住宅与现有的联排住宅相连，从而形成一系列庭院空间。“社区脊柱”是进出场地的首要道路，并将场地与附近的公园联系起来。汽车使用在此处将受到限制；除了工程维护车辆与应急车辆，其他所有车辆将不能进入或穿过场地。实质上，该方案证明了零碳住宅可以在用地相对紧张的城市场地内实现，而不必为了追求长时效综合使用而在建筑特色与生活成本方面做出妥协。

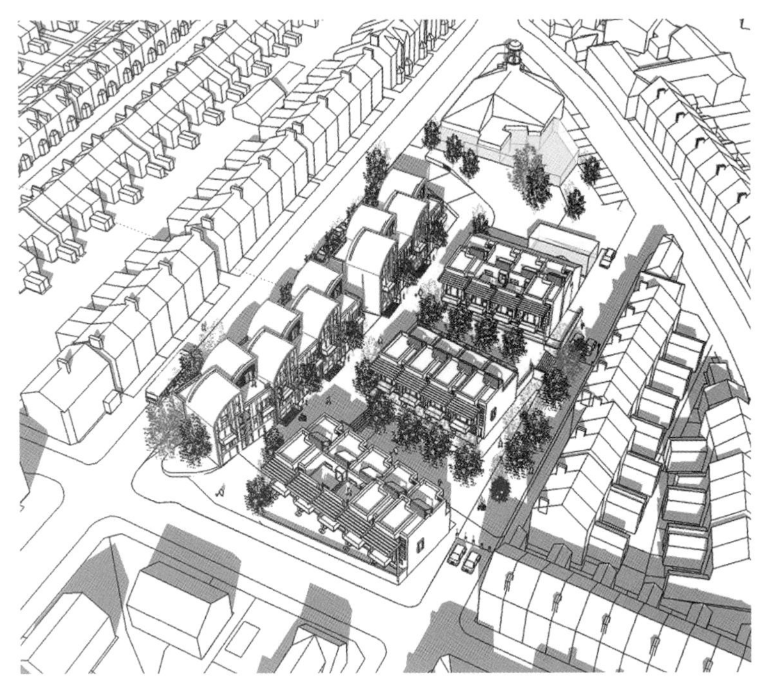

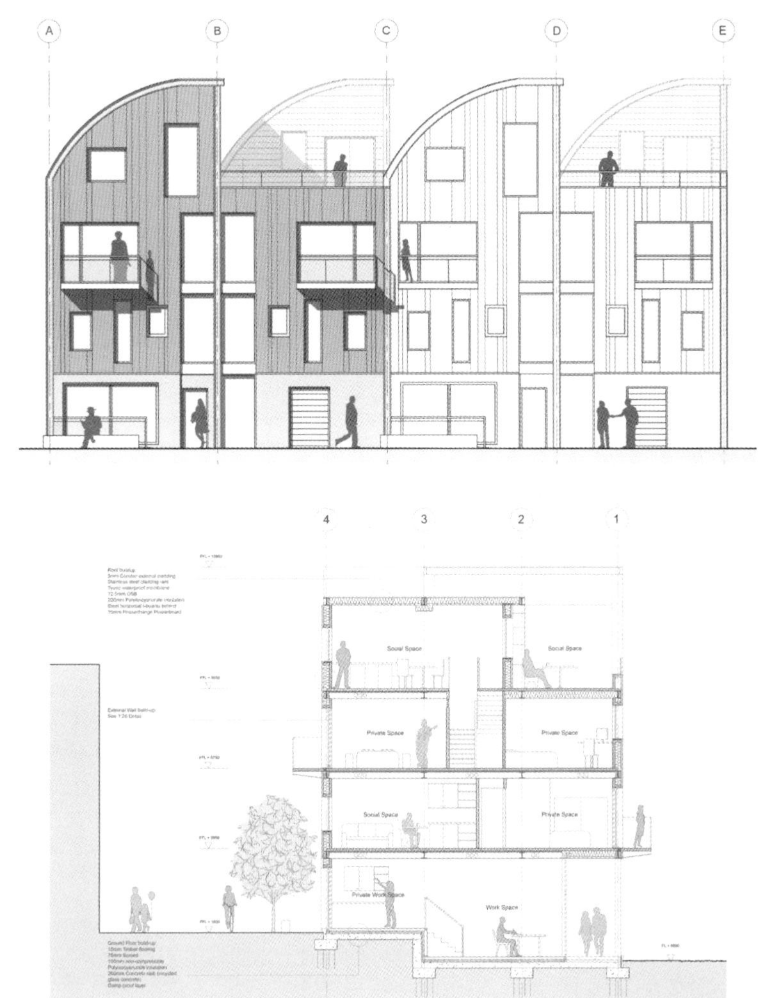

- Section

- Street Section

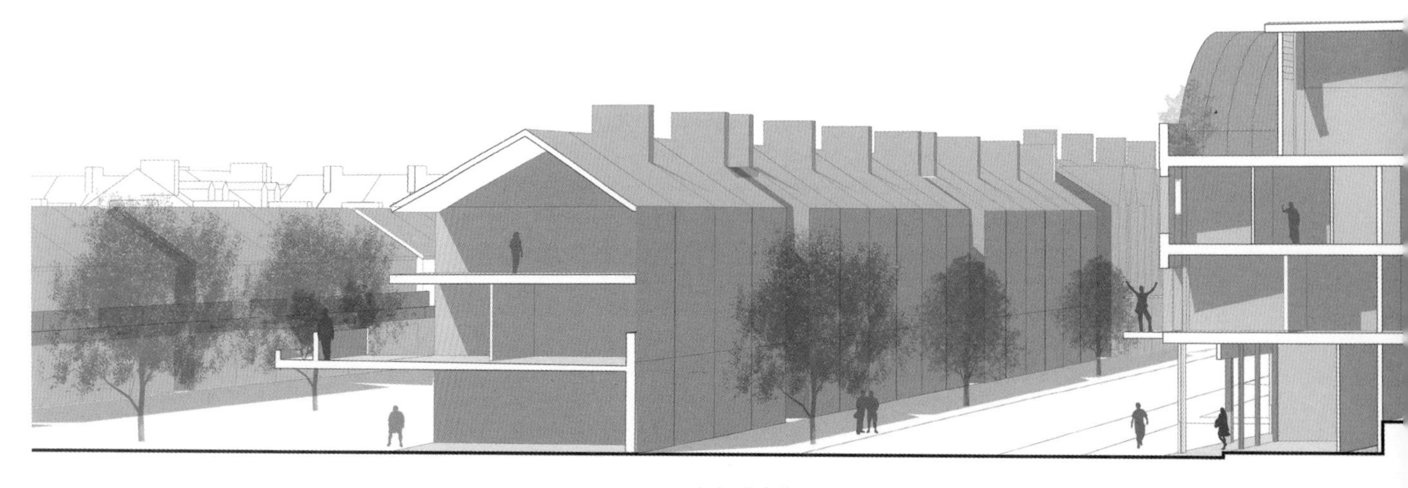

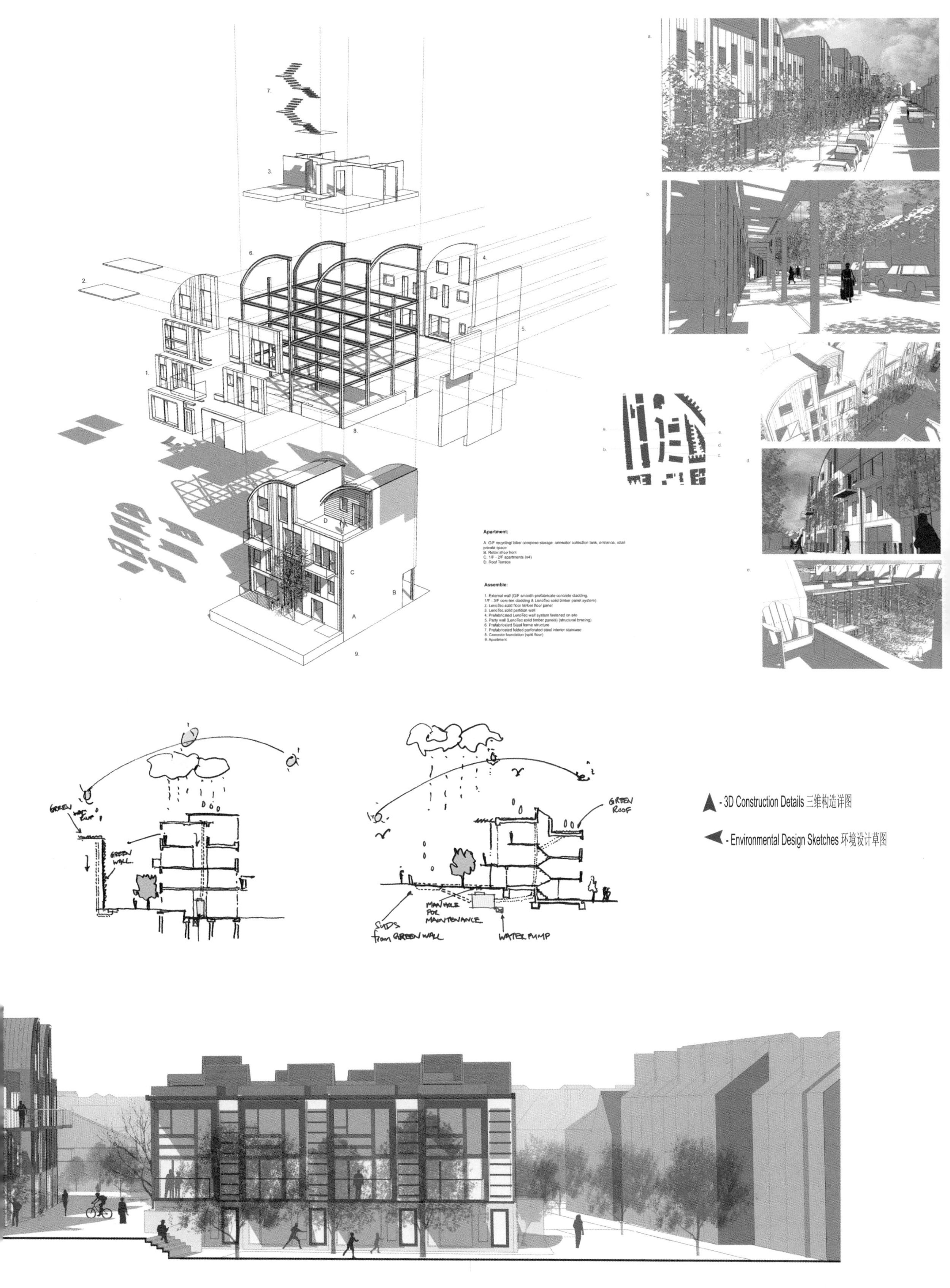
Apartment:
A. G/F recycling/ bike/ compose storage, rainwater collection tank, entrance, retail private space
B. Retail shop front
C. 1/F - 2/F apartments (x4)
D. Roof Terrace
Assemble:
1. External wall (G/F smooth-prefabricate concrete cladding, 1/F - 3/F core-ten cladding & LenoTec solid timber panel system)
2. LenoTec solid floor timber floor panel
3. LenoTec solid partition wall
4. Prefabricated LenoTec wall system fastened on site
5. Party wall (LenoTec solid timber panels) (structural bracing)
6. Prefabricated Steel frame structure
7. Prefabricated folded perforated steel interior staircase
8. Concrete foundation (split floor)
9. Apartment
GREEN WALL
GREEN ROOF
MANHOLE FOR MAINTENANCE
SUDS from GREEN WALL
WATER PUMP
- 3D Construction Details 三维构造详图
- Environmental Design Sketches 环境设计草图

DESIGNERS-设计人: Jindong Wu, Jun Aso

Zero Carbon Housing Scheme in Meadows, Nottingham

The aim of this project was to come up with a good master plan of the site which creates a strong sense of community; a place where all the residents know each other, enjoy each other's company and work together to make their community - a better place, and also a lower carbon emission community. There is a natural order in the forest as each tree is competing for an access to the solar radiation; this is where the design concept comes from. This we believe is exactly the same as we would like every house to have access to direct solar gain which provide significant amount of warmth. We keep the spacing between the houses large enough so that the direct solar gain can be received even in the winter months. The unwanted solar gain during the summer is killed off using movable shutters which allow natural ventilation to take place even when the occupant is not present as they are lockable. The builds are to be prefabricated from timber off site in order to reduce the waste on site.

零碳住宅设计，诺丁汉梅多斯区

这个设计的目的是为了营造一个人们能够相互交流和认识的和谐社区，同时能够实现低二氧化碳排放的目标。设计的概念来自于树林当中树木对于阳光的渴望，而小区当中的建筑也像树林中的树木一样需要阳光来保持室内温暖。设计尽量保持了房子之间的相互距离，使得小区内的建筑在冬天都能得到有效的阳光照射。在夏天的时候，可移动的遮阳格栅可以减少太阳辐射防止室内过热又不影响室内的自然通风。所有的建筑将使用木材在工厂中预制成型，从而减少材料的浪费和对基地环境的影响。

- Ground Floor Plan 底层平面图

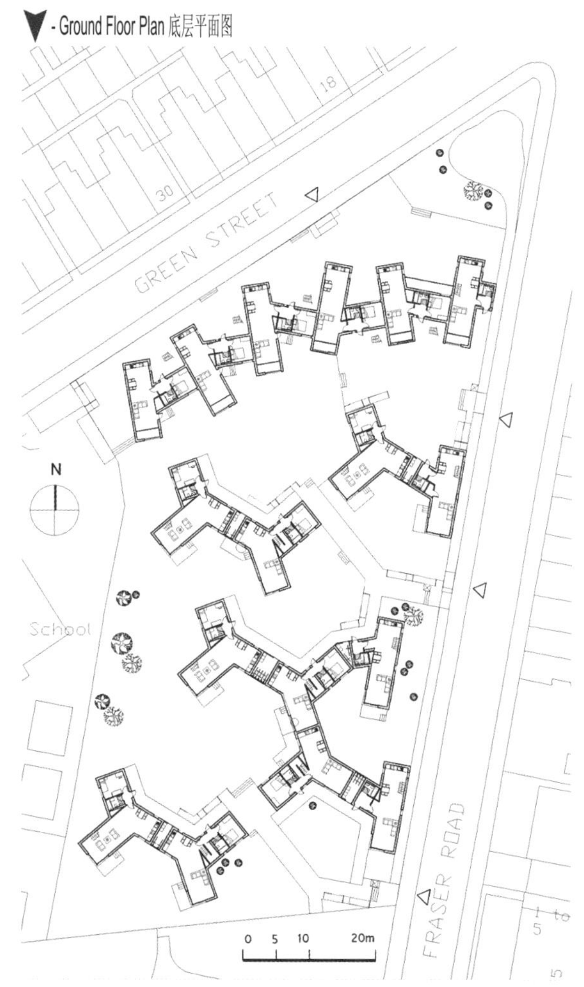

- Site Plan 场地总平面图

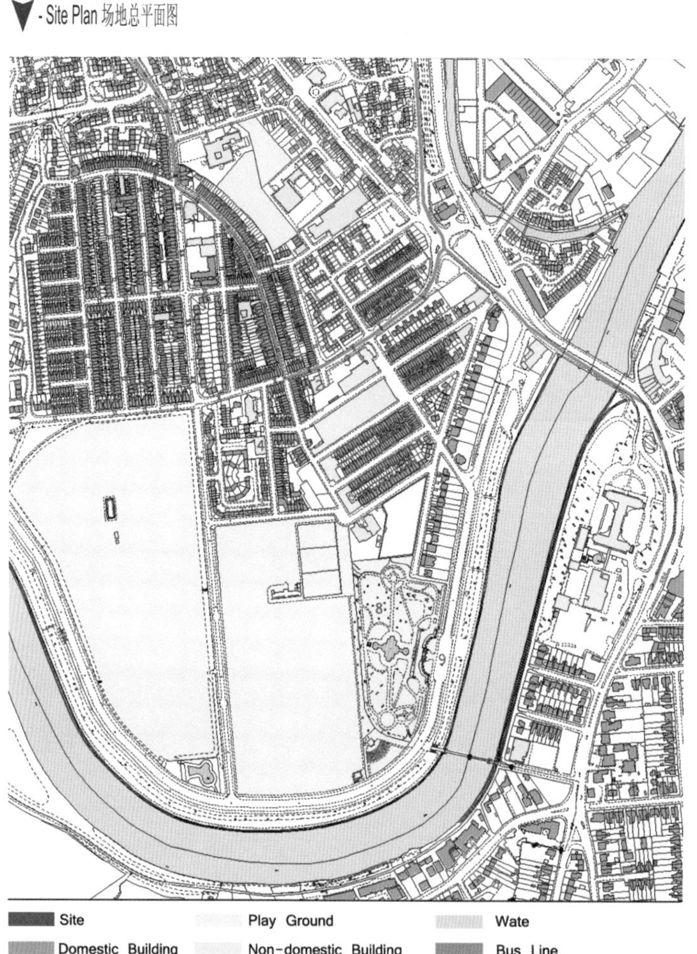

- Site Model 场地实体模型

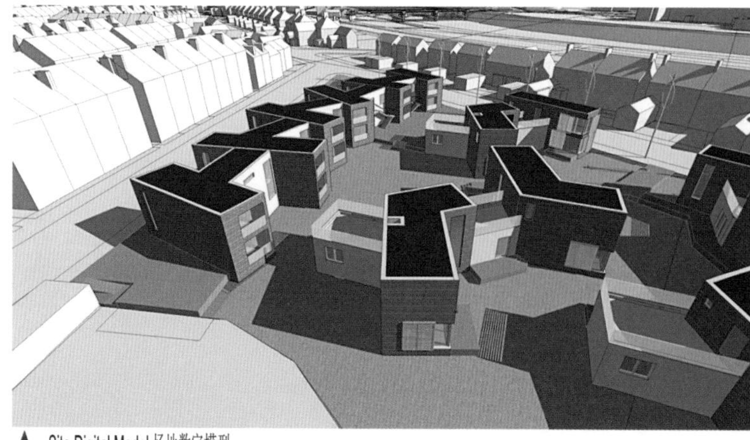

- Site Digital Model 场地数字模型

TUTORS - 指导教师: *Michael Stacey, Swinal Samant*

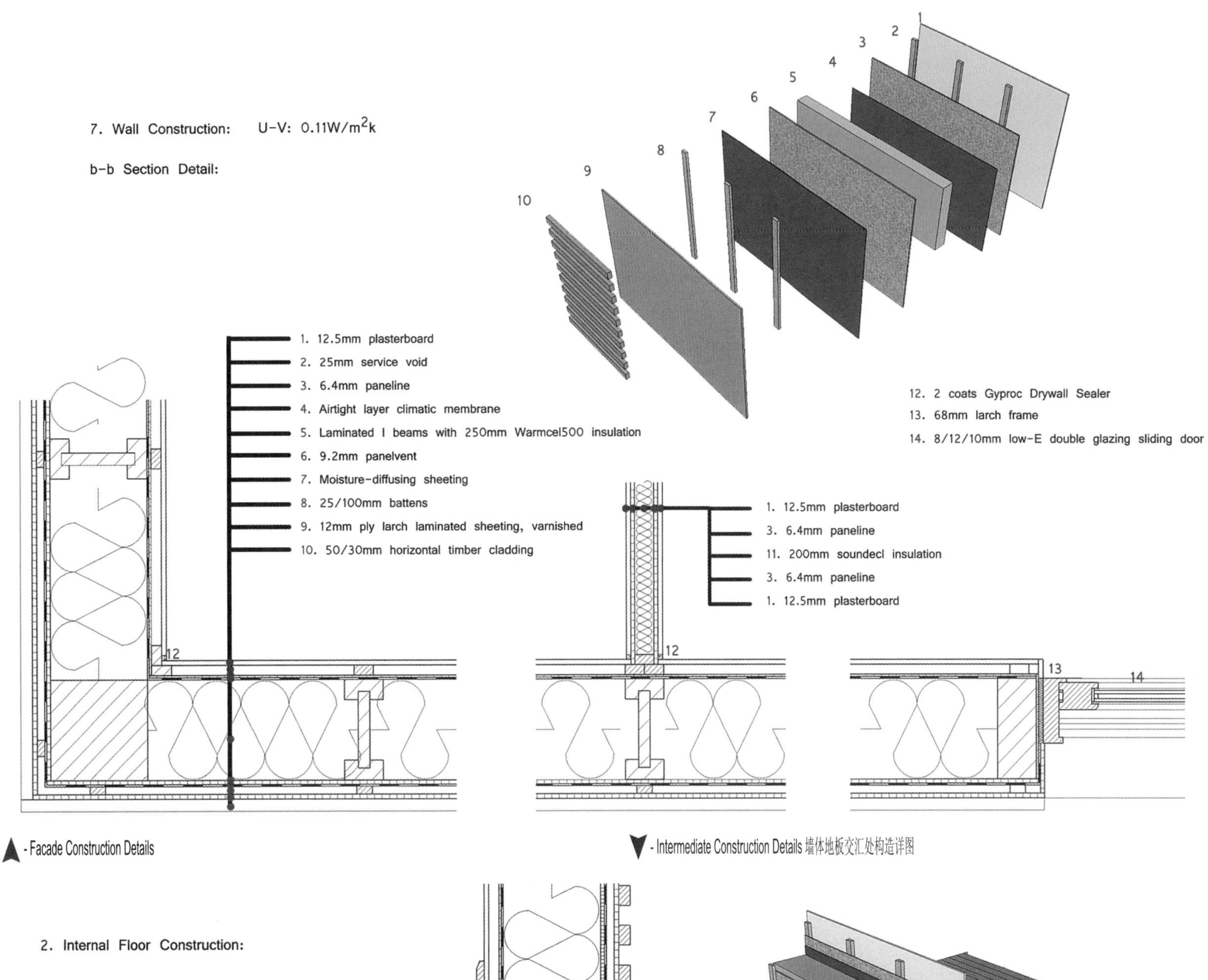
7. Wall Construction: U-V: 0.11W/m²k
b-b Section Detail:
1. 12.5mm plasterboard
2. 25mm service void
3. 6.4mm paneline
4. Airtight layer climatic membrane
5. Laminated I beams with 250mm Warmcel500 insulation
6. 9.2mm panelvent
7. Moisture-diffusing sheeting
8. 25/100mm battens
9. 12mm ply larch laminated sheeting, varnished
10. 50/30mm horizontal timber cladding
12. 2 coats Gyproc Drywall Sealer
13. 68mm larch frame
14. 8/12/10mm low-E double glazing sliding door
1. 12.5mm plasterboard
3. 6.4mm paneline
11. 200mm soundecl insulation
3. 6.4mm paneline
1. 12.5mm plasterboard

- Facade Construction Details

- Intermediate Construction Details 墙体地板交汇处构造详图

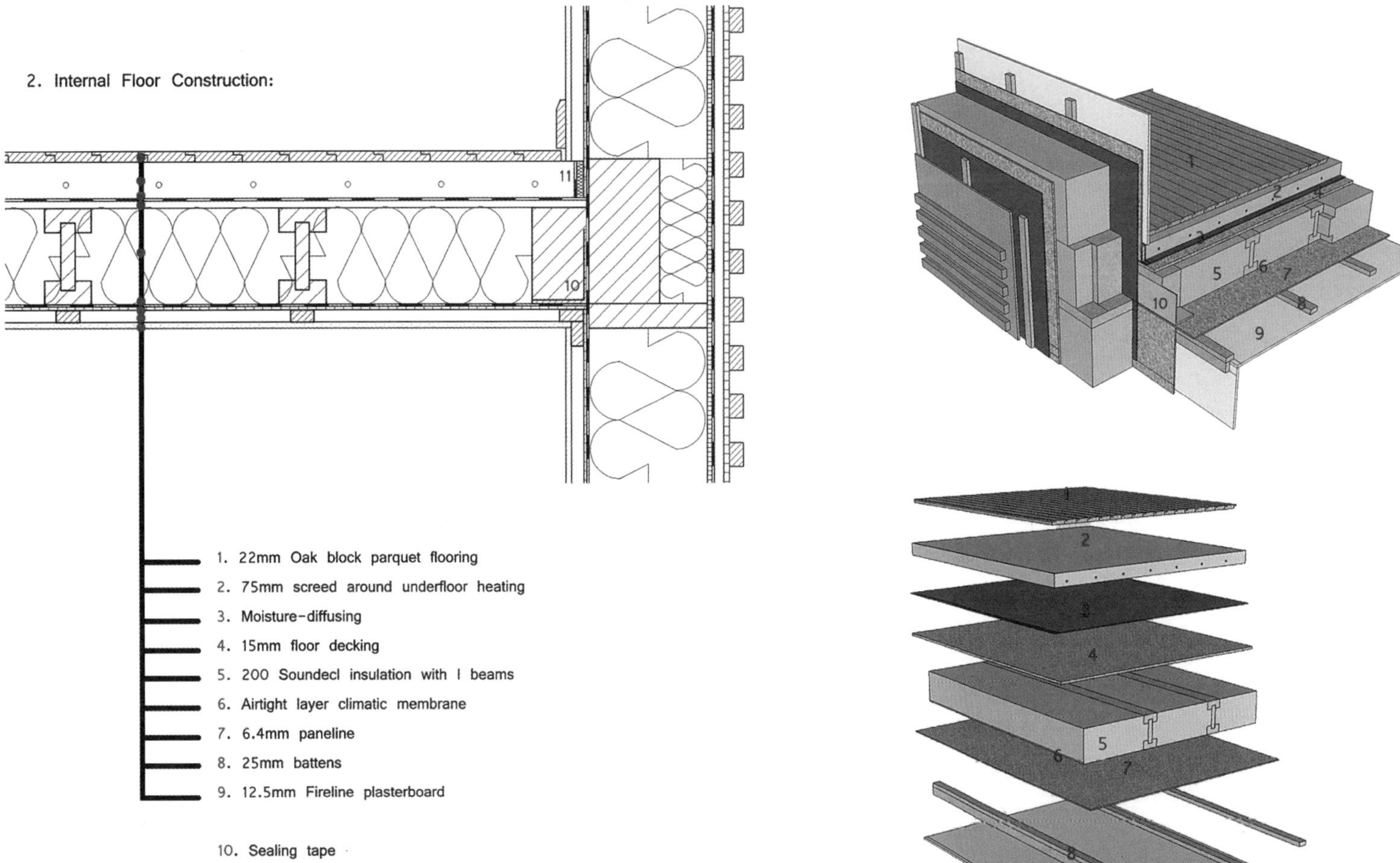
2. Internal Floor Construction:
1. 22mm Oak block parquet flooring
2. 75mm screed around underfloor heating
3. Moisture-diffusing
4. 15mm floor decking
5. 200 Soundecl insulation with I beams
6. Airtight layer climatic membrane
7. 6.4mm paneline
8. 25mm battens
9. 12.5mm Fireline plasterboard
10. Sealing tape
11. Acoustice decoupling

—— URBAN DESIGN
—— 城市设计篇

Research and teaching Urban Design has been a strength of the University of Nottingham for the past two decades. The Urban Design Research Group focuses on the study of the public realm, place making, place branding, urban regeneration, conservation, sustainable urbanism, history and theory of urban design, and waterfront regeneration. The group addresses the design, management and interpretation of urban settings of different scales and geographical contexts, ranging from urban buildings, to public spaces and towns through to entire global cities and their regional contexts. Research of staff members is closely linked to studio teaching.

The projects shown in the next pages show a variety of scales of urban interventions. One might argue that architecture and urban design should not simply be understood as designs at different scales. The process of urban design tends to differ from that of architecture in a number of ways. It foregrounds strategy rather than form. It aims at identifying urban potentials rather than highly articulate solutions. It seeks to identify and bring together a broad range of stakeholders rather than working for a single client. Often it does not deal with a contained site, and instead explores the interrelationship of urban networks – spatial, social and political - across scales.

However, in *The Architecture of the City*, Aldo Rossi argued that the city should be considered as architecture. He not only referred to architecture as the visible sum of its parts, but as the construction of the city over time. For Rossi, architecture not only embodies the city's collective memory, but is an active catalyst in its process of change and transformation. Accordingly, it is architecture's catalytic role in the urban process that renders it "urban". Urban processes are always already linked to spatial and formal transformations, but also to a broad set of negotiations across disciplines, addressing a broad range of questions.

Some of the projects can be understood as Rossi's catalysts, urban buildings that respond to the urban context, but also contribute to the larger urban dynamism. For example, "The Final Act" celebrates a darker side of the city in designing an urban abattoir. The form and expression of the project is manipulated to address the constraints of the urban site. Public routes through the building weave the project into the urban context. The sheer scale of the "Apitherapy Centre" turns it into an urban architecture. The "Urban Dairy Farm" provides a large scale urban infrastructure, responds to the history of the city and brings a manufacturing process back into the city.

Other projects, such as "Lincoln Brayford Pool" or "Basford Urban Framework Study" are developed as a framework for urban development. The former developed linkages and synergies between the university and the city by consolidating Brayford Pool North and Brayford Pool East as distinct quarters within the city. This project was also developed further as a professional consultancy for Lincoln City Council and the University of Lincoln. "Basford Urban Framework Study" was developed with the help of Nottingham City Council. The framework proposed a range of urban strategies to harness, propel and interlink existing and new urban potentials – ranging from the enhancement of street life, a range of public space interventions, reprogramming existing buildings, the promotion of programmatic synergies, to the identification of potential development sites.

在过去的二十年中，研究和教授城市设计已经成为诺丁汉大学的一个强项。城市设计研究组着眼于公共领域，场所营造，区域标识化，城市再生，保护，可持续发展都市化，城市设计历史和理论，以及滨水地区复兴。这个小组旨在对不同尺度的，不同地理文脉的城市布置进行设计，管理和解释，其范围涵盖城市建筑，公共空间和小镇，以至全球化的城市与区域文脉。小组成员的研究工作同时也与教学工作紧密相连。

在随后篇幅中展示的各个项目均体现了不同尺度下的城市介入。我们认为建筑设计与城市设计不应该被简单地理解为仅在尺度上有所不同。其实在设计过程的很多方面，两者差异显著。城市设计注重策略，而非形式。它旨在标识出城市的潜力，而不是特别实际具体的解决方案。它希望能在广泛的范围内确定并聚集与之相关的人群，而不是单单为一个客户工作。城市设计时常不是处理一个封闭的基地，而是透过不同的尺度去探索存在于空间，社会与政治体系中的城市网络关系。

然而，在《城市建筑》一书中，阿尔多•罗西则认为城市应该被看作是一种建筑。他强调建筑不仅是所有可见部分的总和，也是随着时间推移而进行的城市建设。对于罗西而言，建筑不仅体现了城市中那聚合起来的记忆，也是城市变革和转型中活跃的催化剂。因此，正是建筑这种在城市化进程中所扮演的催化剂角色，使城可谓之“城”。城市化进程总是与空间和形式的转型相联系，且同时也涉及多学科间的协商与合作，从而可以在广泛的领域内探讨问题。

本篇中的一些方案即可被理解为罗西所说的催化剂——城市建筑，它们不仅回应城市的文脉，也更强烈的激发了城市活力。例如，方案“最后一幕”通过设计位于城市中的屠宰场来强调了城市阴暗的一面。该项目的形式和表达手法均源于城市基地的种种限制条件，而穿街走巷的公共道路则将其与城市脉络编织在一起。垂直生长的“蜂疗中心”看上去更像是一座城市建筑，而“城市奶牛场”则展示了一个大规模的城市基础设施，与这个城市的历史相呼应，同时也将工业生产重新带回了城市。

其他项目，例如“林肯布雷福德湖”或“巴斯福德城市框架研究”，也均被视为针对城市化发展框架的研究。前者通过将布雷福德湖北部和东部巩固为不同的城市区域，推进了大学和城市间的联系和协同合作。此外该项目还被作为林肯市政府与林肯大学的专业参考而得到了深入发展的机会。“巴斯福德城市框架研究”的发展也得到了诺丁汉市政府的协助。这个框架提出了一系列城市发展策略——从强化街道生命力，系列化公共空间介入，重塑现有建筑功能，推动各种功能整合，到确定潜在的发展用地——进而来配合，推动及联系现有的与新兴的城市潜力。

—— **Dr. Katharina Borsi**
卡特琳娜 · 波尔西 博士

The Head of the Urban Design Research Group
城市设计研究组主任

DESIGNER-设计人: Dana Halasa

Sneinton Market Master Plan, Nottingham

The creative industries district is a destination for local, national and international productions, exhibitions and events, which ideally showcases emerging digital and new media work. Clustered around there are the media café hub, interactive exhibition spaces, an experimental interactive hanging space, a parade ground and outdoor terraces ideal for multi-media performance, live performance, game screenings, festivals and seminars. Other facilities include the enterprise centre, computing studios for animation, virtual reality, visual arts and design technology, media labs for science and technology activities, engaging young people to explore the potential for careers in science and developing effective links between science and education at all levels will be a central thread in Nottingham's Approach. Hence creating public understanding of science appreciating the benefits science brings in.

斯奈顿市场总体规划，诺丁汉

巧妙地借助当代数字和媒体技术，这个富有创意的工业区成了举办地区级、国家级，甚至是国际级产品展销会和活动的重要场所。这里拥有媒体咖啡厅、互动展览空间、实验性的互动悬挂空间、一个大展场，以及一处室外露台，可为各种多媒体表演，现场表演，游戏屏幕，节日庆典和学术研讨会提供非常理想的举办场所。这里的其他设施还包括企业中心、计算机动画工作室、虚拟现实、虚拟艺术和设计技术，为科学和技术服务的媒体实验室，从而不仅鼓励年轻人去发掘自己投身科学领域的未来职业发展机遇，而且可在各个层面上建立起科学与教育的有效联系。这些都将成为诺丁汉未来发展的主线，使全民科普为当地发展带来利益。

- Master Plan 总平面图

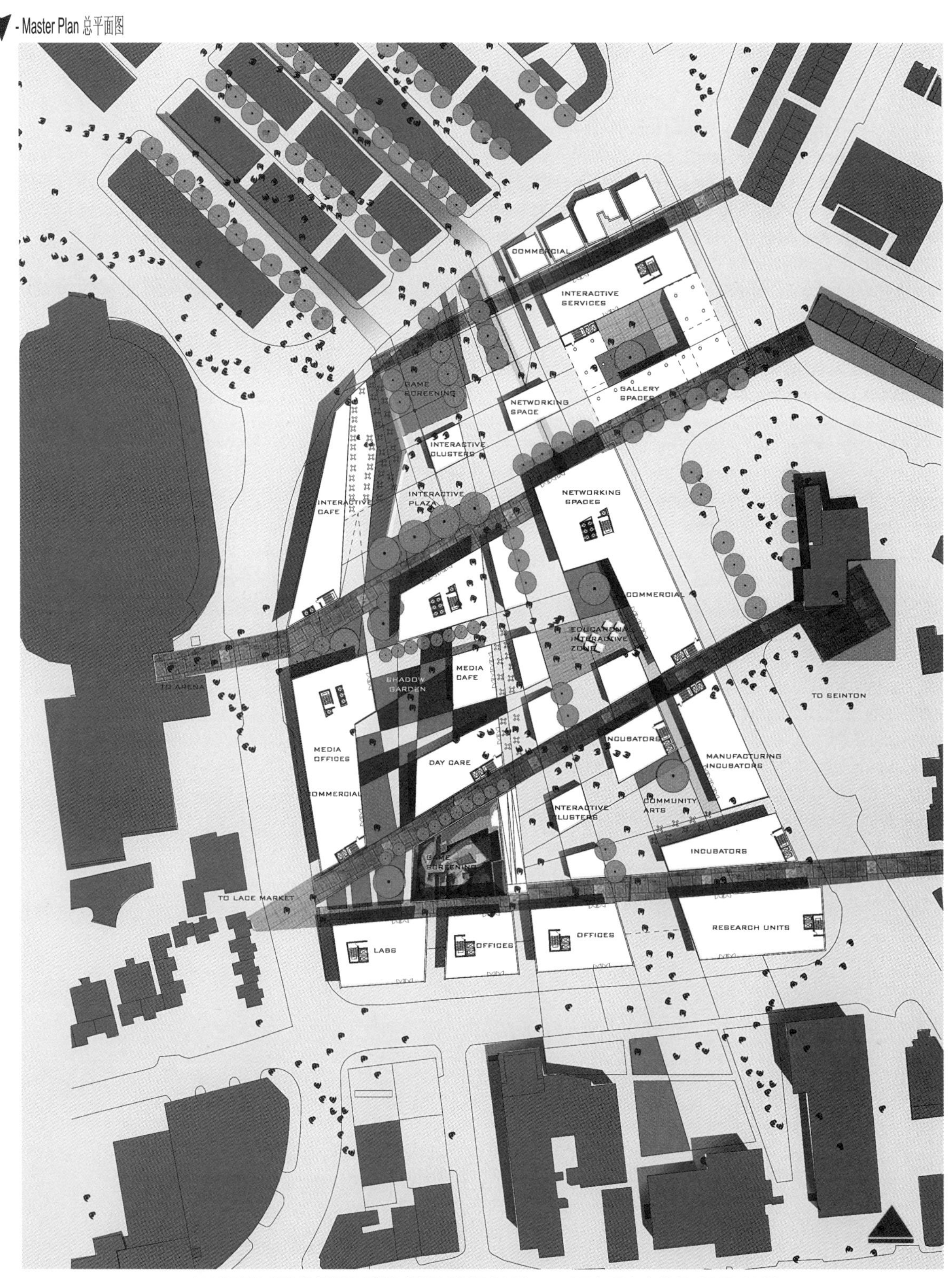

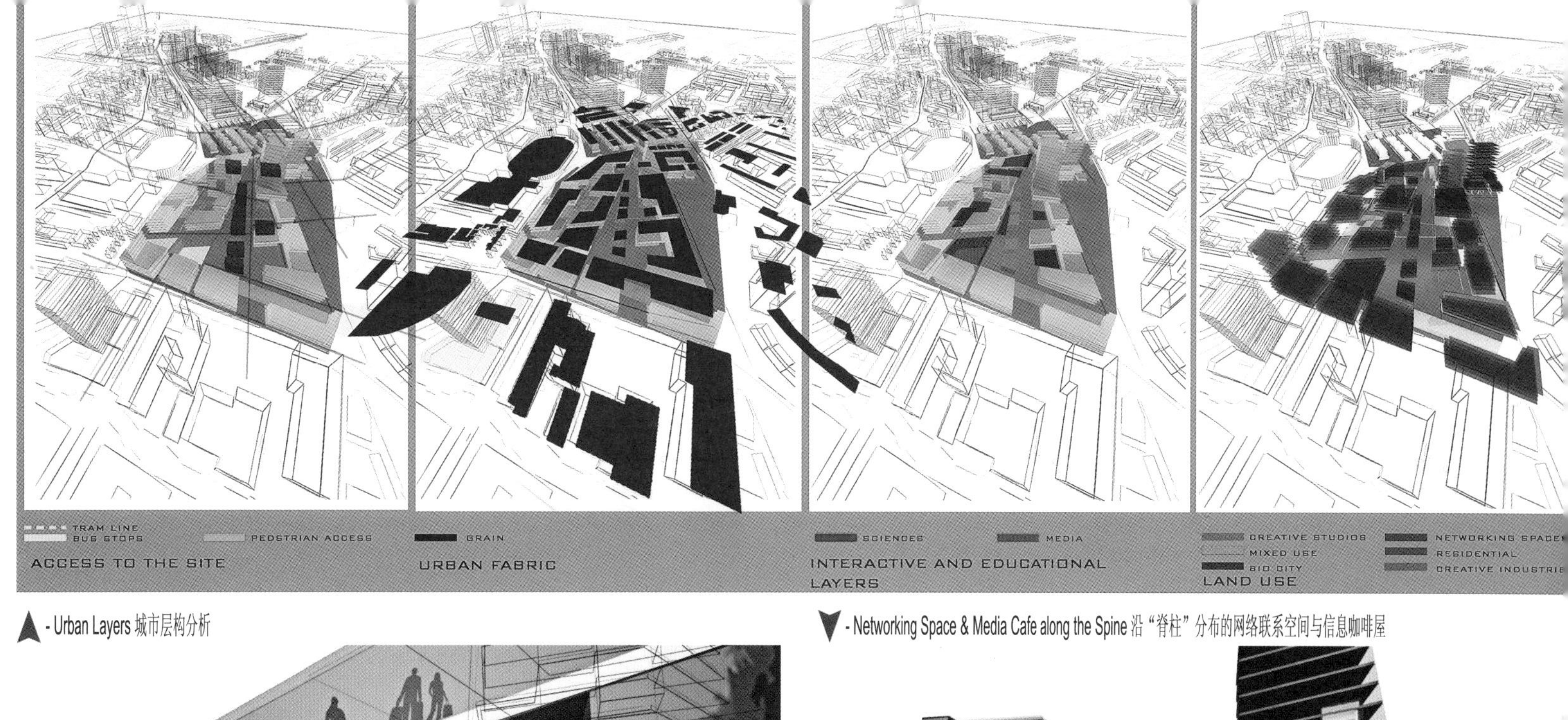

- Urban Layers 城市层构分析

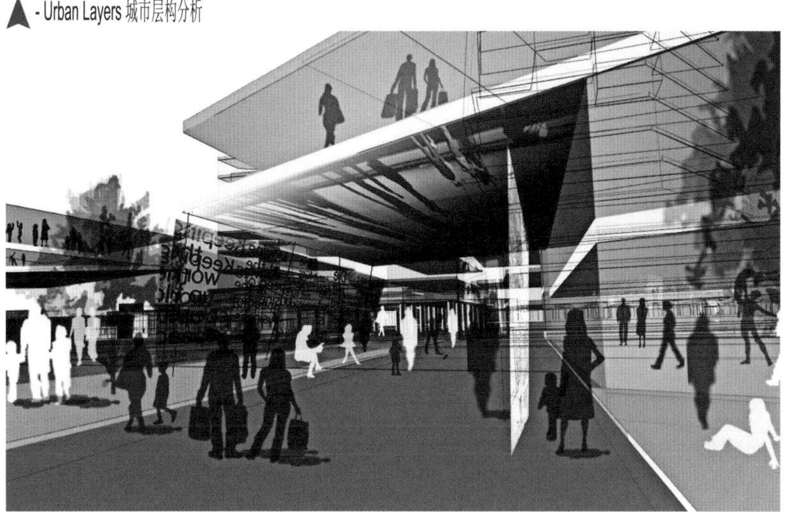

- The Interactive Plaza 互动广场

- Sectional Perspective of Main Public Space 主要公共空间剖透视图

- Networking Space & Media Cafe along the Spine 沿“脊柱”分布的网络联系空间与信息咖啡屋

- Section in Mixed Use Tower 综合高层剖透视图

- Section along the Spine 剖透视图－沿“脊柱”剖切

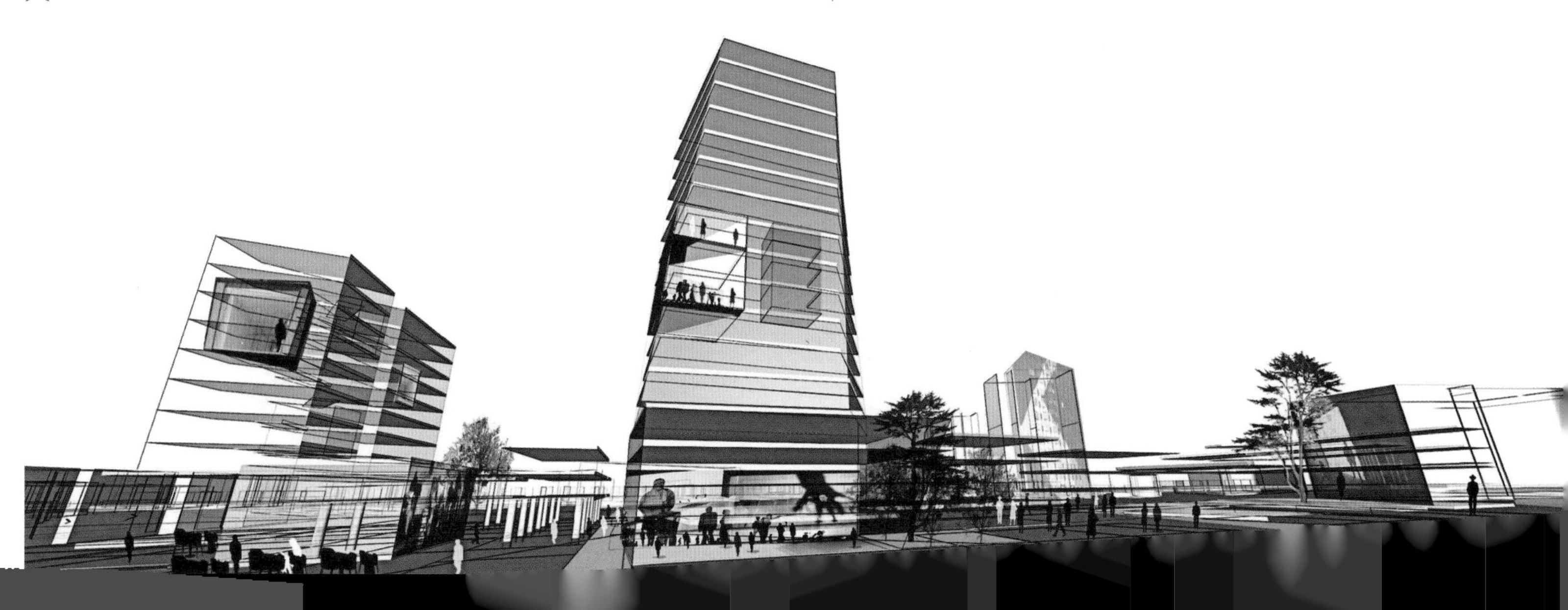

083

DESIGNERS-设计人: David Baggaley, Alina Toosy, Lenard Wong

Lincoln Urban Design Masterplan

Lincoln is a cathedral city lying in the east midlands. The city used to be a flourishing settlement for the Romans due to its namesake, "lindon", which means "the pool". The pool in Lincoln is a natural lake formed from the widening of the River Witham. Here a port was built. Today, the Brayford Pool is a quiet area, next to the heart of the city, but disconnected from the high street, cathedral and castle. To the south of Brayford Pool lies the main campus of the University of Lincoln. The northern edge is overlooked by a cinema, a few bars and restaurants. To the west, a concrete overpass acts as the endpoint. The eastern edge of the pool is overlooked by residential and office blocks. The main objective of the masterplan is to link Brayford Pool to the rest of the city of Lincoln. The aim is to bring an economic growth for the city whilst incorporating elements to keep the existing residents involved in the overall scheme.

林肯城市设计总体规划

林肯是一座位于英国东米德兰地区的天主教之城。罗马人曾经在这里进行过蓬勃的发展建设，其旧名"林顿"正是由原古罗马语"水塘"演变而来的。林肯的这个"水塘"是一个由威特姆河冲刷形成的自然湖，这里还曾经修建了码头。如今，幽静的布雷福德湖比邻城市中心，遥望着远处的中心商业街，大教堂和城堡。布雷福德湖的南部坐落在林肯大学的主校区内；湖的北边正对着一家电影院和一些餐馆酒吧。西部边界则以一架混凝土立交桥标明；而居住区和办公区主要分布在湖东岸。总体规划的主要目的是将布雷福德湖与林肯市其余的区域相连，以此帮助现有的居民参与到整个发展计划中，同舟共济，推动林肯市的经济增长。

▼ - Masterplan, top-left. Lincoln Pocket, Bottom-right. Culture District Waterside 总平面图，左上方为林肯角，右下方为滨水文化区

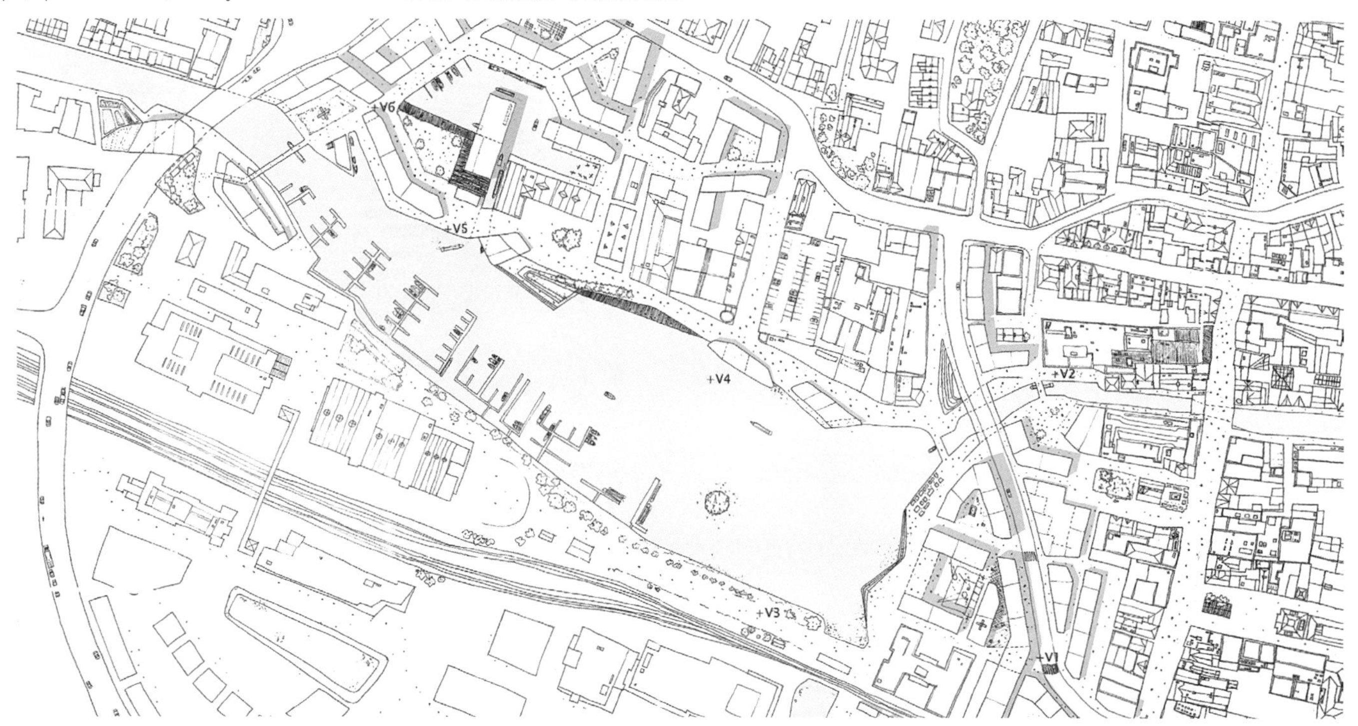

▼ - Life in Lincoln 林肯城日常生活场景

▼ - Lincoln Panorama 林肯全景

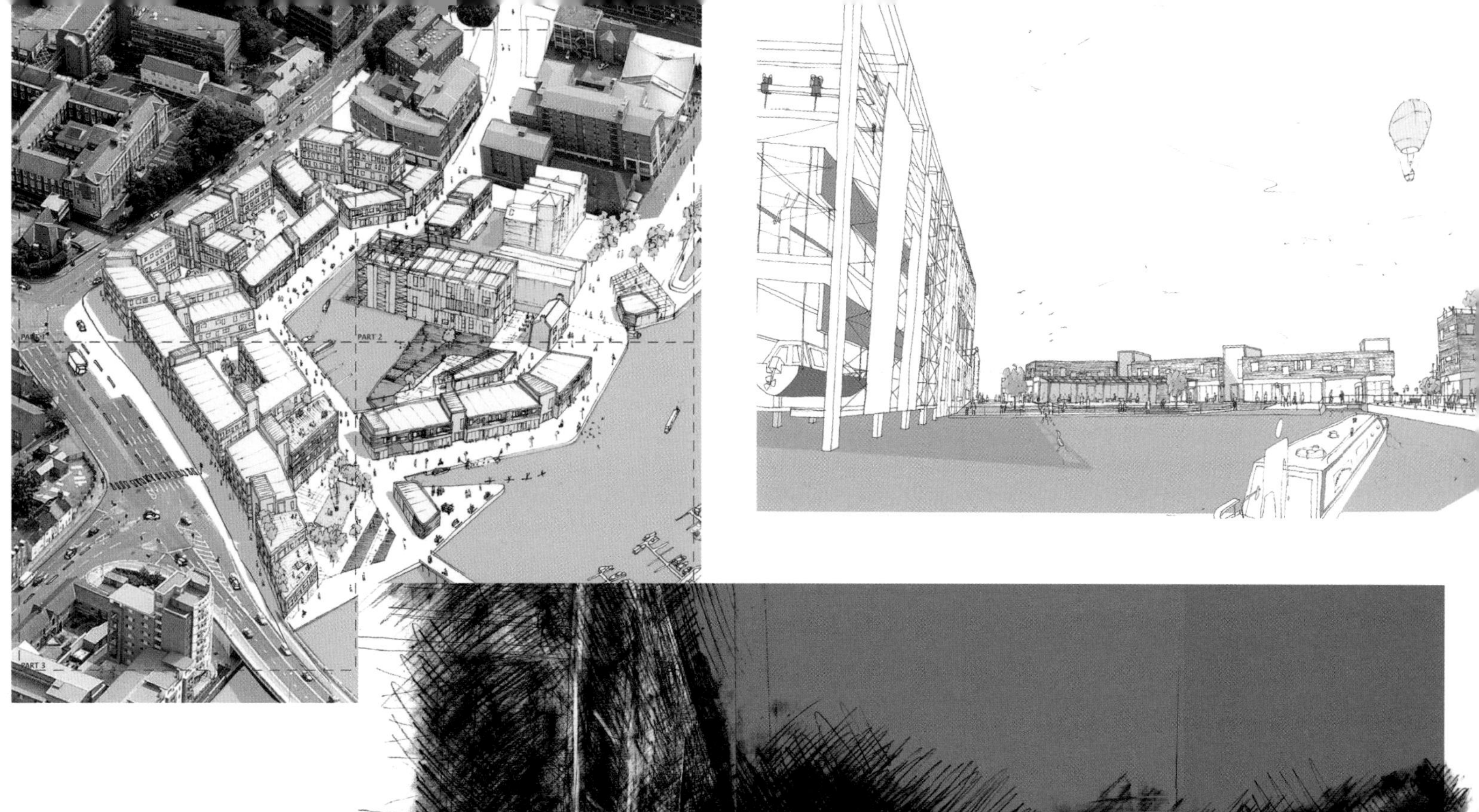

◄ - Media Square 媒体广场

▼ - Culture District Waterside Plan 滨水文化区平面图

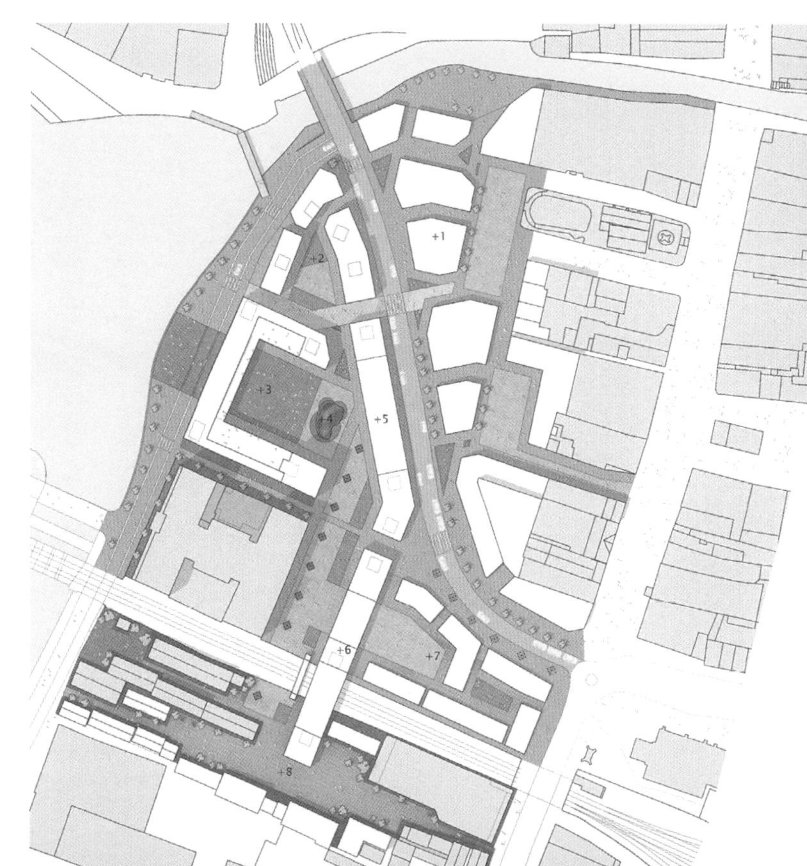

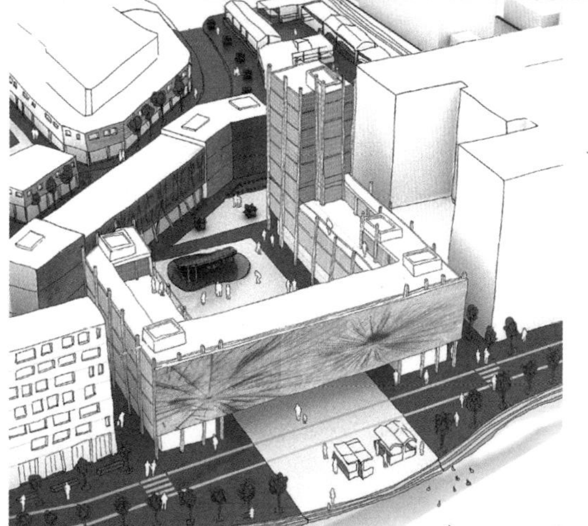

◄ - Media Square Bird View 媒体广场鸟瞰图

DESIGNER-设计人: Sam Johnson

Basford Urban Framework Study, Nottingham

Basford encompasses Nottingham's suburb districts of New Basford and Old Basford, which lies approximately 3 kilometres NNW of Nottingham's city centre. Basford houses 15,113 people which are 5% of the population of the greater Nottingham metropolitan area. The physical area of Basford is also 5% of that of Nottingham. In the current economic climate it is hard to make any assumptions about the implementation of any large scale development project. Our analysis and interventions are focused on a local scale and flexibility is a key theme throughout. The intention is to identify the areas which need connecting, embracing and resurrecting to create a network of small scale catalysts which will trigger the continued development of the area. In keeping with the theme of rhythms, these catalyst interventions can be seen as accents or crescendos on a larger scale. The locations for each of the catalysts respond to our research and attempt to embrace the community and strengthen and develop a much needed urban centre.

巴斯福德城市框架研究，诺丁汉

巴斯福德包括诺丁汉郊区的新老巴斯福德两个区域，大约在诺丁汉市中心西北偏北3公里处。巴斯福德共有居民15113人，占大诺丁汉都市区人口总数的5%，而其占地面积也相当于诺丁汉的5%。在当前的经济气候下，很难想象去实施什么大规模的发展项目。于是我们的分析和介入主要集中在当地小环境，而灵活性则是贯穿始终的要点。方案主要目的是发现那些需要联系，接纳和复苏的区域，进而去形成一个由多个催化区域构成的网络，去激发整个地区的持续发展。通过保留主题旋律，这些催化区域的介入好比一曲更大乐章中（更大区域中）的变音和高潮。根据我们的研究成果，可以确定出每个催化剂的位置，它们将融入当地社区，共同加强与发展一处急需的城市中心。

▼ - Master Plan 总体规划图

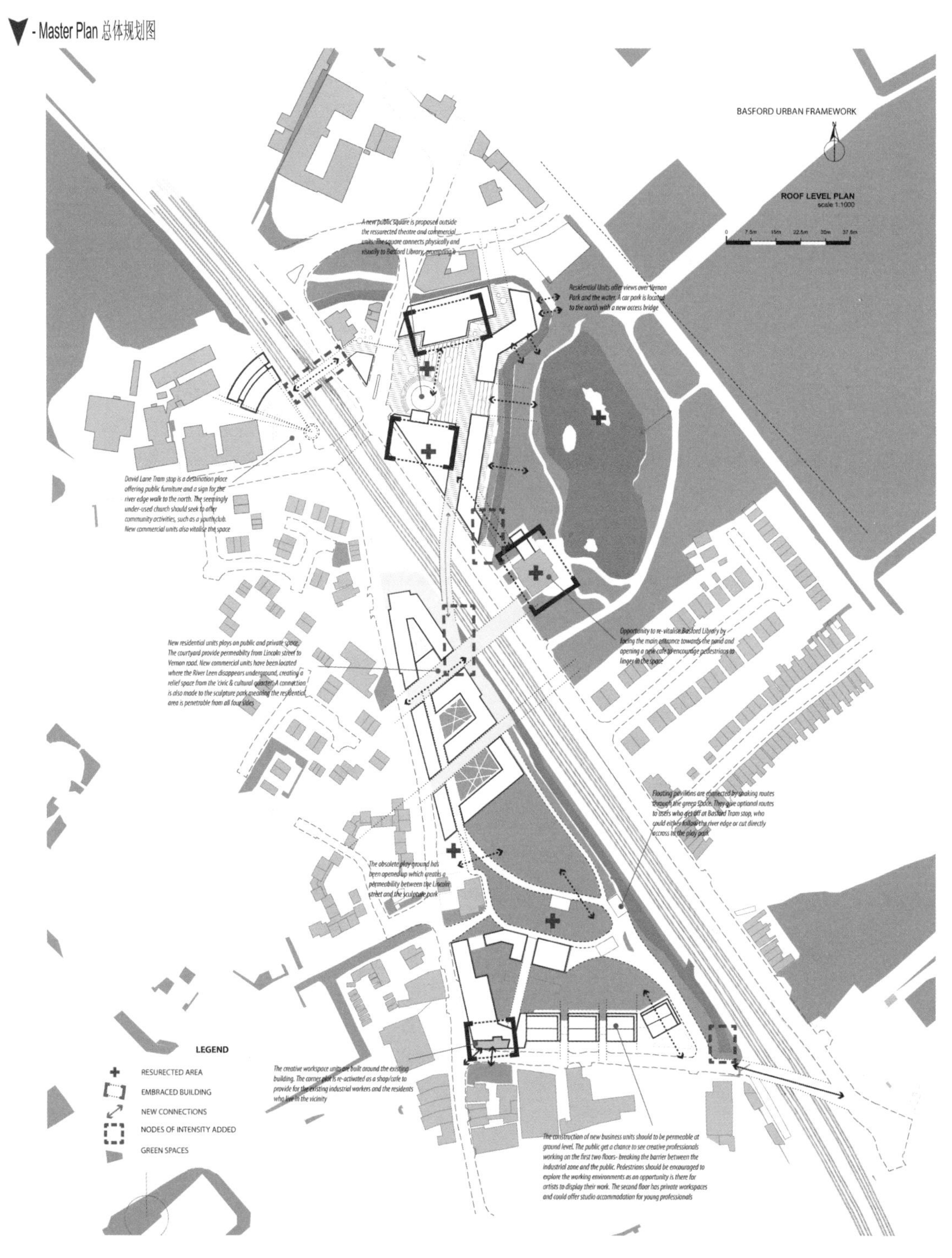

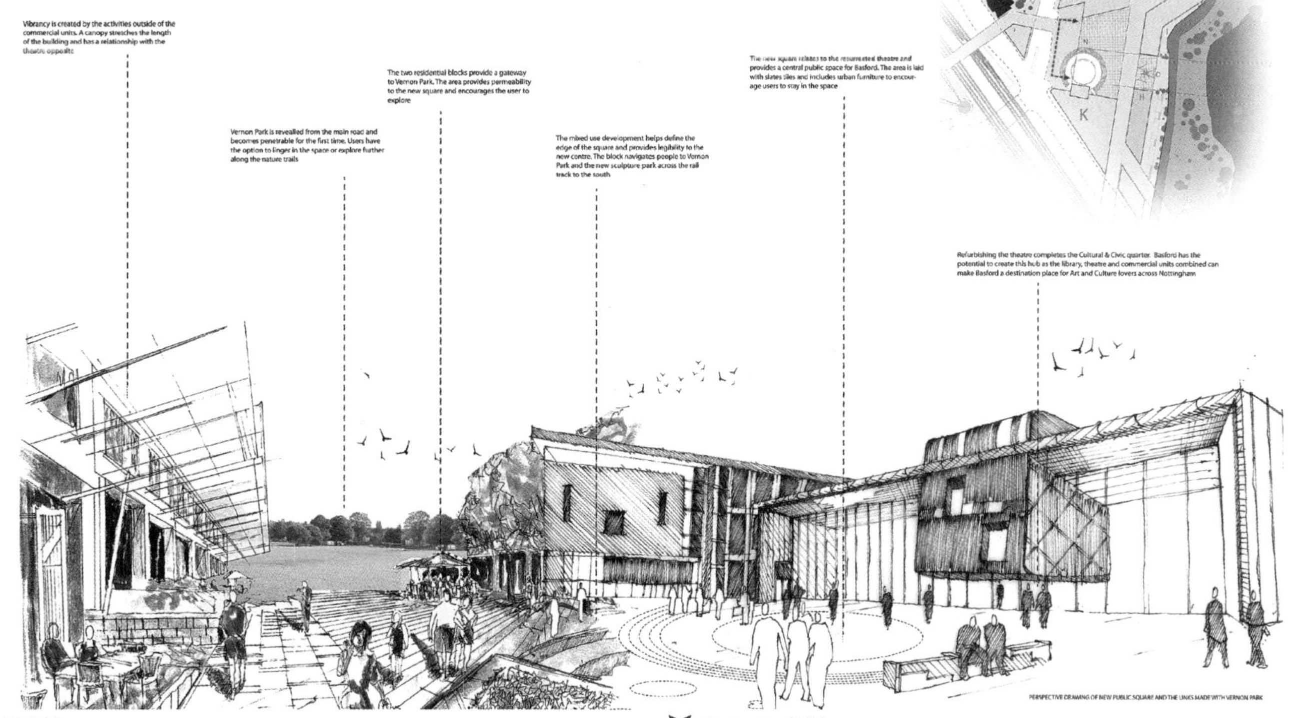

▲ - New Public Square新公共广场

▼ - Site Overview 鸟瞰图

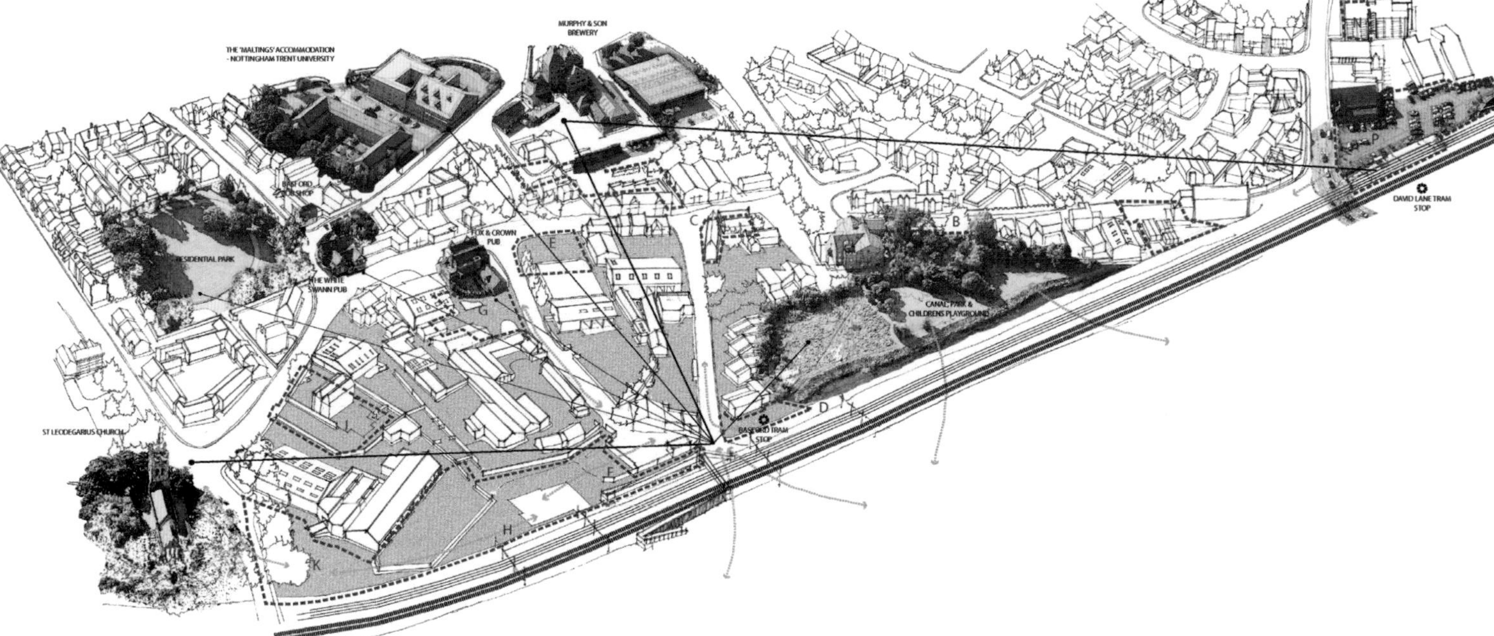

▼ - Creative Workspace Units 创新工作室单元透视图，Team Work 团队成果：Andrew Tindale, Jenny Bachelor, Sam Johnson

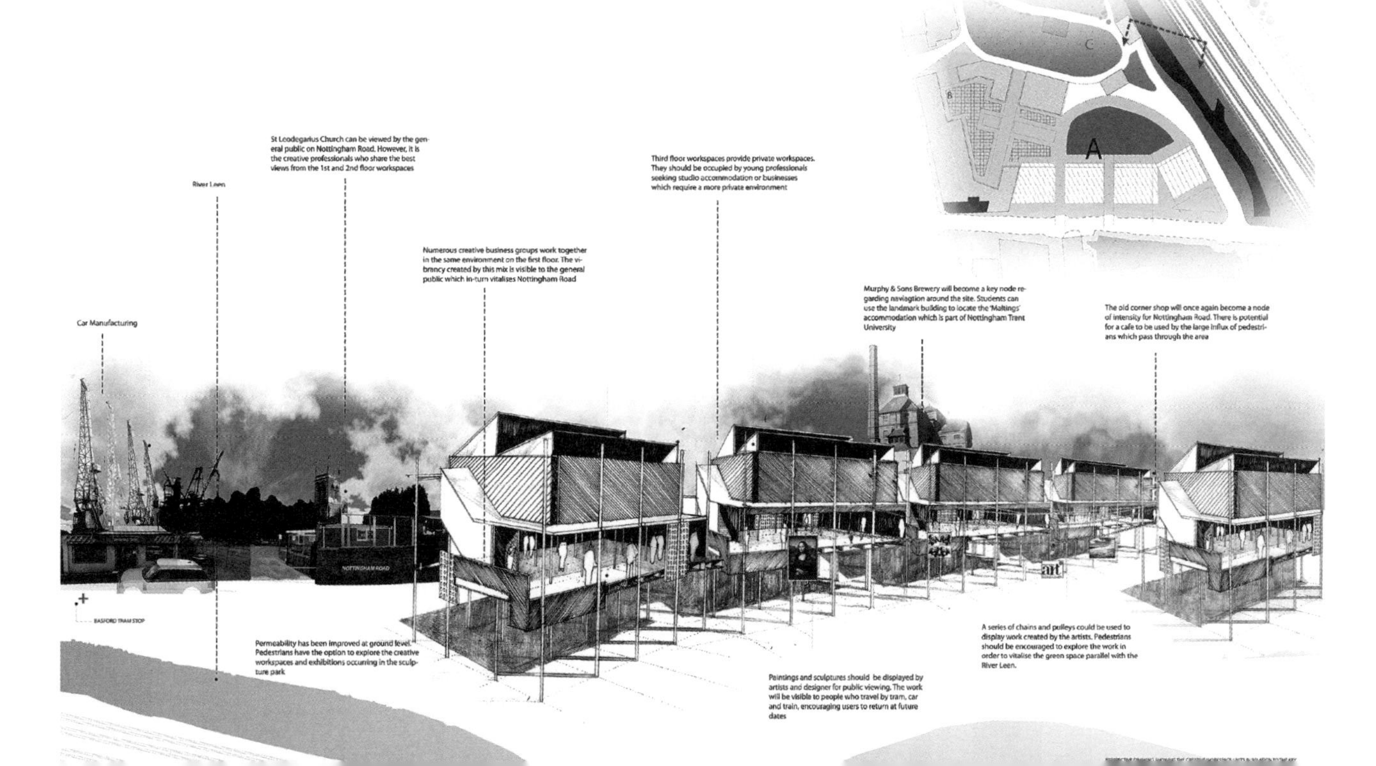

DESIGNER-设计人: Farzad Zamani Gharaghooshi

TangGu Station Urban Regeneration, China

TangGu South Station is one of the oldest railway stations in China constructed in Qing Dynasty. It was built in 1888 as the terminal of Kaiping Mining Railway. The station buildings have a long history and are well-preserved today. Nowadays there are two old station buildings. One of them is timber structure in English style and the other one is a two-story building with sign "TANG KU" on the wall. The British architects, engineers and their works in China especially TangGu South Station is of great value in technological and cultural exchange between China and Britain in the history. This project based on the information above. The aim of the design was to preserve and renovate the industrial buildings and environment, as a respect to Chinese culture and to create an urban area which could be interesting, economic and unique. In this project, the main concept was "back to the tracks" physically, functionally and conceptually. It means that this project proposed to create an atmosphere which belongs to the past, and remind the memories of industrial age.

塘沽火车站城市再生计划，中国

塘沽火车站始建于清朝时期，是中国最古老的火车站之一。它建于1888年，当时被作为开平煤矿铁路的终点站，现在，这些珍贵的历史都被完好地保存了下来。这里共有两栋老建筑，一座是具有英式风格的木构建筑，而另一座是墙上带有"塘沽"标志的两层建筑。那些曾经在中国工作过的——特别是在塘沽工作过的英国建筑师和工程师们在历史上为中英两国的技术与文化交流曾做出了巨大的贡献。基于上述的调研信息，我的设计希望能够在保护现有遗址的同时，实现整个工业建筑和环境的再生，以此来表达对中国文化的敬意，也同时为市民提供一个多彩而繁荣的城市空间。这个项目力图在物质，功能和概念上实现方案的主要概念——"回归轨道"，并最终营造出追忆过去工业时代的氛围。

- Master Plan - Land Use 总体规划 - 土地利用

- Master Plan - Green Spaces 总体规划 - 绿带空间

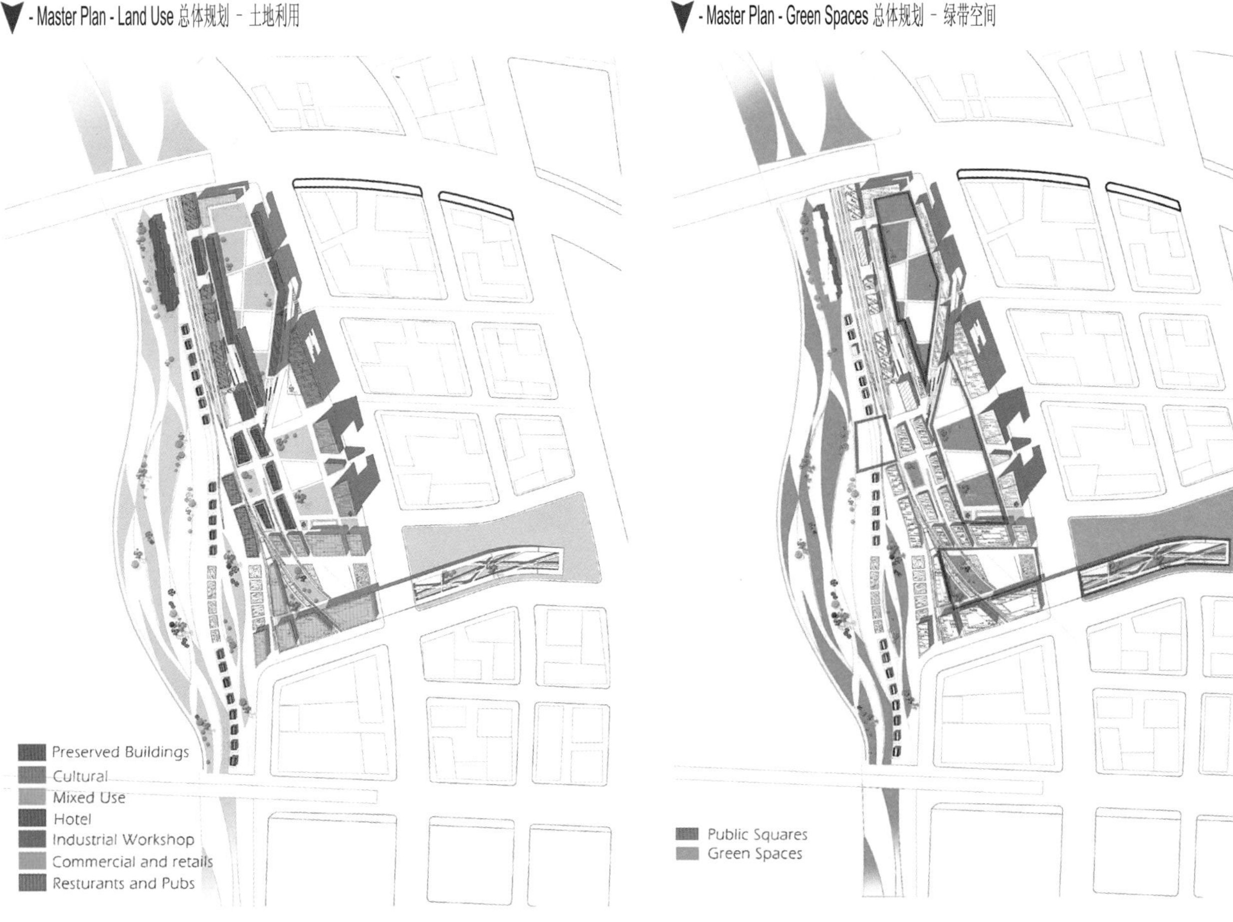

- Functional Strategies 局部功能策略研究

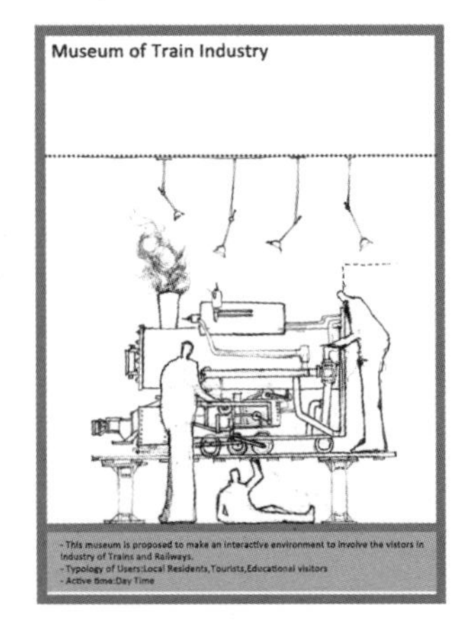

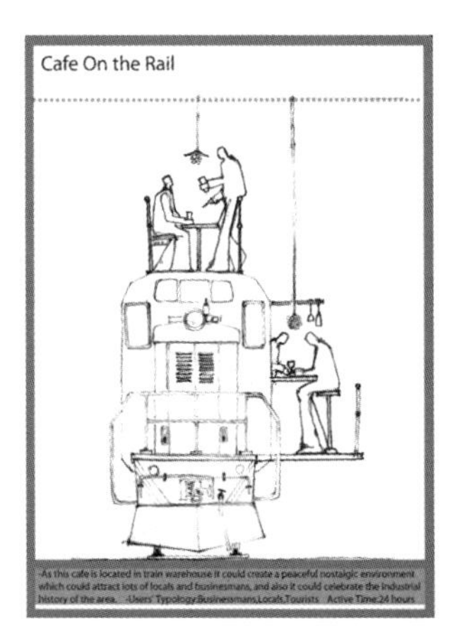

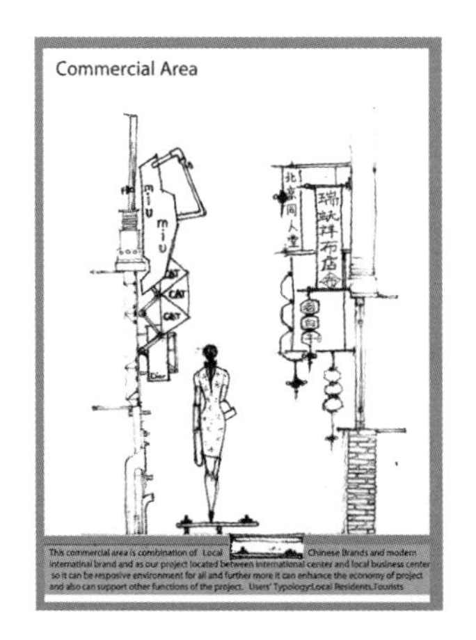

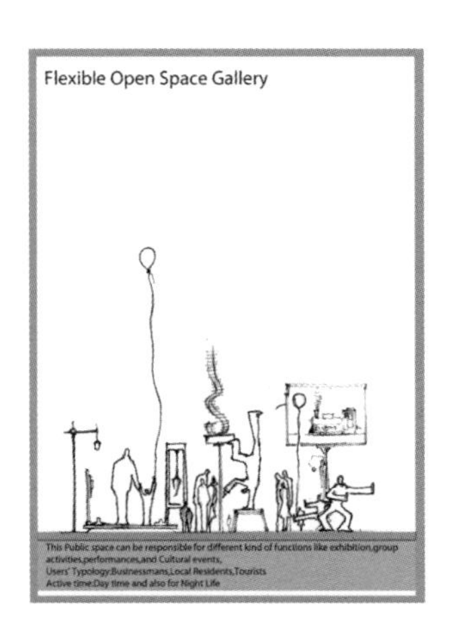

▲ - Construction Details 细部构造

▲ - Construction Framework 建筑构造意象

▲ - Preserved Area 保护区域

▲ - River Front View 滨水区景观

▼ - West Elevation 西立面

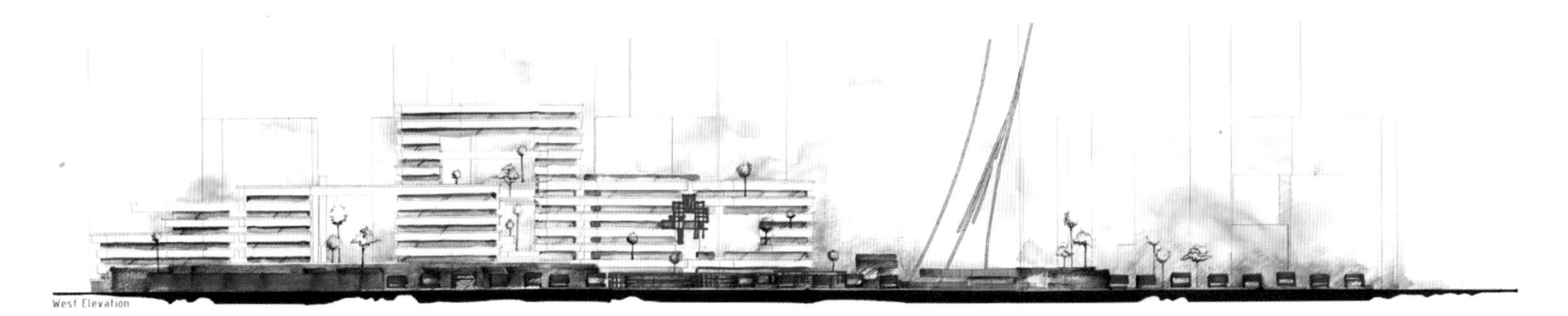

DESIGNER-设计人: Delvis D Anes

TangGu Station Master Plan, China

The Tanggu Station is located Tianjin, China. The train station is located near to the Haihe River and in the centre of the CBD of Tianjin. The site was an old industrial area. Today the old industrial site is under construction soon to be the new CBD of Tianjin. Tanggu Station have historical heritage divided between the station and port. The station built in April 1888 and then rebuilt in 1907 is composed of six buildings the station, the tower of control, the warehouse for the trains and other warehouse. Tanggu Station is the end point of Tang Xu extension railway, but today it is used just for cargo. After learning about the history of Tanggu Station and having the opportunity of visiting, I have concluded that in my site proposal I want to create a place not just for culture and entertainment, but also a place where people can remember what that the Tanggu Station used to be and what it is today. I want to keep an abstract memory of the place.

塘沽火车站总体规划，中国

塘沽火车站位于中国天津市中央商务区中心，紧邻海河。这里曾经是老工业区，而今新中央商务区的建设即将在这里破土动工。这个火车站始建于1888年4月，1907年重建了一次，共有6栋建筑组成——控制塔，停放火车的仓库，以及其他仓库。在火车站和码头之间分布着很多塘沽火车站的历史遗址。塘沽火车站是唐胥铁路扩展路线的终点站，不过现在只用于货物的运输了。通过对塘沽火车站进行实地考察和历史研究，我总结出一下设计要点：将来这里应不仅能丰富人们的文化和娱乐生活，又能将塘沽火车站的过去和今天深深刻入公众的记忆中，进而营造一个令人流连忘返的场所。

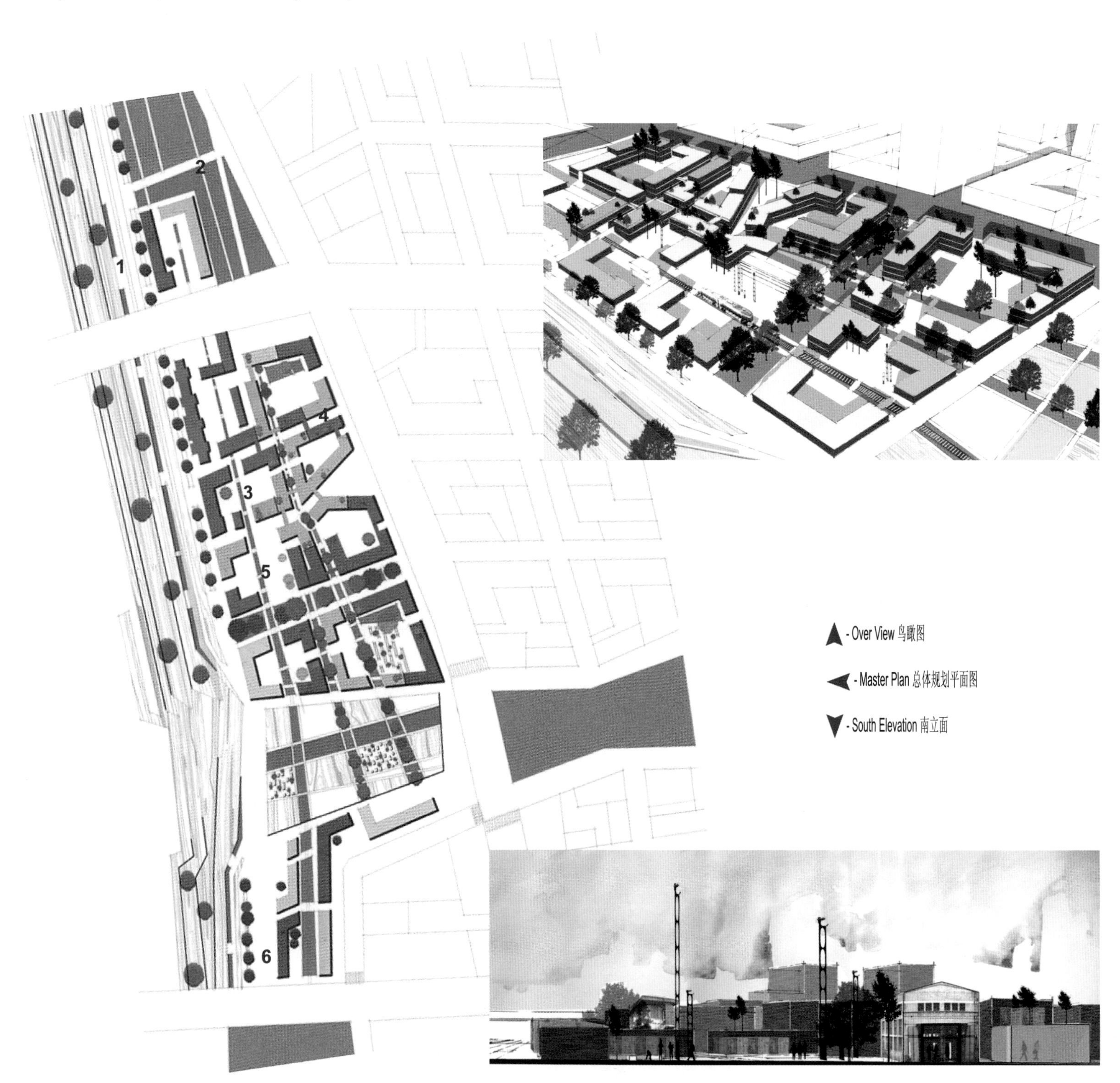

- 1. Walking on Water 亲水步道

- 2. Train Arch 机车拱门

- 3. TangGu Plaza 塘沽广场

- 4. The Bridges 天桥

- 5. TangGu Open Theatre 塘沽露天剧院

- 6. TangGu Main Entrance 塘沽主入口

- West Elevation 西立面

087

DESIGNER-设计人: Yamen Badr

TangGu Station Master Plan, China

Yesterday around one hundred years ago, trains, railways and Tangku train station had appeared within the site; theoretically this appearance was the representation of china to reach the world, the attempt to become international. Today throughout the twenty first century, with massive developments such as SOM master plan, the site with all its industrial and historical value is under the threat of globalization. Nevertheless contrast represented by black and white bring beauty to each, the same apply for TangGu station project, while the memories of the station will bring historical and industrial value to the surrounding area, as well the site will consider the surroundings skyscraper as opportunities rather than obstacles. This report is partly the story of the site between yesterday and tomorrow.

塘沽火车站总体规划，中国

在一百多年前的昨天，火车、铁路和塘沽火车站在此出现。理论上讲，这些新事物的出现可以被理解为中国希望与世界接轨，实现国际化的一种姿态。而二十一世纪的今天，面对大规模的发展，例如SOM设计的总体规划，这个地区所具有的工业和历史价值已经开始面临全球化的威胁。其实，黑与白，这两个极端颜色只要搭配的巧妙，也可以相得益彰，同样的道理也可以运用在塘沽火车站项目上。关于火车站千丝万缕的记忆能提升周边地区的历史和工业价值，火车站的存在也能为周围的摩天大楼带去发展机会而不是障碍。于是这份设计报告也将成为该区域从过去迈向将来的故事中的一部分。

▼ - Analysis 场地综合分析

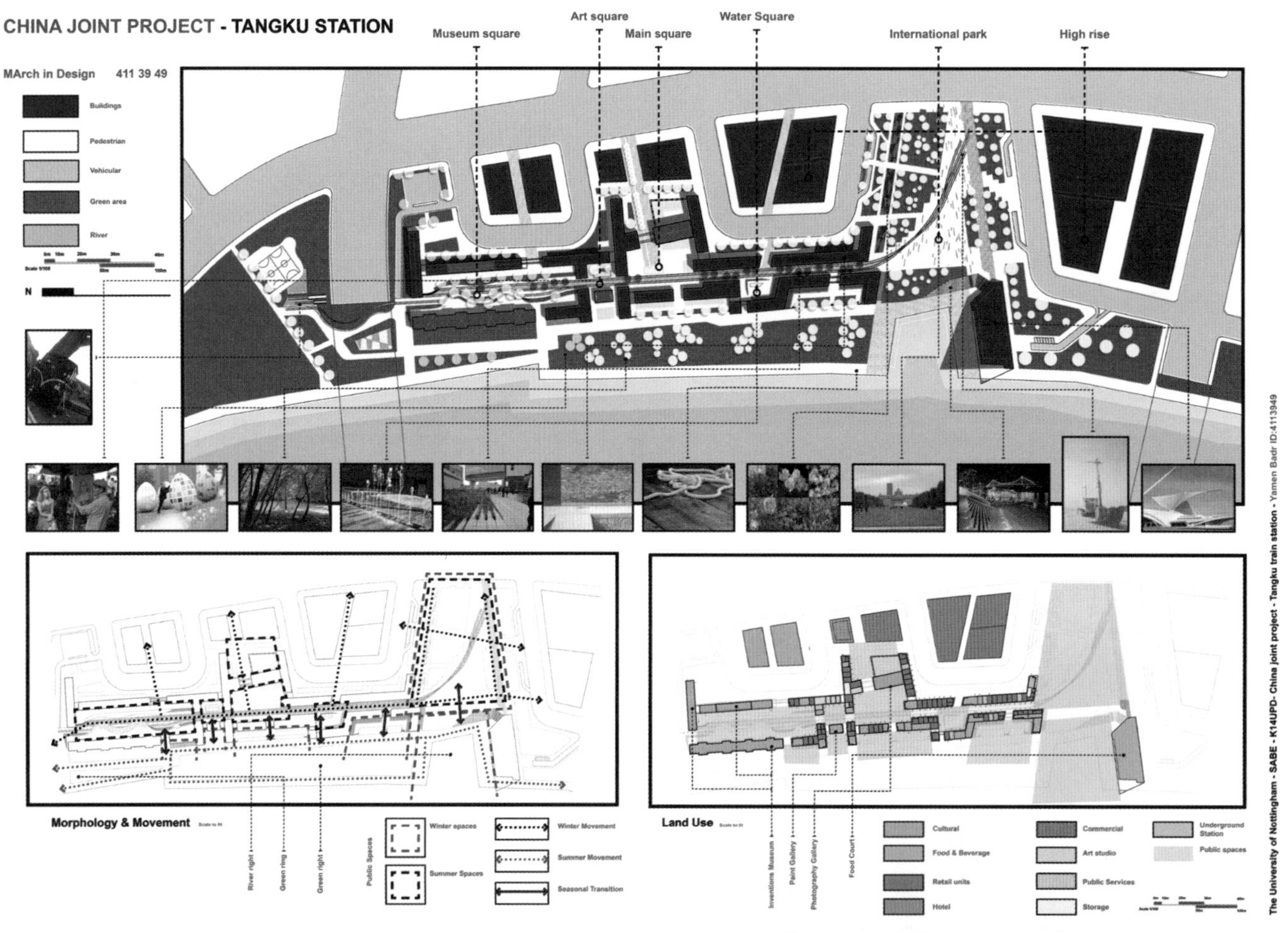

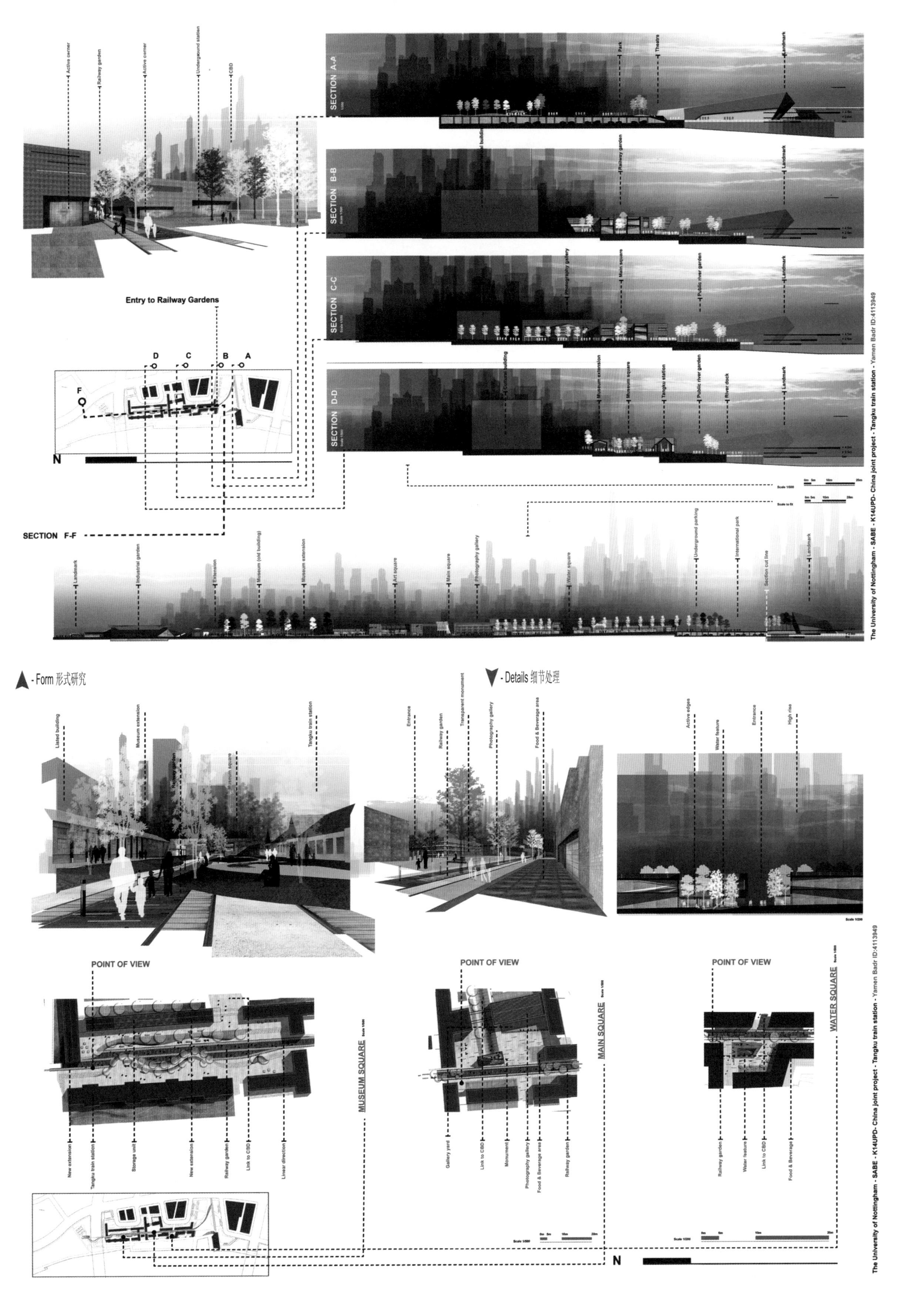
Entry to Railway Gardens
SECTION A-A
SECTION B-B
SECTION C-C
SECTION D-D
SECTION F-F
Landmark
Industrial garden
Extension
Museum (old building)
Museum extension
Art square
Main square
Photography gallery
Water square
Underground parking
International park
Section cut line
- Form 形式研究
- Details 细节处理
POINT OF VIEW
MUSEUM SQUARE
MAIN SQUARE
WATER SQUARE
New extension
Tangku train station
Storage unit
Railway garden
Link to CBD
Linear direction
Gallery yard
Monument
Food & Beverage area
Water feature
Food & Beverage
Entrance
Transparent monument
Active edges
High rise
Listed building
Museum square
The University of Nottingham - SABE - K14UPD- China joint project - Tangku train station - Yamen Badr ID:4113949

DESIGNER-设计人: Eric Cheung

Heterogeneous Interface: Continuity of Identity with Inclusive Diversity

The city of Jingdezhen situated in the JiangXi province in China is undergoing rapid and planned change. Despite an enviable history (the Porcelain City), it faces the danger of losing its actual identity to typical approaches to planning resulting in gentrification, homogenous slab block housing and disneyfication of existing production hubs. In the face of imminent erasure of the historic city centre and its replacement with an 'acceptable urban face' for external investors, the project attempts to provide an alternative and open ended approach for regeneration, through the development of digital design tools that incorporate historic and new patterns and spaces as layers, strengthening and continuing the possibilities of authentic social and cultural networks that make up the true identity of the city. The tools are designed on the basis of concepts related to space and pattern identity, and the implicit relationship between spatial configurations, time, processes of self organisation and social networks/cultural habits. The aim is to retain spatial heterogeneity and hence urban potentialities into unknown but influenced future topographies.

异质界面：多样化可识别性的延续

坐落在中国江西省的景德镇正经历着飞速的发展和变化。尽管有着举世瞩目的历史（瓷都），景德镇却面临着城市发展中典型的身份缺失问题，最后导致了中产阶级化，均质的板式住房和现有生产中心的迪士尼化。历史悠久的城市中心即将消失，而取而代之的则是一种为外来投资者设置的“可接受的城市面貌”，面对这一问题，该项目试图提供一种可替代的，开放式的更新方式，通过采用数字化设计工具，结合新旧模式和空间层次，强化和延续真实社会和文化脉络的可能性，并以此来构建起整个城市的真实身份。该工具的设计基于两大方面：空间与模式可识别性，以及空间形态、时间、自我组织过程、社会网络与文化习惯等元素之间的固有关系。其目的是保留空间的异质性，并借此将城市的发展潜力设置在一种不可预料，却可对未来产生影响的拓扑变化之中。

▼ - Site Analysis 基地分析

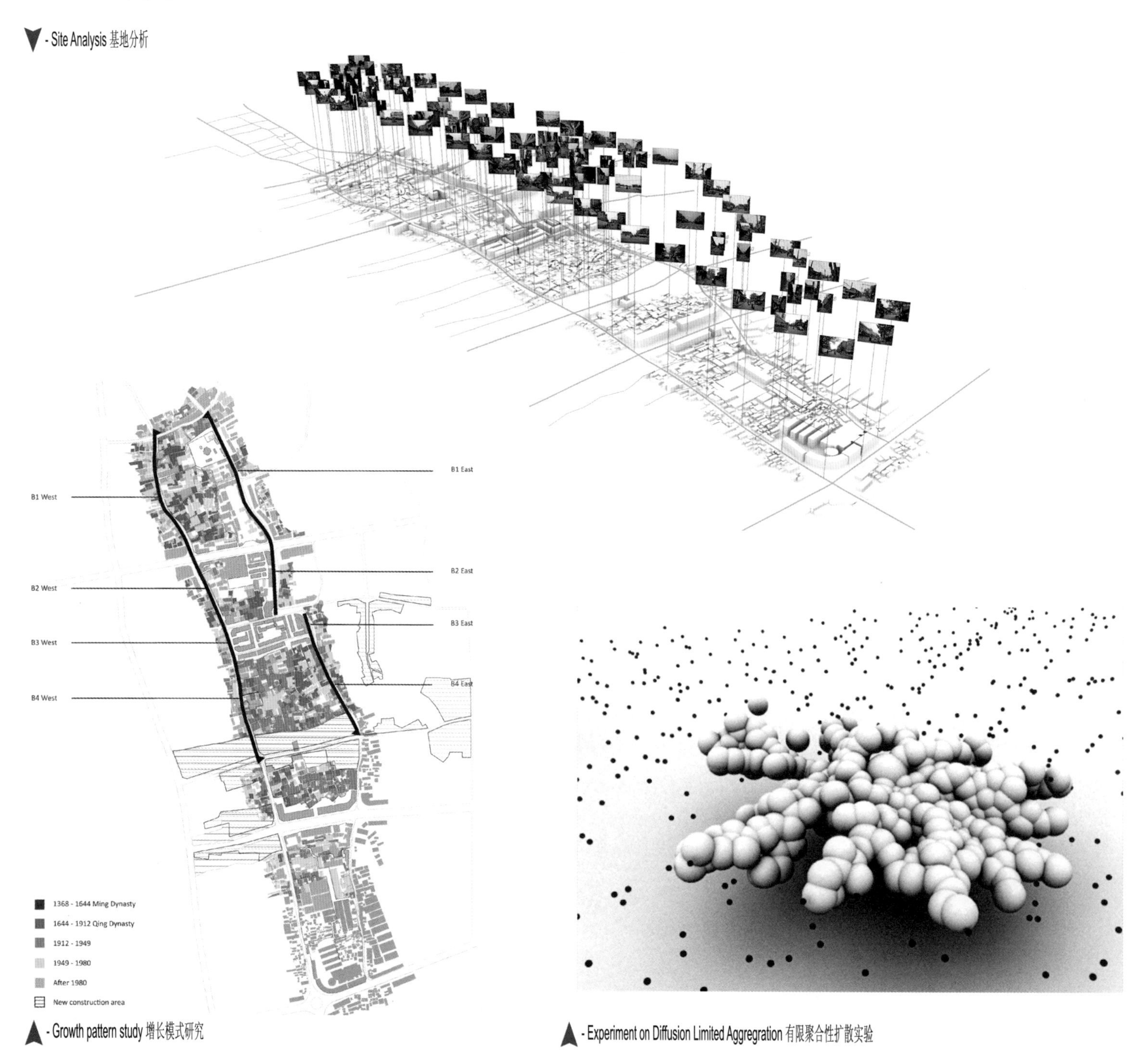

▲ - Growth pattern study 增长模式研究

▲ - Experiment on Diffusion Limited Aggregration 有限聚合性扩散实验

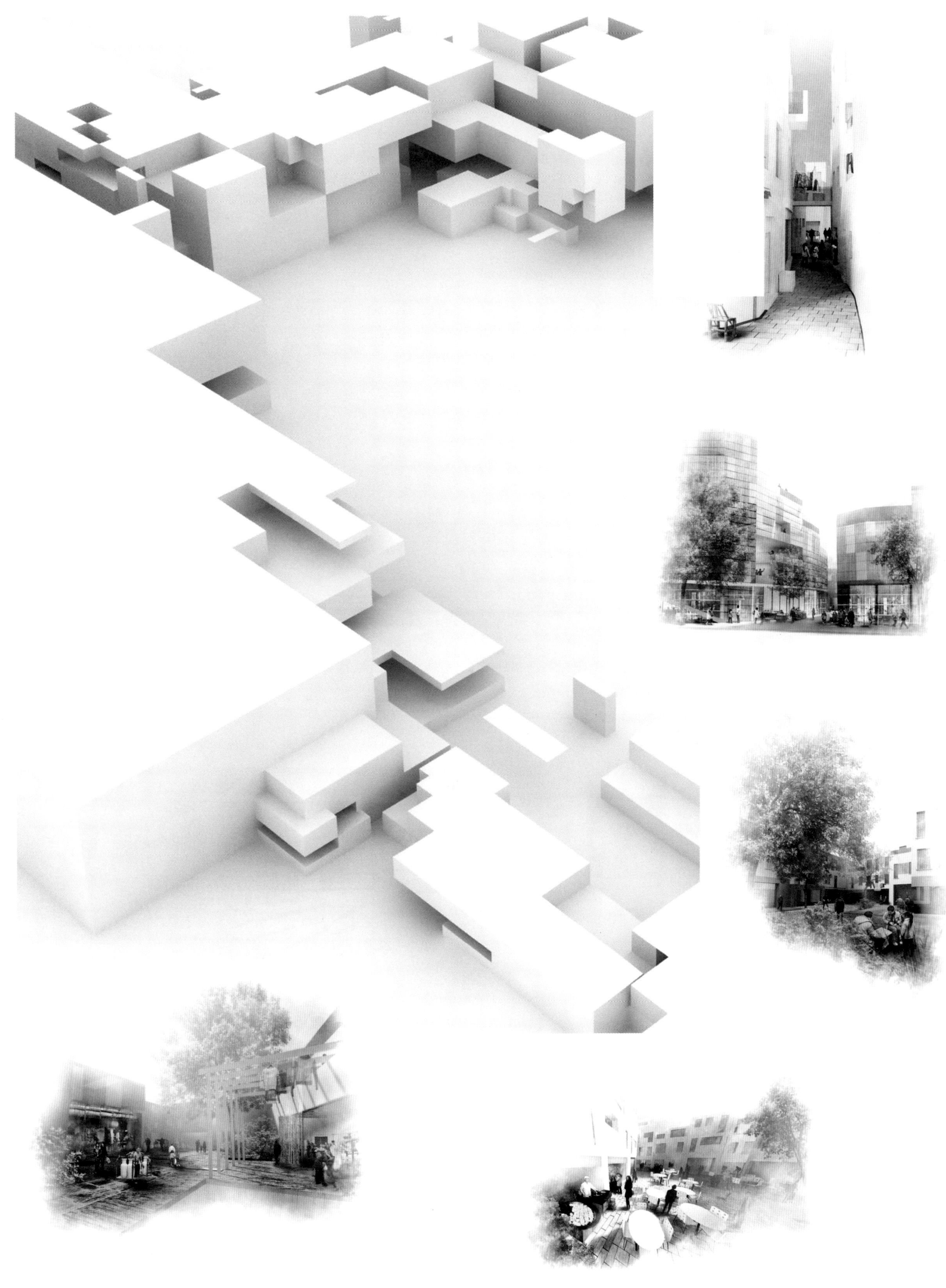

▲ - Cellular model of Spaces and City Scenes 细胞式空间模型与城市场景意象

DESIGNER-设计人: Jonathan Pick

Addressing Imbalance: Scenarios for an Inclusive City

As Mumbai strives forward in its mission to achieve 'Global City' status, the urban topography of the city is undergoing change at an accelerating rate. However, the changes do not incorporate the needs of the majority of the population, leading to instances of large scale gentrification meanwhile encouraging unacceptable conditions for the existing and increasing population living in slums. With recent examples of private developments following the economically lucrative trends of luxury flats and exclusive offices, even when occupying land previously designated for public amenities and affordable housing, there is a need to imagine a different city for the real people of Mumbai. The project attempts to address the incorporation of the eastern waterfront of the seven island city (a strip approximately 14km long) currently owned by port authorities, back into the city of Mumbai. The complex set of changing conditions and uses are explored as temporal events while newly invented urban tools attempts to re-stitch the urban patterns and typologies of the historic city centre to a newly designed waterfront through an understanding of the processes by which the urban fabric of Mumbai hardens from the informal to the formal.

解决失衡：一个包容的城市意境

当孟买正努力朝着"世界性都市"这个目标前进的时候，整个城市的地形也正经历飞速的变化。然而，这种变化与大多数人的需要并不吻合，它既造就了大规模的中产阶级，也同时导致了越来越多的人居住在贫民窟中这种难以忍受的恶劣条件之中。最近的几个私营开发的例子都是唯利是图的豪华公寓和专属办公楼，甚至占用了原本用来建设公共设施和经济适用房的土地，那么，为真正的孟买人民设计一个完全不同的城市就变得极其必要了。这个项目希望将位于东部海滨的七个，目前属于孟买港口管理部门的岛镇结合起来（长约14公里的城市带），重新融入孟买城。一系列由持续变化的现状与用途所构成的复杂性被视为暂时的事件，而新发明的城市设计工具则希望通过对孟买的城市肌理从非正式到正式的硬化过程之理解，将古老的城市中心的模式和类型与新设计的滨水区缝合在一起。

- Road Network Development 路网规划发展图

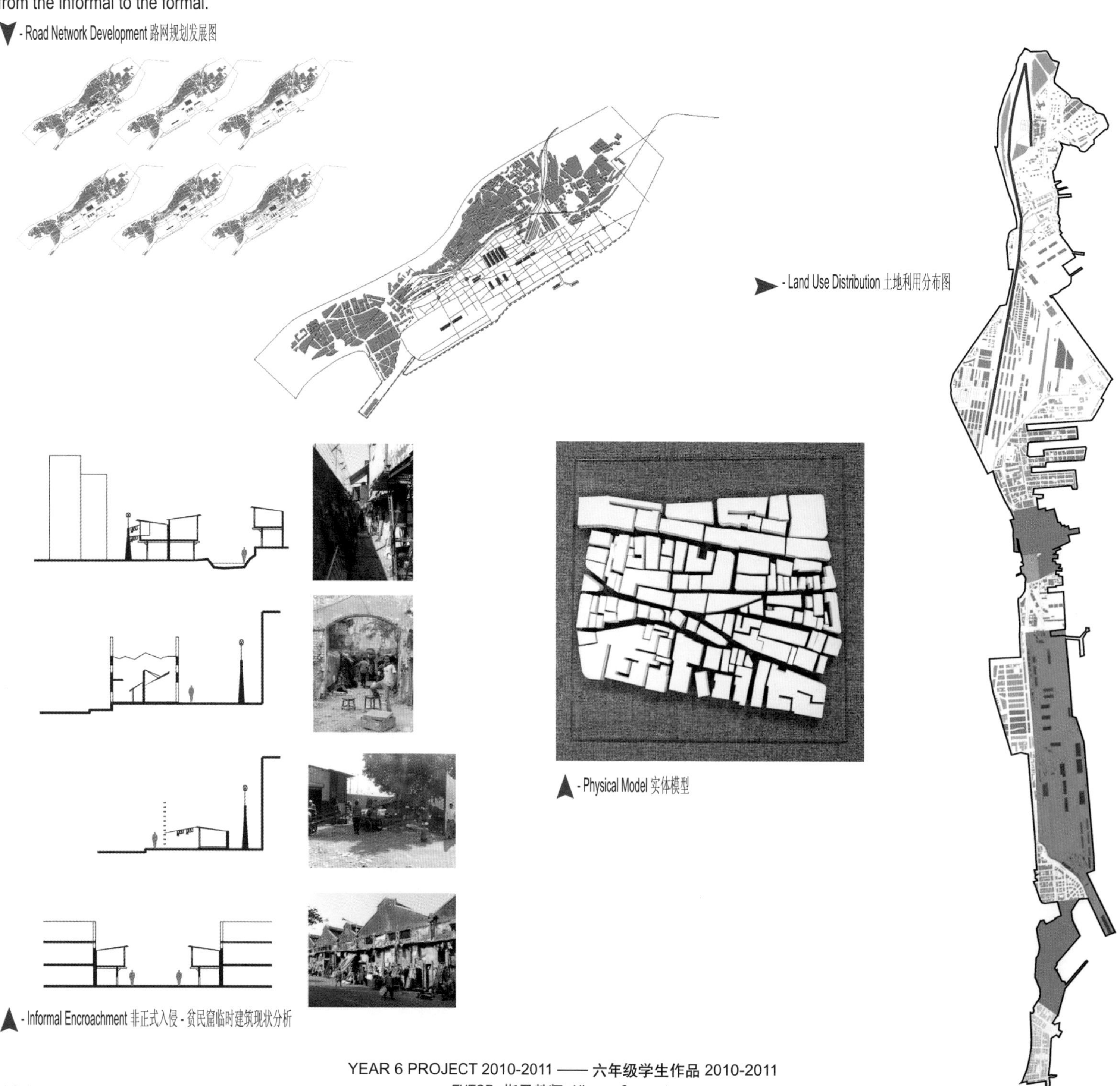

- Land Use Distribution 土地利用分布图

- Physical Model 实体模型

- Informal Encroachment 非正式入侵 - 贫民窟临时建筑现状分析

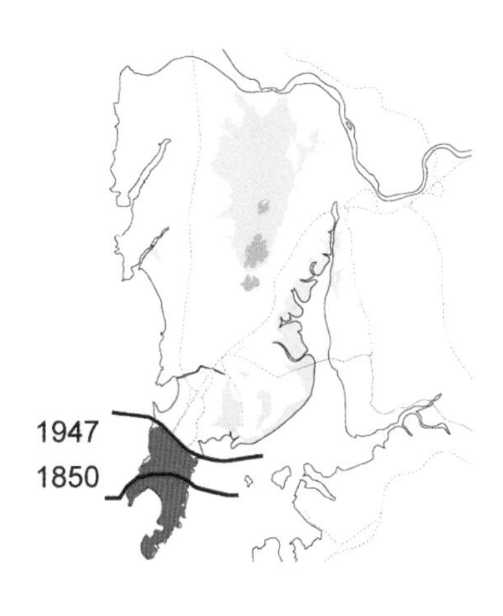

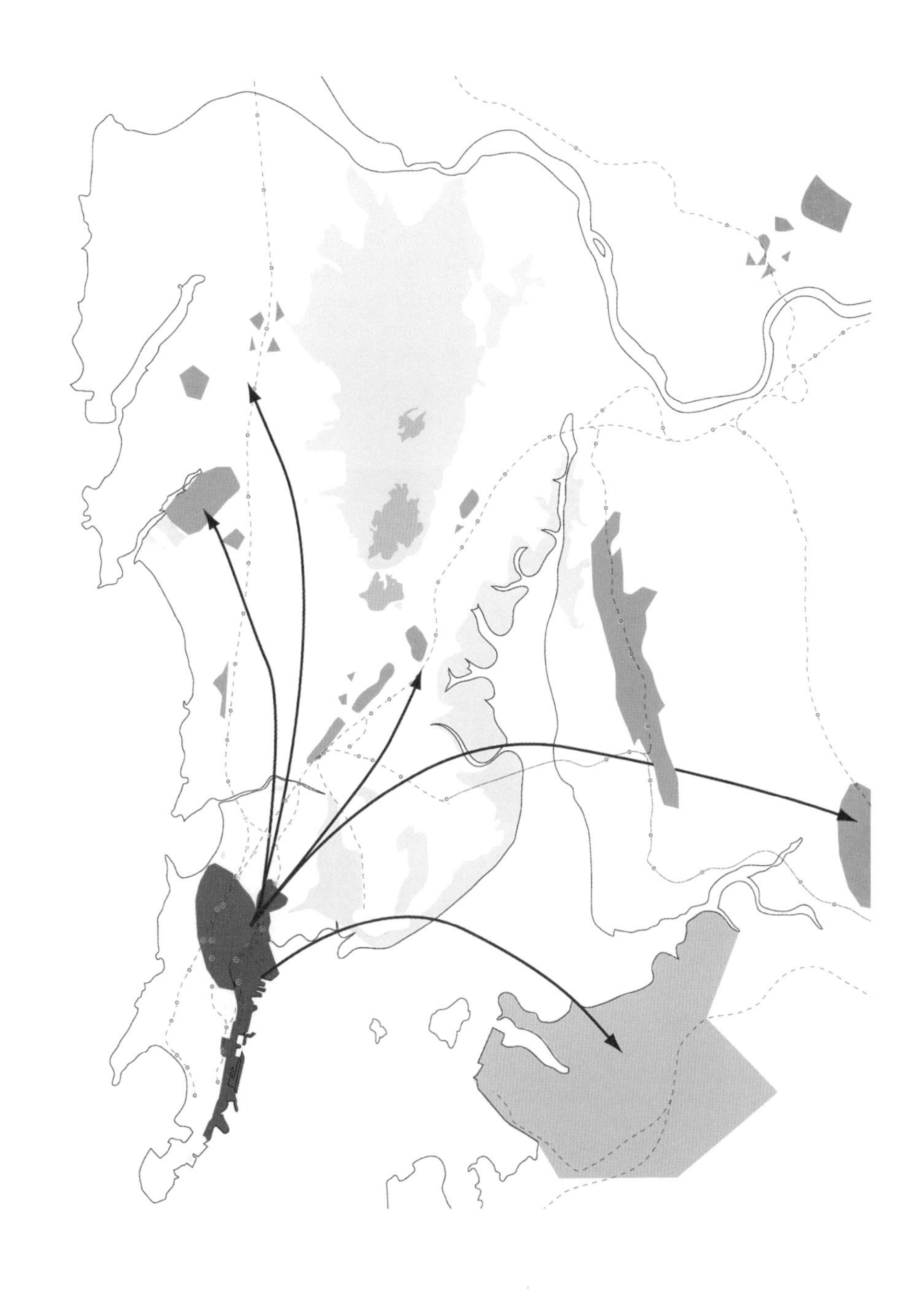

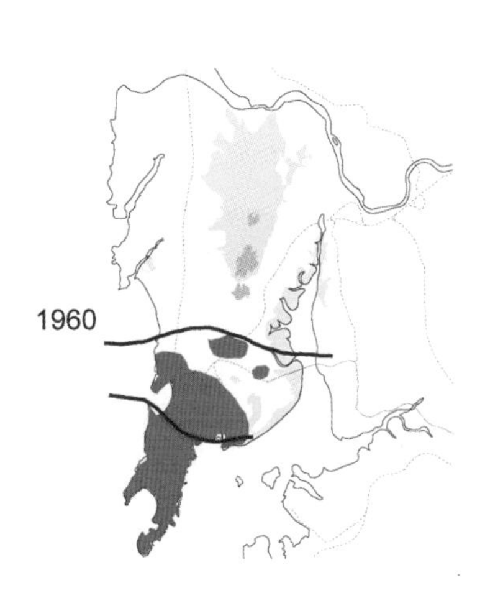

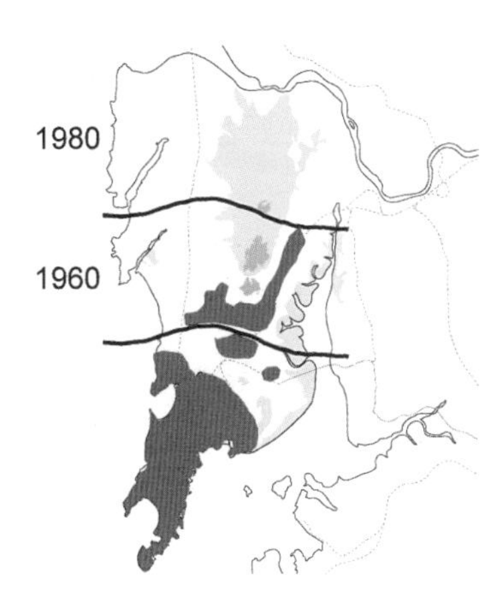

Industrial growth and shift

Legend

- Mill lands
- Mumbai port
- New port activity
- New industrial activity
- Lakes
- Parks & Mangroves
- Industrial activity over time

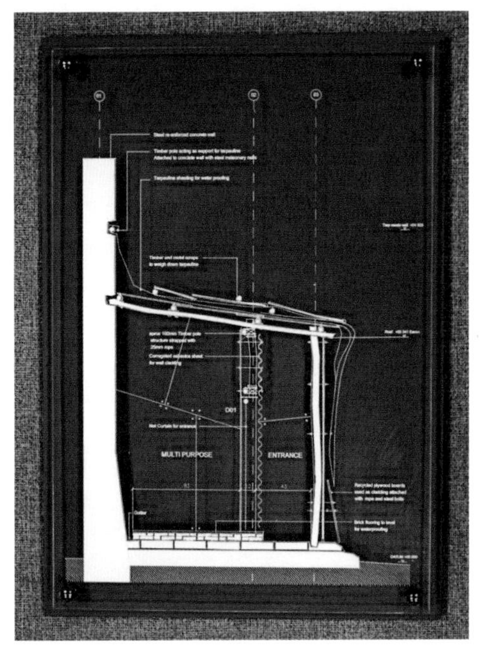

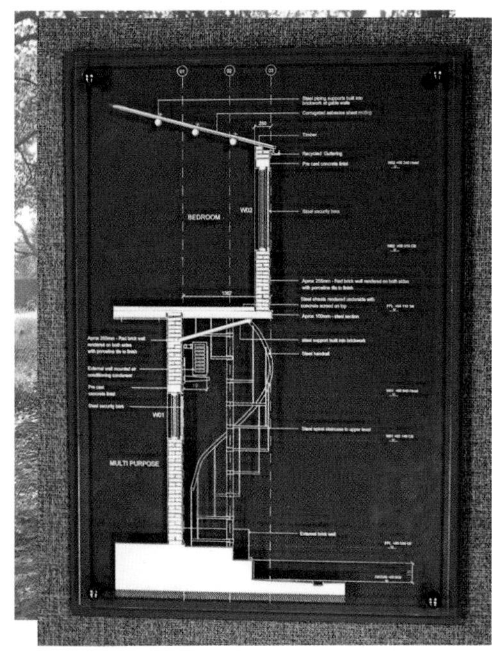

▲ - Industrial Growth and Shit, Physical Model Sections 工业增长与转变分析图，剖面实体模型

090

DESIGNER-设计人: Katherine Coleman

Necropolis

In order to create a meaningful fragment that enhances experience and alters the dialogue of the space, a small scale piece of urban furniture led to the development of a necropolis. Where an overlap of activity once extended into the city of the dead, life and death coexisted through the juxtaposition of the cemetery and the social city. These spaces are today becoming a more separated; a stagnant society of the dead which has lead to the conception of the Necropolis, intended to readdress and reintroduce the notion of urban burial to establish an important quarter within the city where the living and the dead choreographically come together.

死之卫城

为了创造一处有意义的，可强化空间体验以及改变空间对话的片段，一个小尺寸的城市家具设计引申出对“死之卫城”的诠释。在这里，一种重叠的行为延申入死者的城市之中，通过城市与墓地的并置，生与死实现了共存。而在今天，这些地方却变得更加割裂，孤立；一个凝滞的死亡世界，不但导致了“死之卫城”的概念，并且意在重新解释，介绍在城市安葬的老传统，从而在城市中进一步建立一处重要的区域——在那里，生者与死者共舞共栖。

▲ - Experiential Models 实验性模型

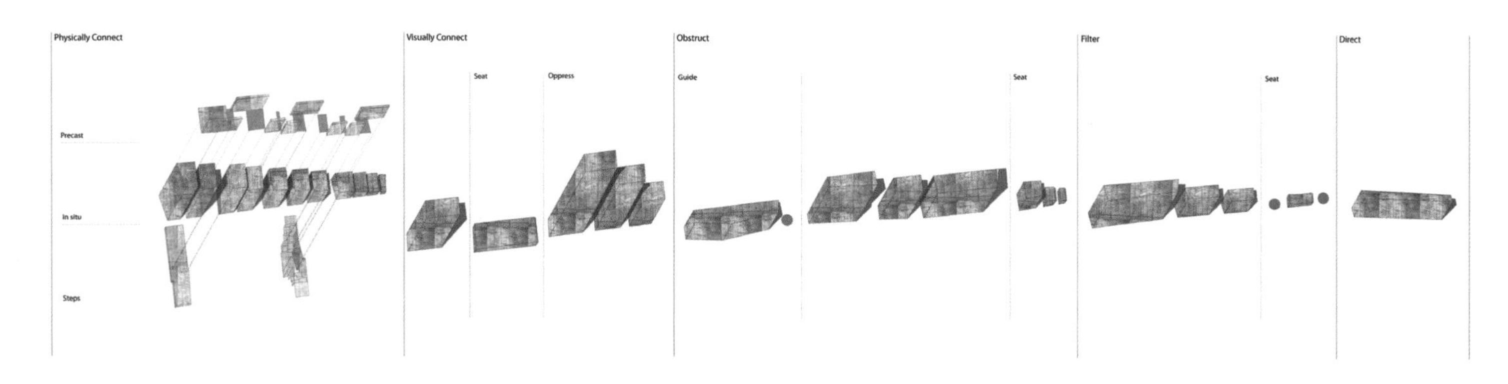

▲ - Abstract Site Study 抽象性基地分析

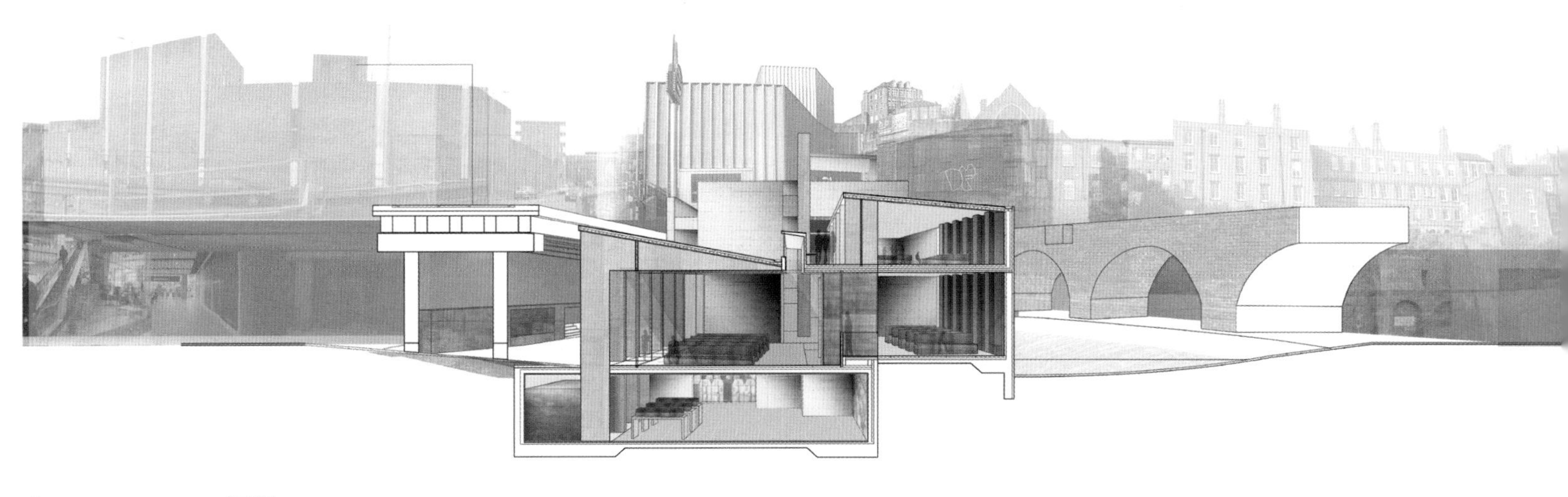

▲ - Perspective View on Section 剖透视图

▲ - Plan 平面图

091

DESIGNER-设计人: James Reynolds

Topographical Incisions

This project considers the idea that the city is made up of multiple fragments, and suggests that our 'readings' of the city are really a mixture of endless encounters and memories whose combination and re-collection provide the basis for a meaningful understanding of our environment. The scheme is situated in Nottingham and directly addresses the impact of Maid Marian Way on the Western edge of city centre. A sequence of interventions and incisions into the city fabric announce a new succession of public routes, separated from the dominant patterns of vehicular movement, and help to establish a symbolic agenda of reclaiming public ground from this chaotic part of the city. A vertical scrap yard is used to reline one of these cuts taken through an existing car park.

因地制宜的切入

该方案将城市视作由许多碎片所构成，并相信我们对城市的解读的确是对发生在每时每刻的，无休止的偶然相遇与记忆的一种混合印象。而这种混合与再收集则为充分理解我们的周围环境架构起了基础。方案定位于诺丁汉，主要研究位于城市中心区西边缘的玛丽安女士大道所造成的影响。沿着大道，一系列介入与割裂深入城市机理之中，展示了对公共道路系统的崭新延伸，并同时将其与主要机动车流线相隔离。这一设计有助于在这片嘈杂混乱的城市环境中建立重塑公共空间的象征性表述，而一处现有停车楼则被改造为一所立体废料场，用来隐喻那穿过停车楼的诸多割裂中的一条，重新编制其机理中的乱线。

▲ - City Plan 城市总平面图

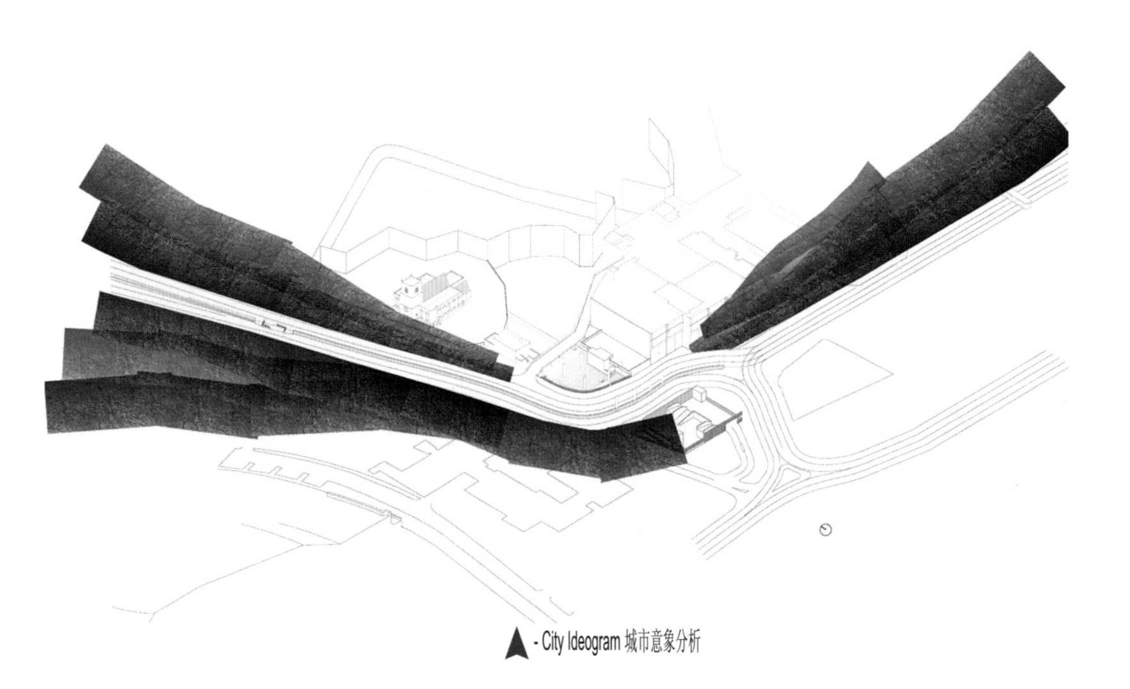

▲ - City Ideogram 城市意象分析

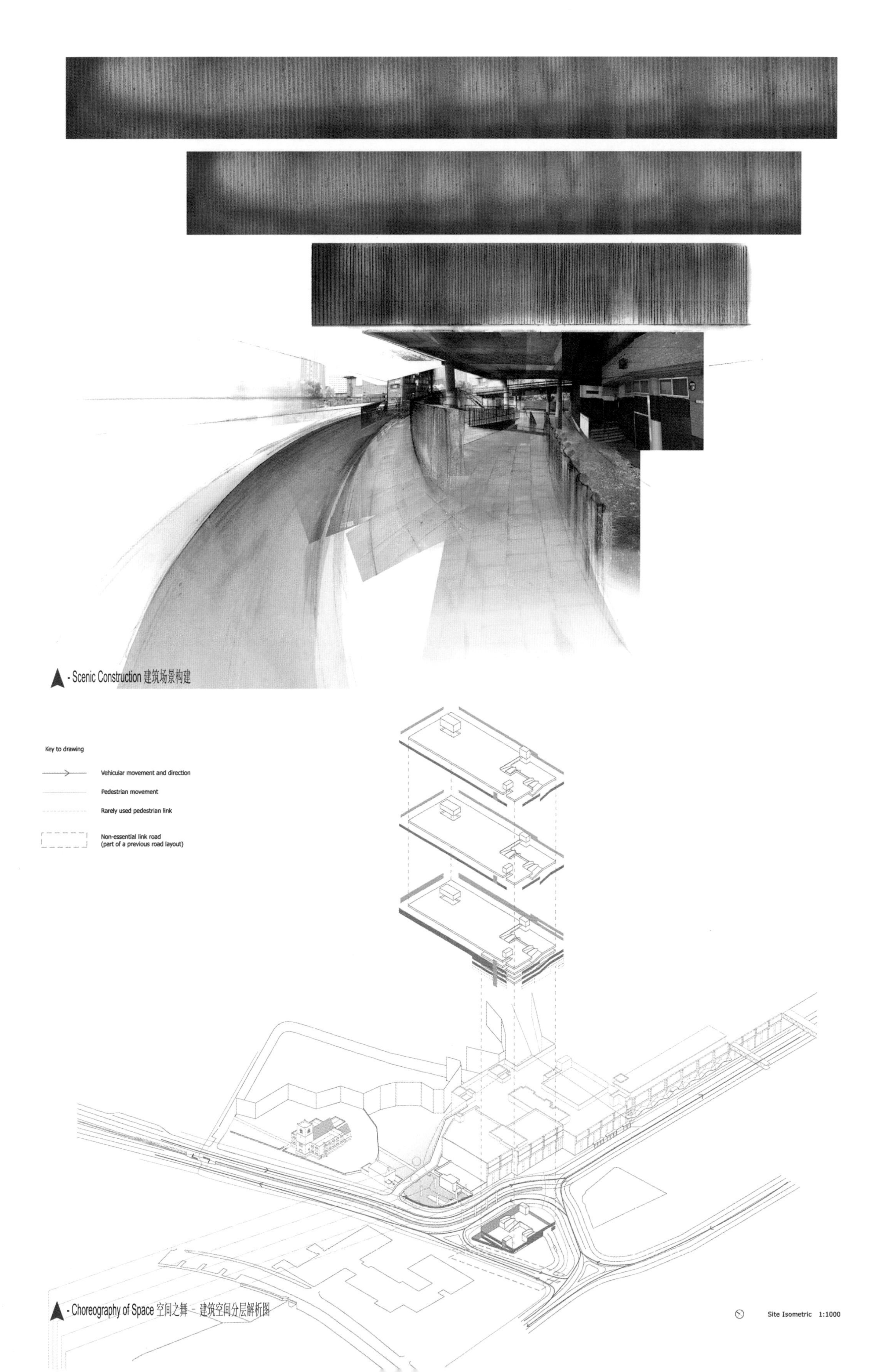

- Scenic Construction 建筑场景构建

- Choreography of Space 空间之舞 - 建筑空间分层解析图

092

DESIGNER-设计人: Nicholas Lowe

Urban Lobby: Layered Threshold

Nottingham rests on a sandstone escarpment and its topography is a palimpsest of many layers of infrastructure. The urban lobby is intended to strengthen this condition, providing a forum that rediscovers lost infrastructures, both industrial and cultural, and connects them to the new. Two new pieces of infrastructure are laid upon the old. The first is a concrete table top laid on top of the old viaduct. The second is a hood constructed from steel plate boxes which encloses and shelters the concourse. These include a hotel, restaurant, outdoor cinema and natural history museum.

城市门厅：层叠的入口

诺丁汉坐落在一处砂岩石壁之上，这里那层层叠起的地质结构就像是城市中相互重叠的多层基本元素。城市门厅就意在加强这种状态，为重新发现遗失掉的那些基本元素——工业的或是文化的，提供了一处讨论空间，并将它们与新的元素相连接起来。两层新的基本元素被架构在老的上面；第一层是架构在轻轨高架桥之上的一块混凝土"桌面"。而第二层是一个由钢板盒子构成的"兜帽"结构体，围合遮掩着这个综合体。这里包括一所宾馆，一所饭店，室外电影院与一处自然历史博物馆。

- Collection of Models 工作模型

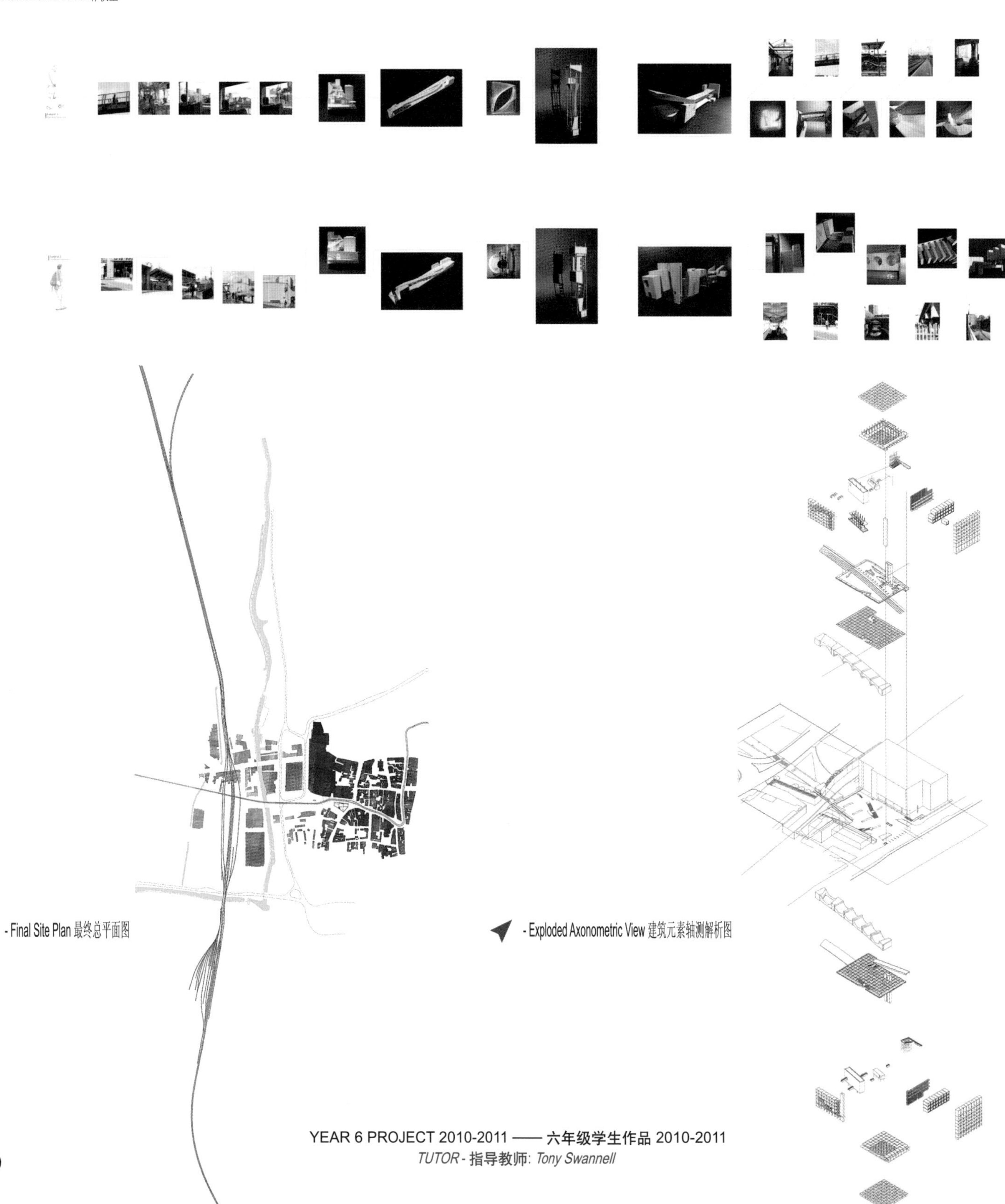
- Final Site Plan 最终总平面图

- Exploded Axonometric View 建筑元素轴测解析图

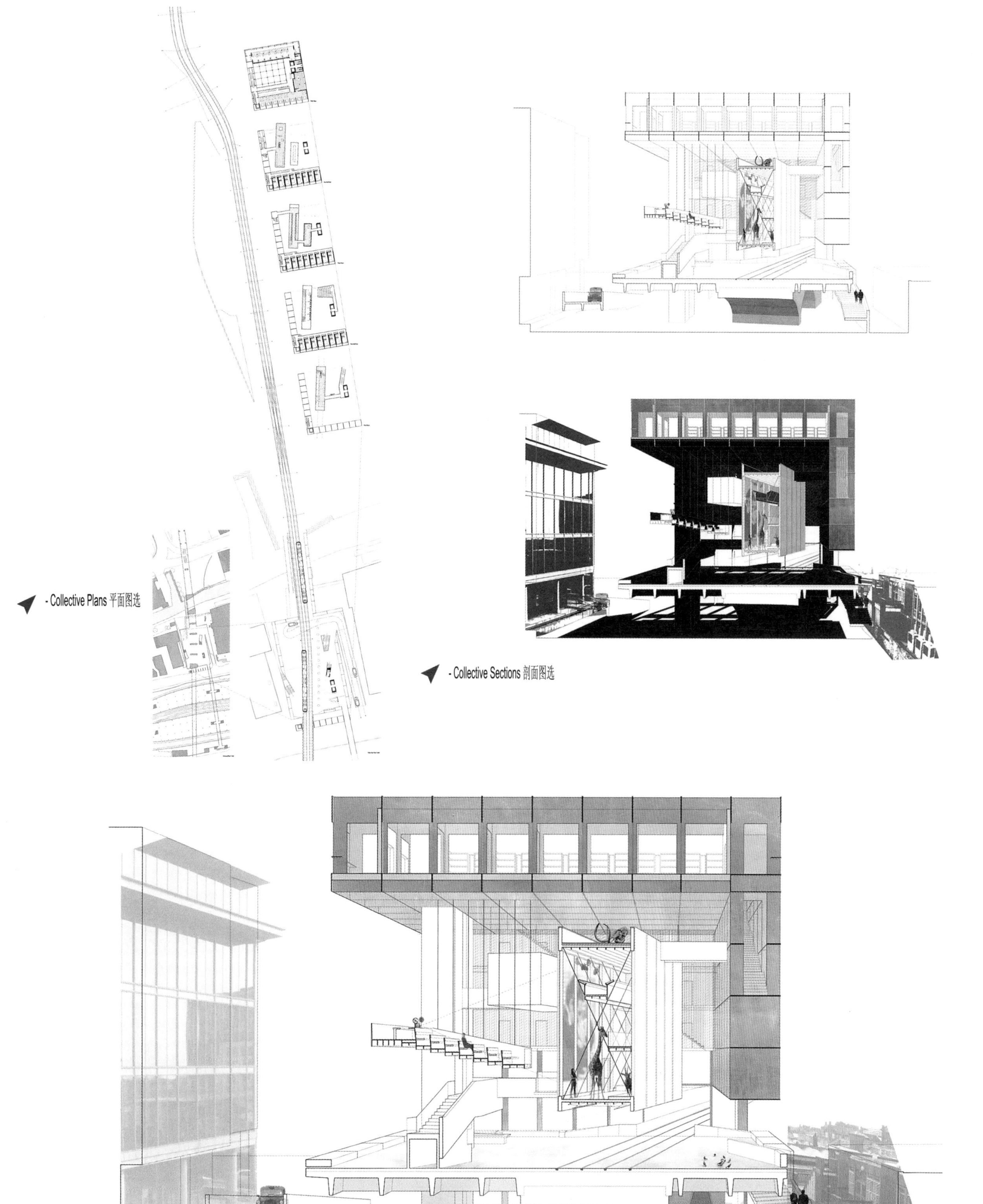

- Collective Plans 平面图选

- Collective Sections 剖面图选

- Foyer, Urban Table Top and “Hood” 门厅，城市“桌面”与“兜帽”

DESIGNER-设计人: Martin Leow-Clifford

Popham Street Galleries: Nottingham's Existing Masterplan

Through historical and material explorations, I have looked at the material importance of buildings, and their thresholds to neighbouring structures of a different time. The realisation that framing these material meetings and thresholds is a more valuable way of preserving an old degenerate area, and it leads me to believe in keeping the old, not replacing it. The design proposal sets out a resistance to big city regeneration schemes. By utilising existing buildings within a site, a site deemed to be either flattened or overshadowed within a set out master-plan, one can design a historical reclamation and save the future of the site. The idea behind the Reclamation of the Narrow Marsh is to create a new urban threshold between the lost community and the city. By instigating a series of architectural interventions I have transformed the collection of buildings on Popham Street into an annex of the Nottingham Contemporary, displaying sculpture, and housing gallery related functions such as studio's and workshops.

波帕姆街画廊：诺丁汉现有的总体规划

在对历史和材料的研究中，我发现材料对于建筑的重要性，以及其对毗邻的，建于不同时期结构的联系过渡作用。我意识到，重新组织这些材料性对话与联系过渡，是保护没落街区的一种极有价值的手段，它使我相信，应该去保留旧有遗存，而不是轻易将其取而代之。该设计方案设想了一股针对大规模城市更新的抵抗力。在一处看上去既非常低矮平缓，由被现有总体规划过度控制，只能苟延残喘于其阴影之下的场地上，通过利用基地现有的建筑，我们可以重新呼唤其历史价值，进而拯救此地的未来。对“狭窄沼泽”（地名）的重新开发主要希望在遗忘的社区和城市之间创造出新的城市界面。通过实施一系列的建筑介入，我使位于波帕姆街的建筑群转变成为诺丁汉当代艺术馆的延伸，从而可以用来展览雕塑，也可用作与画廊相关的功能，如艺术家工作室与创作室等。

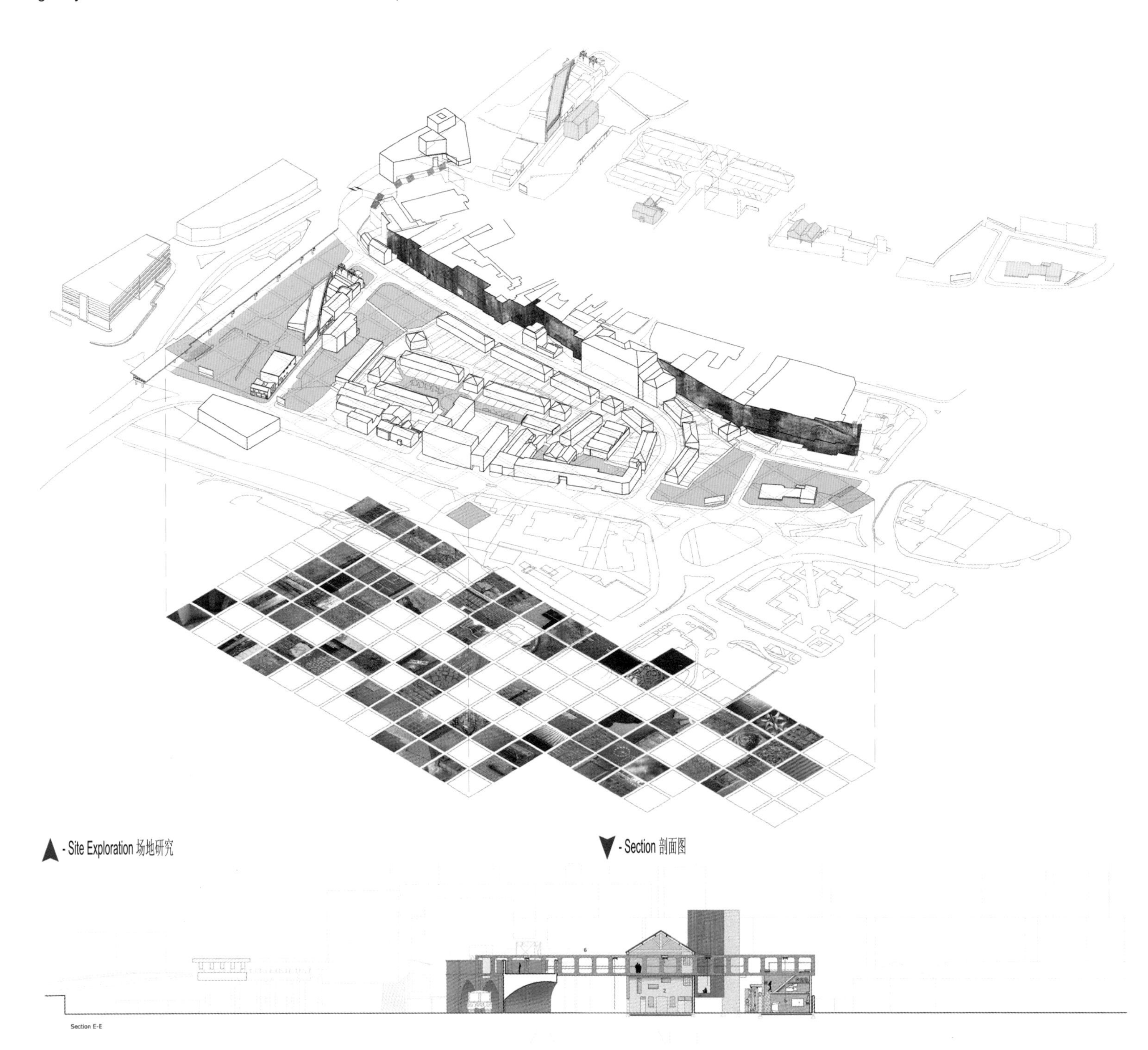

▲ - Site Exploration 场地研究

▼ - Section 剖面图

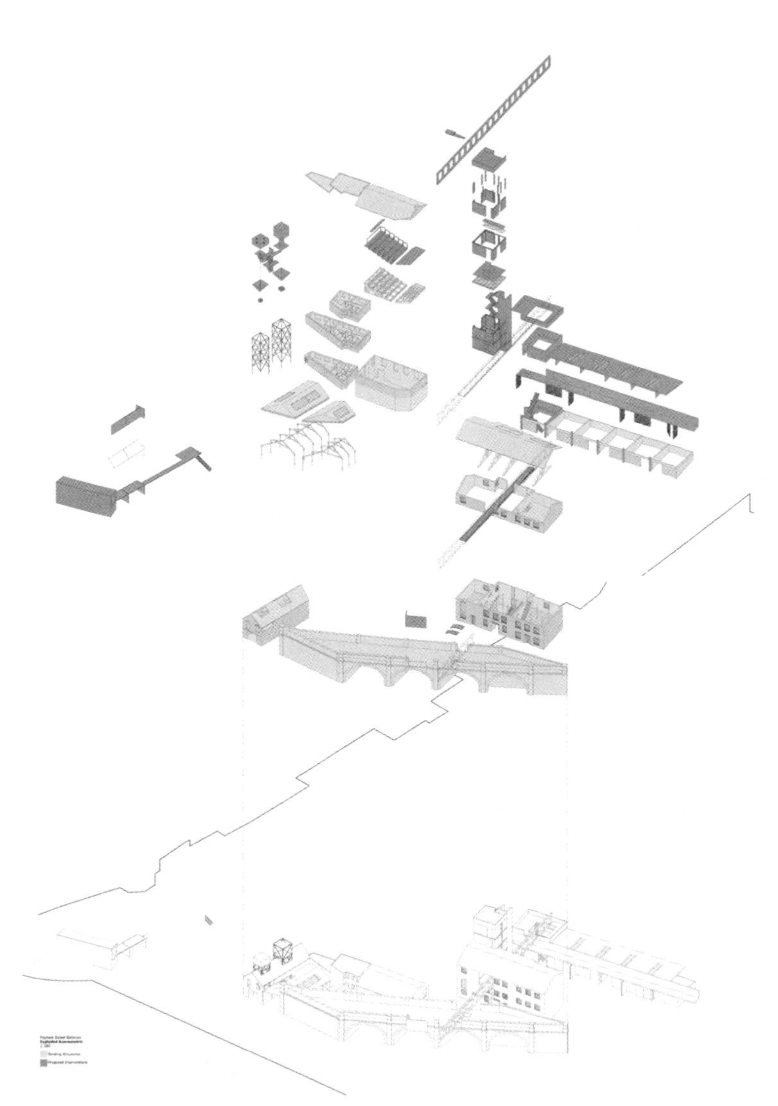

▲ - Exploded Elements 建筑元素分解图

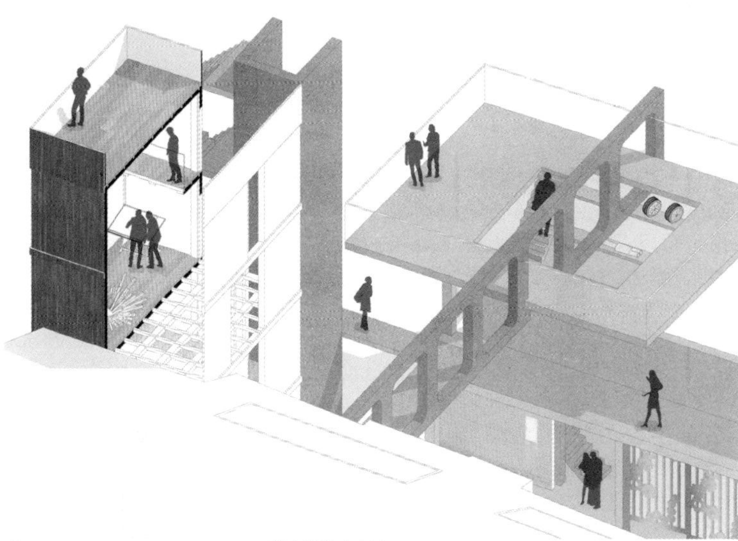

▲ - Studio Tower and Mechanics Entrance 研究室塔楼与自动化入口

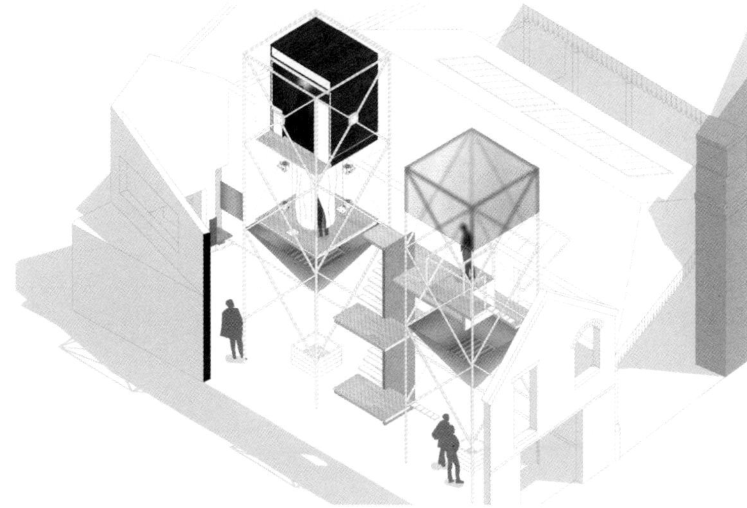

▲ - Maltmill Photography Gallery “老磨坊”摄影展廊

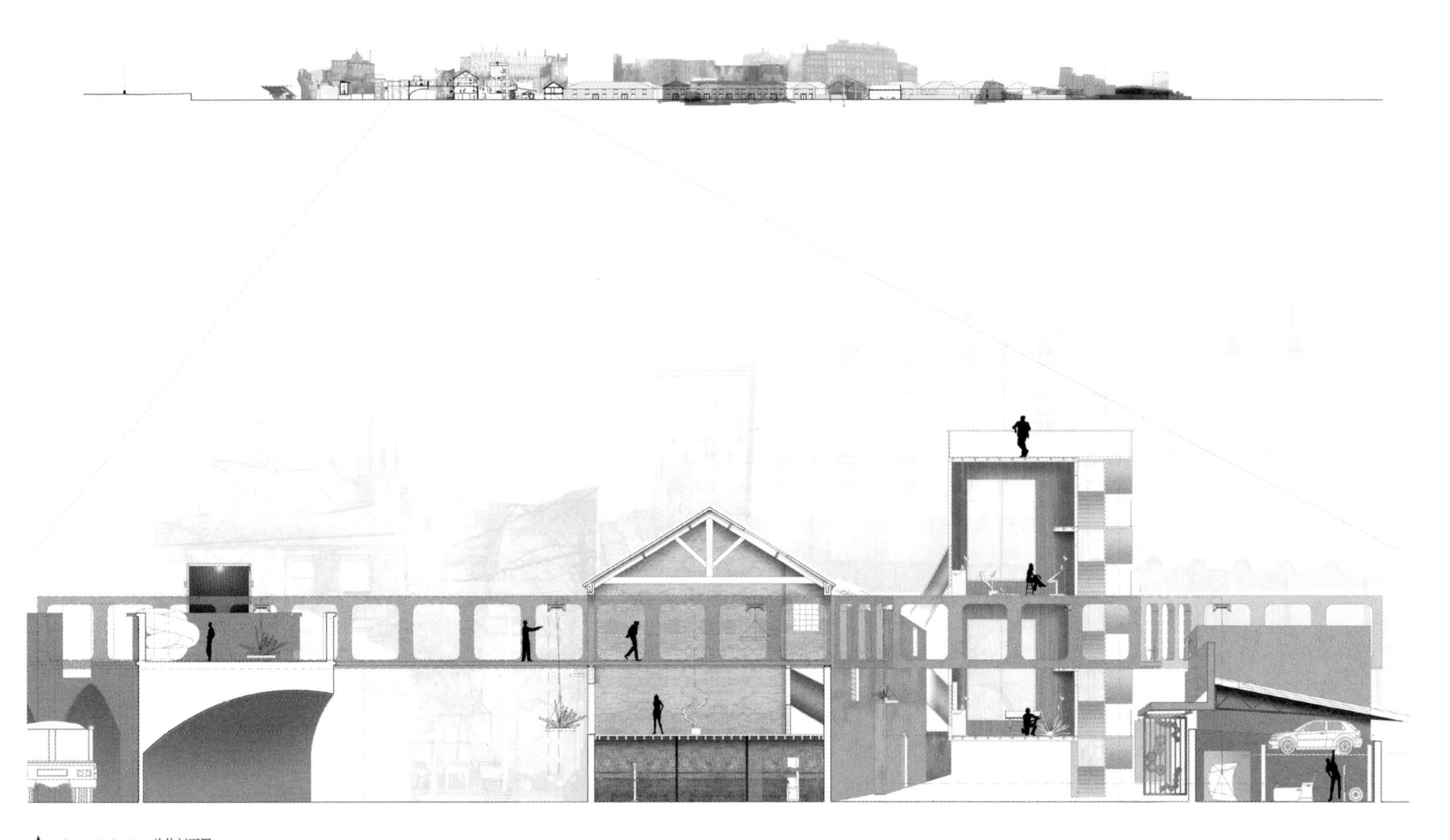

▲ - Synoptic Section 总体剖面图

DESIGNERS-设计人: Anna Michael, Grant Giblett, James Reynolds

The Final Act (Urban Abattoir)

The Final Act celebrates the beauty of death. By Integrating an Abattoir into the urban fabric of Nottingham city centre the aim of the project is to challenge people's preconceptions and highlight the precision, efficiency and beauty that can be found within the process of slaughtering pigs. The form and expression of the project is manipulated to address the constraints of the urban site and provide aspects of performance which extends through the project, which is macabre and more revealing with the aim of recreating the final act of death full of beauty and drama. The architecture per se is both expressive and thought provoking, with dramatic lighting used to create a strong sense of rhythm and by doing so weaving the project into the surrounding urban fabric.

最后一幕（城市屠宰场）

“最后一幕”是为了颂扬死亡的美丽。通过将一所屠宰场整合入诺丁汉的城市肌理之中，本方案旨在展示宰杀猪的过程中体现出的精准，高效和美感，进而来挑战人们一些先入为主的想法。这个项目的形式和表达是基于城市基地的限制而产生的，并且提供纵贯整个项目的表现力。这种表现力令人感到毛骨悚然，但其目的却在于重塑充满美感和戏剧性的最后一幕——死亡。建筑设计本身则既富有表现力，又能发人深省，戏剧性的光照营造出了强烈的韵律感，将其与周围城市肌理交织在一起。

▼ - Site Analysis 基地分析

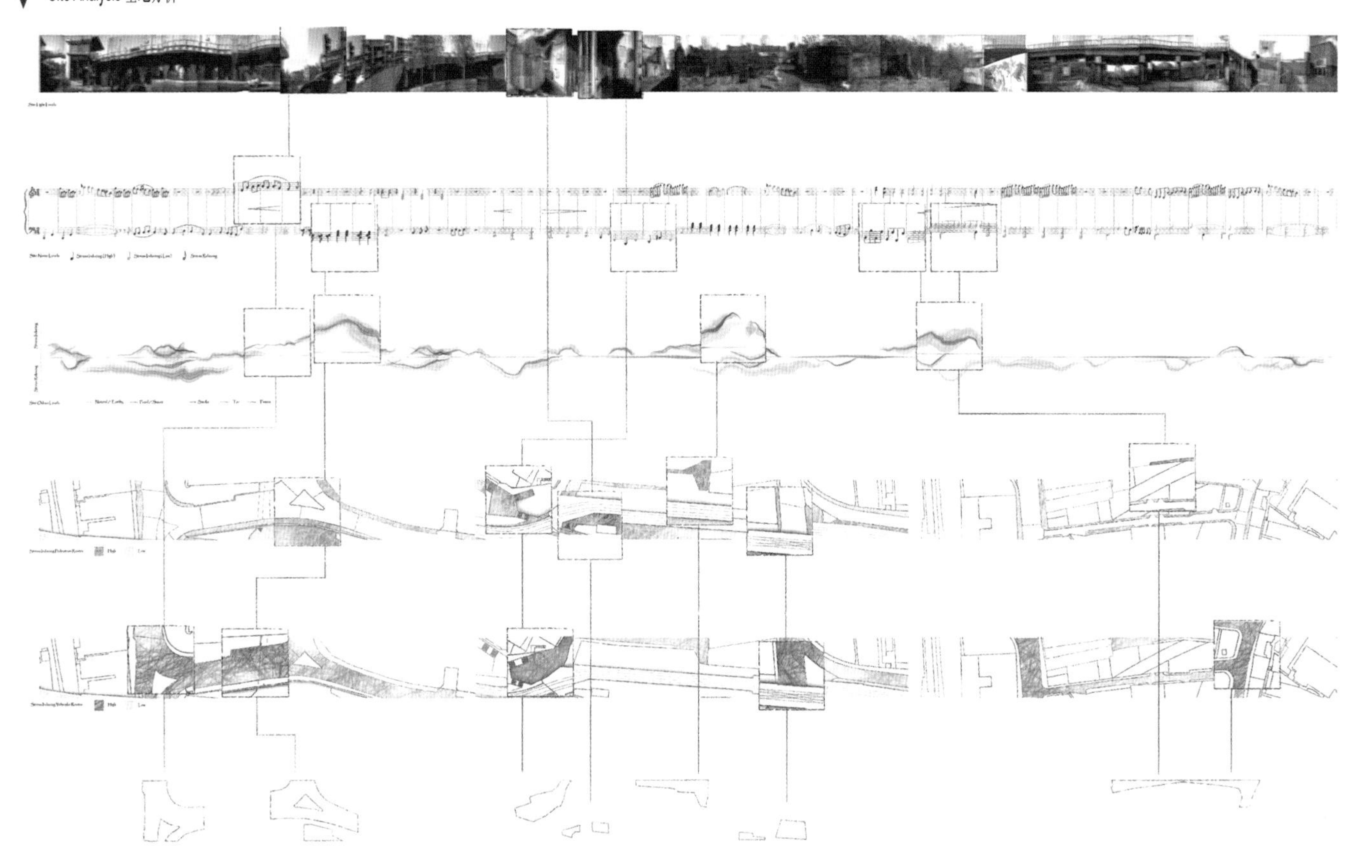

▼ - Site Words 基地语汇概念图

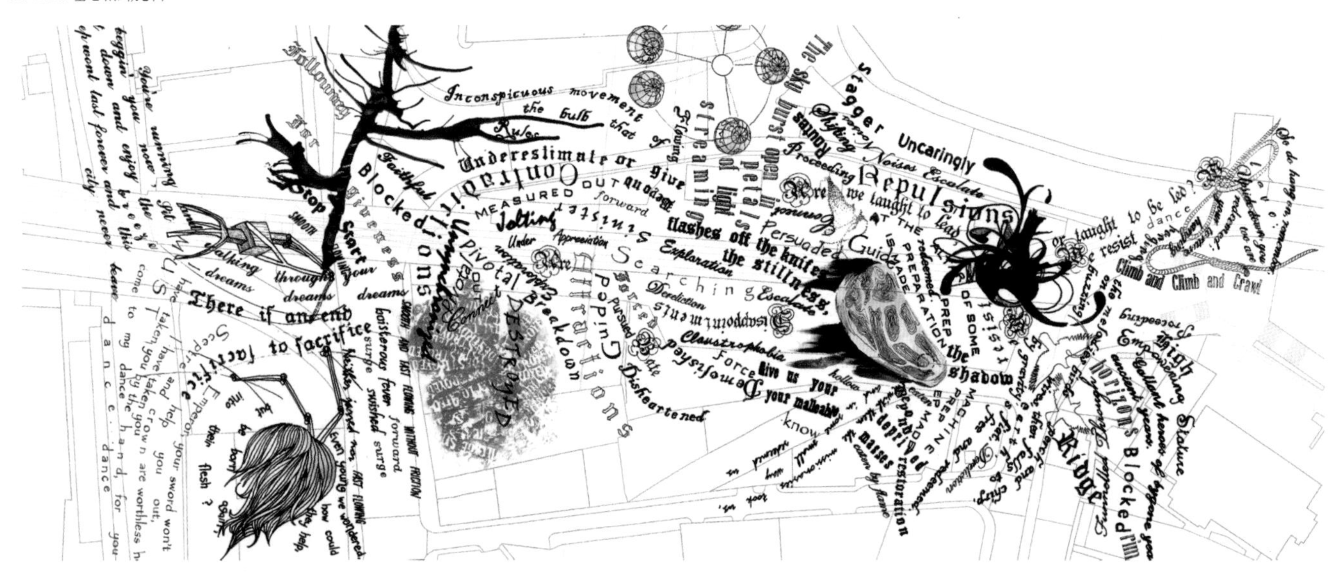

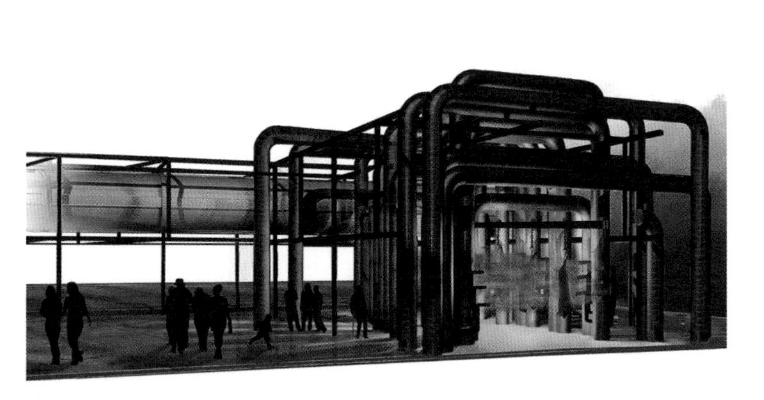

- Singe "灼烧" - 概念效果图

- Steam "蒸汽" - 概念效果图

- Hanging Pigs "悬挂的猪" - 概念效果图

- lairage "入栏" - 概念效果图

- Comprehensive Image 综合表达

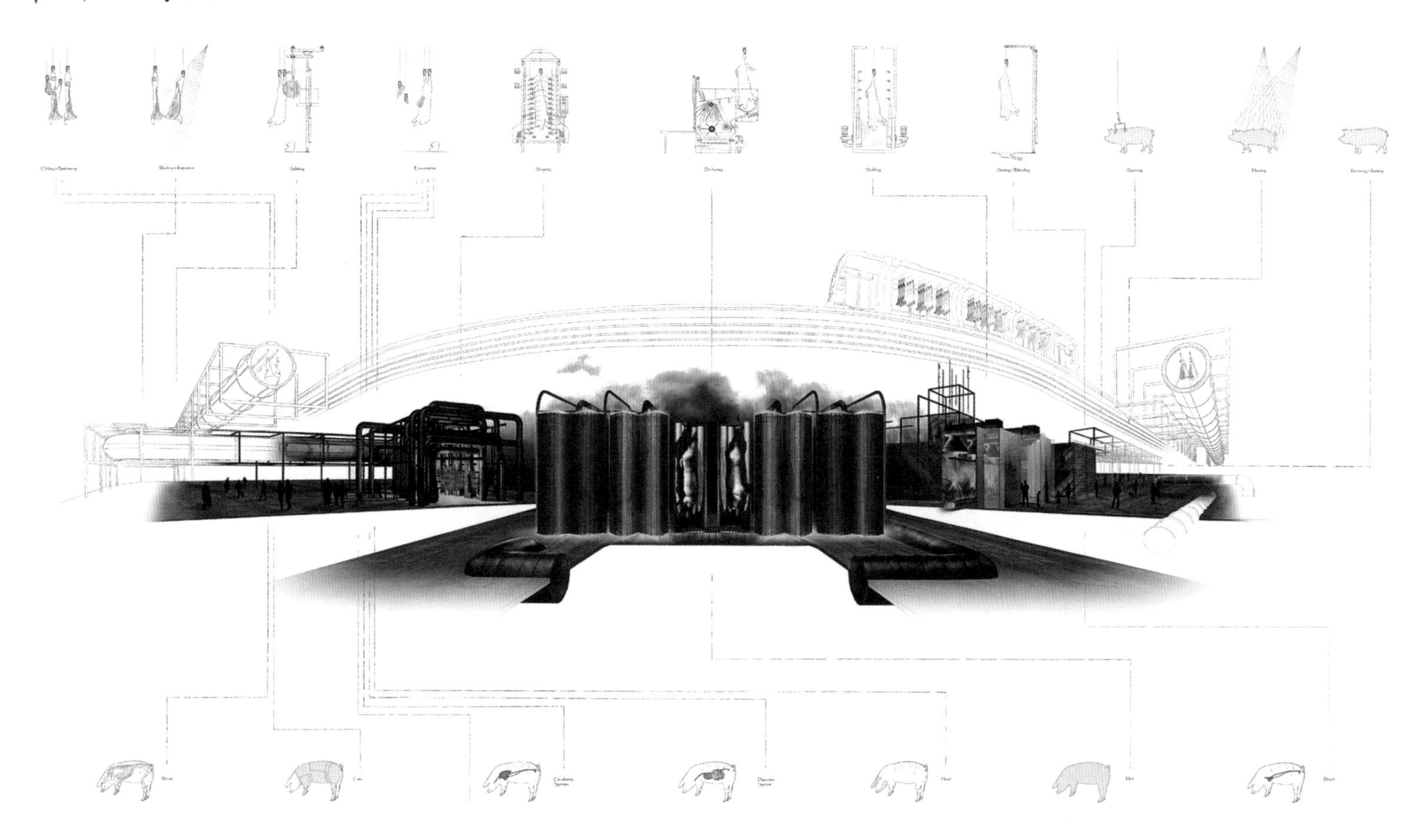

095

DESIGNERS-设计人: Christopher Lee, Harriet Palmers, Xu Xu, Shuo Liu

Apitherapy Centre

Apitherapy is the use of products derived from bees as medicine including venom, honey, propolis, pollen and royal jelly. The Apitherapy Centre aims to remove the traditional purpose of the hive or apiary, to produce honey for consumption, and let the by-products become useful products. Natural medicinal properties are exploited for application of raw materials in an apothecary setting, and in the manufacture of dispensing pharmaceutical products. Where an excess of honey is harvested, the natural chemical properties are used in the corrosion and eventual decay of temporary facades. The building as a whole is designed as a framework of elements that supports a complex network of bespoke mechanical technologies that facilitate the process housed within. A controlled bee hive level transports frames for harvest to an extraction system driven by either supply or demand from the consumer. The project expresses its function through the verticality adopted, which allows the process to use gravity as a generator for its expression.

蜂疗中心

蜂疗是一种利用与蜜蜂有关的产品来治疗的方式，包括蜂毒，蜂蜜，蜂胶，花粉和蜂王浆。蜂疗中心希望将蜂巢或蜂房本身的自然功能消解掉，将其变为可供商业消费的蜂蜜生产，并且让相关副产品变废为宝。这种天然药品被作为原材料探索着应用在药剂配方和制药行业上。当采集的蜂蜜有盈余时，这种天然的化学品将被用来腐蚀，并最终分解掉临时的建筑立面。整个建筑被设计为一个由基本结构元素构成的框架，用来支撑一个复杂的预设化机械技术网络，使得整个流程得以安装进去。根据供需要求，一个可控的蜂房杆架系统会将含有业已成熟，可以进行蜂蜜采集的蜂巢的架子传送到萃取系统中进行采集。此外该项目还可通过利用垂直高度的势能，借助重力来发电支撑各项功能的运转。

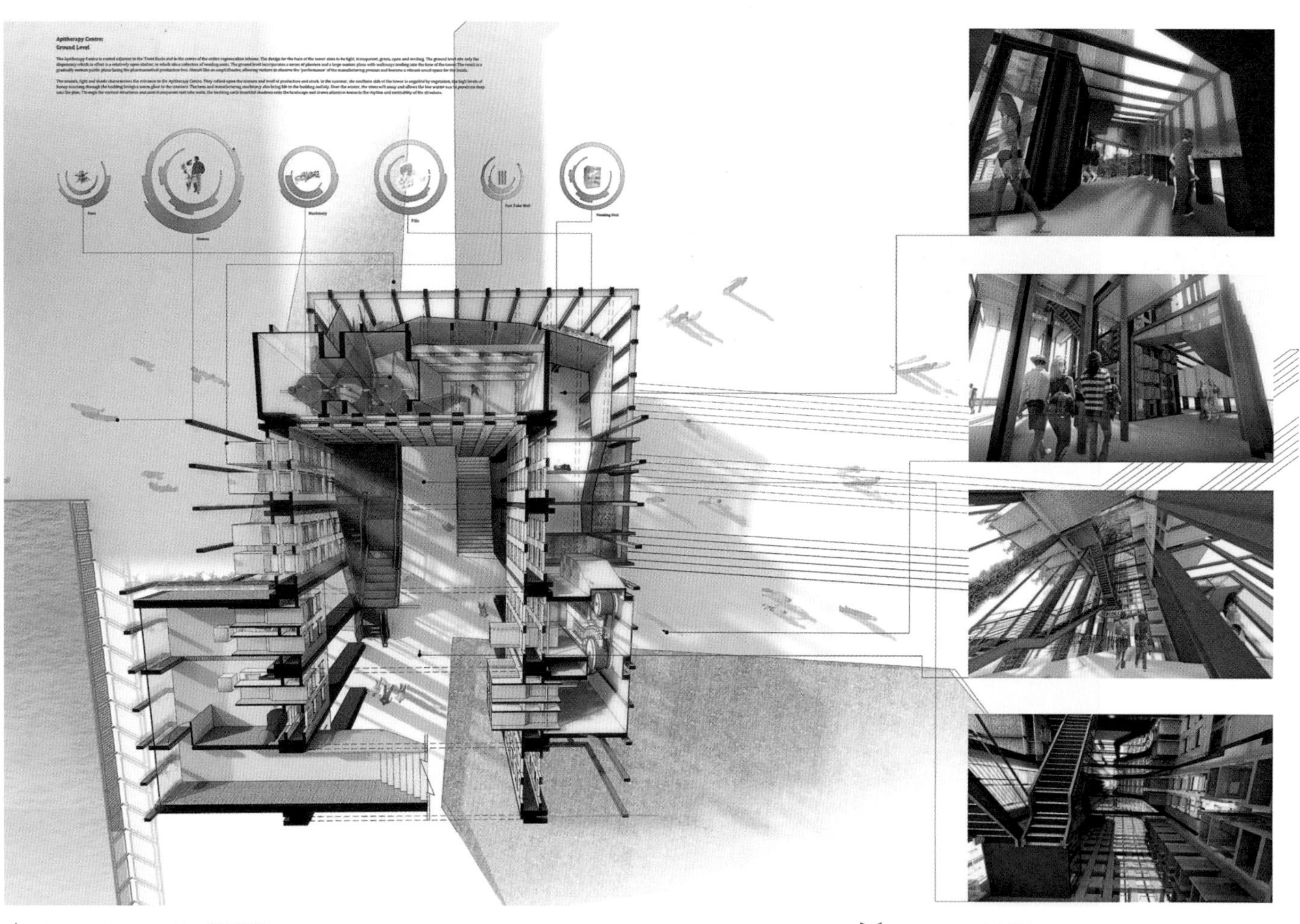

▲ - Groud Level Perspective View 底层透视图

▼ - Concept Models 概念模型

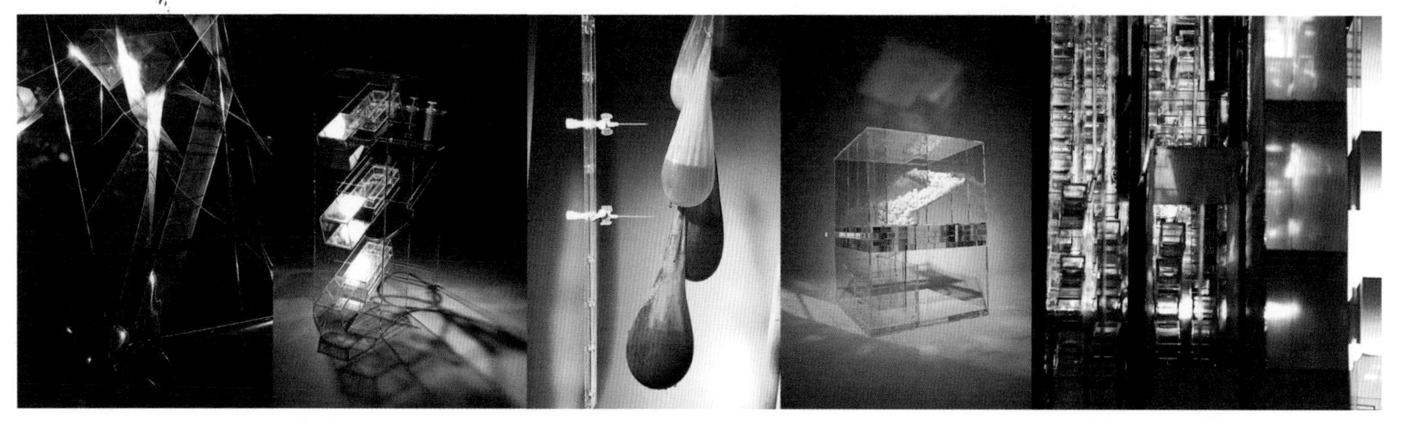

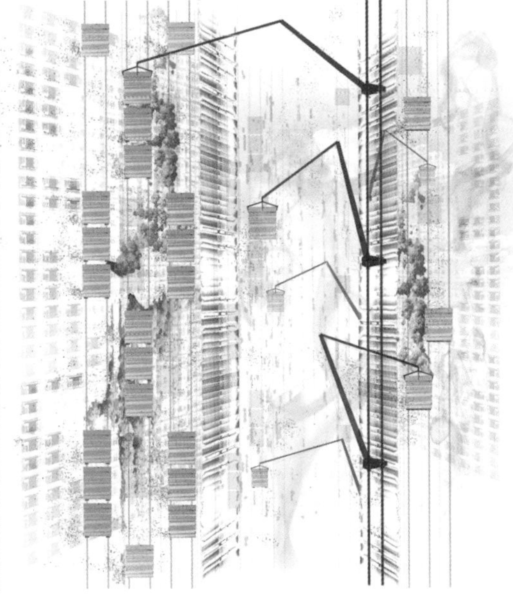

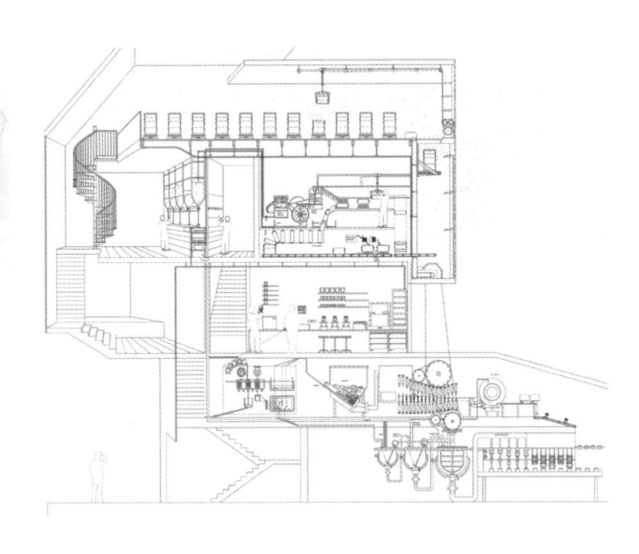

▲ - Mechanical System 机械系统示意图

▲▼▶ - Comprehensive Image and Model 综合表现图与模型

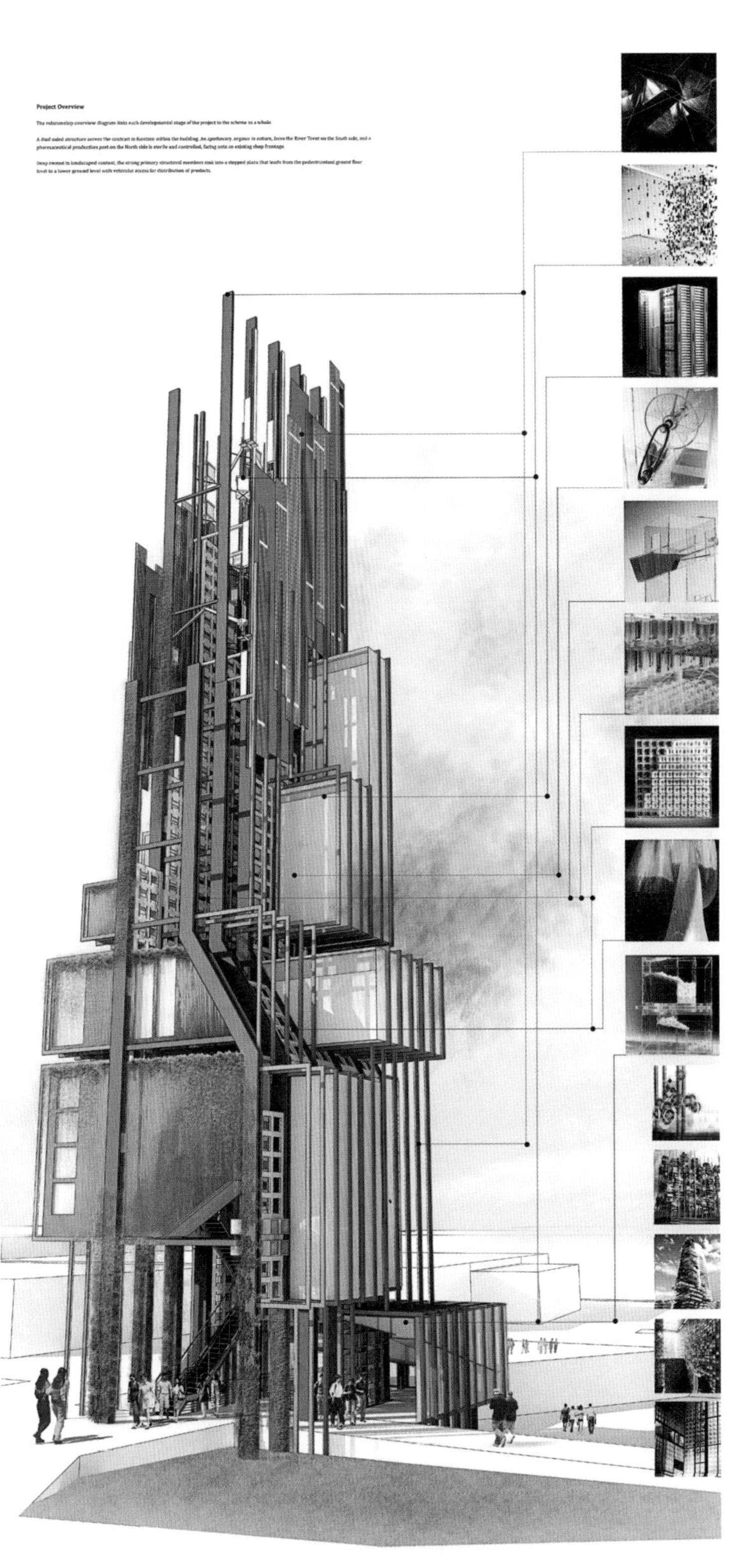

DESIGNERS-设计人: Andrew Tindale, Matthew Holt, Sam Johnson, Nick Emblem

Urban Dairy Farm

The Nottingham City Dairy focuses on reinstating the dairy farm to its historic position within the urban fabric of the city. Initially by exploring the idea of "milk mile"; producing milk within the city, for the city, the aim is to tie the milk production cycle with the demand cycles of the city in a sustainable manner, combining the production of milk with its pasteurization process; refining and closing loops within the cyclical process. The aim is to utilize the by-products of the milk production process, namely manure, in order to produce methane to harness biogas as a means to fuel the process and transport milk to the end users. The building which houses this production process manifests itself as highly intelligent creation which responds directly to the needs to the city as its goes about its daily life. This dominant identity is contrasted by a strong construction ethos which is deeply routed in vernacular farming techniques.

城市奶牛场

诺丁汉城市奶牛场主要是想恢复其在城市肌理中的历史地位。该设计主要源于"喝牛奶一英里赛跑"的概念，为这个城市，在这个城市中生产，提供牛奶。设计的宗旨在于将牛奶的生产周期和城市的需求周期协调在一个可持续发展的模式中，并将生产过程与巴氏高温杀菌流程相结合，形成一处完善的、闭合的循环过程。方案还希望能利用牛奶生产过程中的副产品——粪便，来产生甲烷，并将其作为一种生物天然燃料来为牛奶的生产以及终端运输提供动力。容纳整个生产过程的建筑自身也是一个高度智能化的产品，能够直接对城市日常生活的需求量作出反应。可这一主导性特点却与建筑本身那深深地扎根于农耕技术的普通构筑精神形成了鲜明对比。

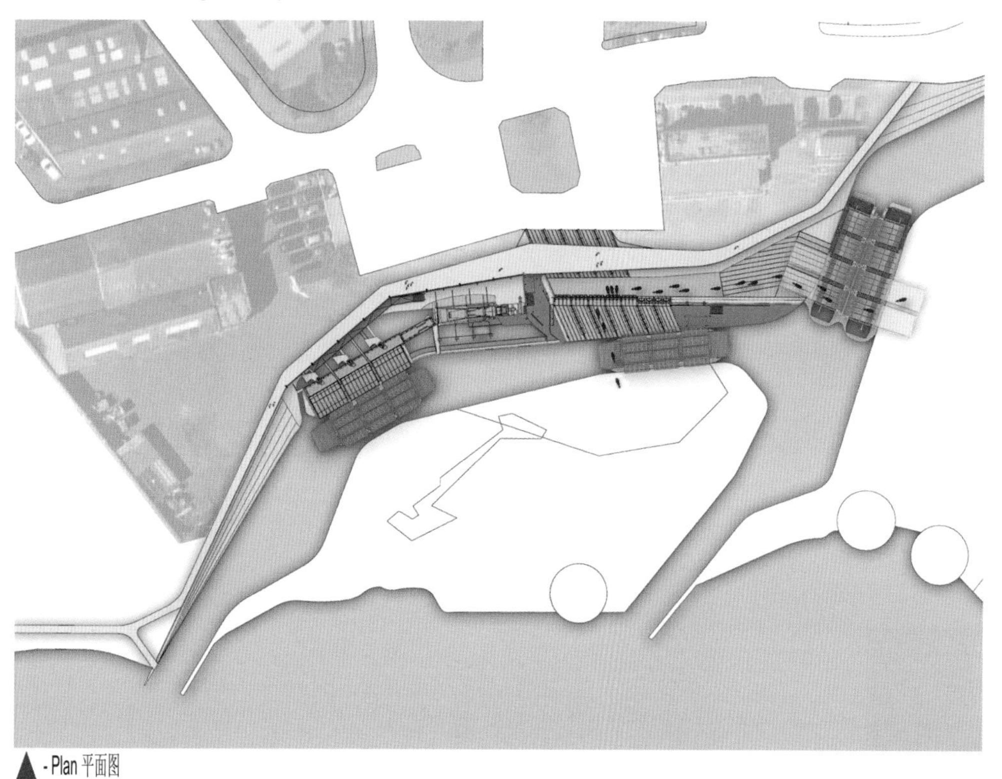

▲ - Plan 平面图

▲ - Cut Through Perspective 剖视图

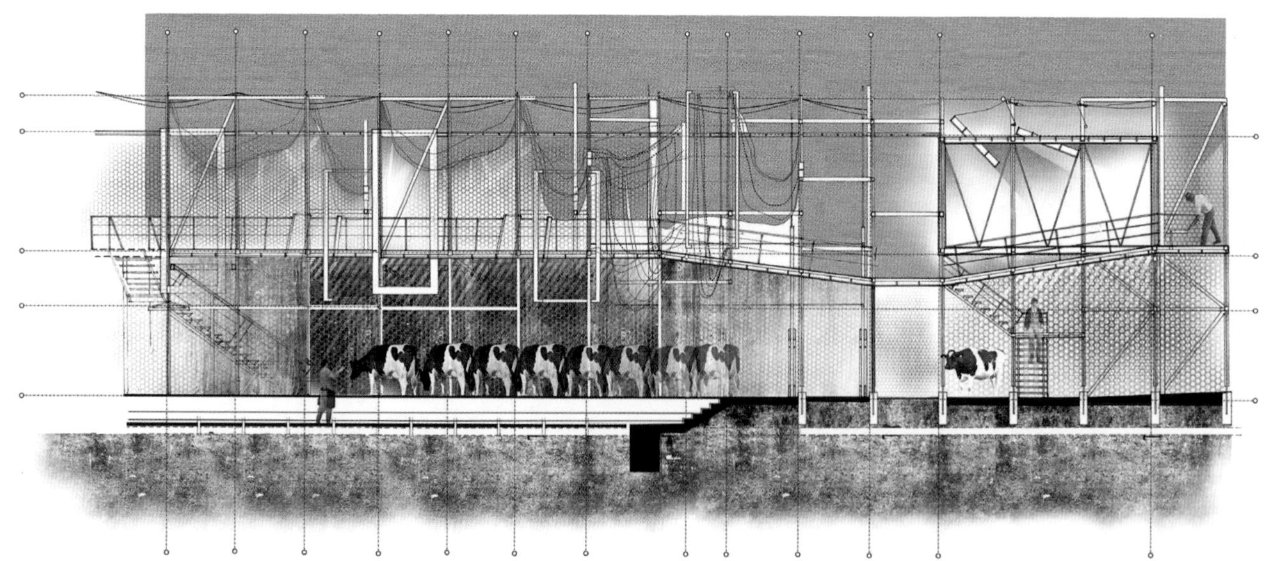

▲ - Parasitic Milk Parlour Section 配套牛奶店剖面图

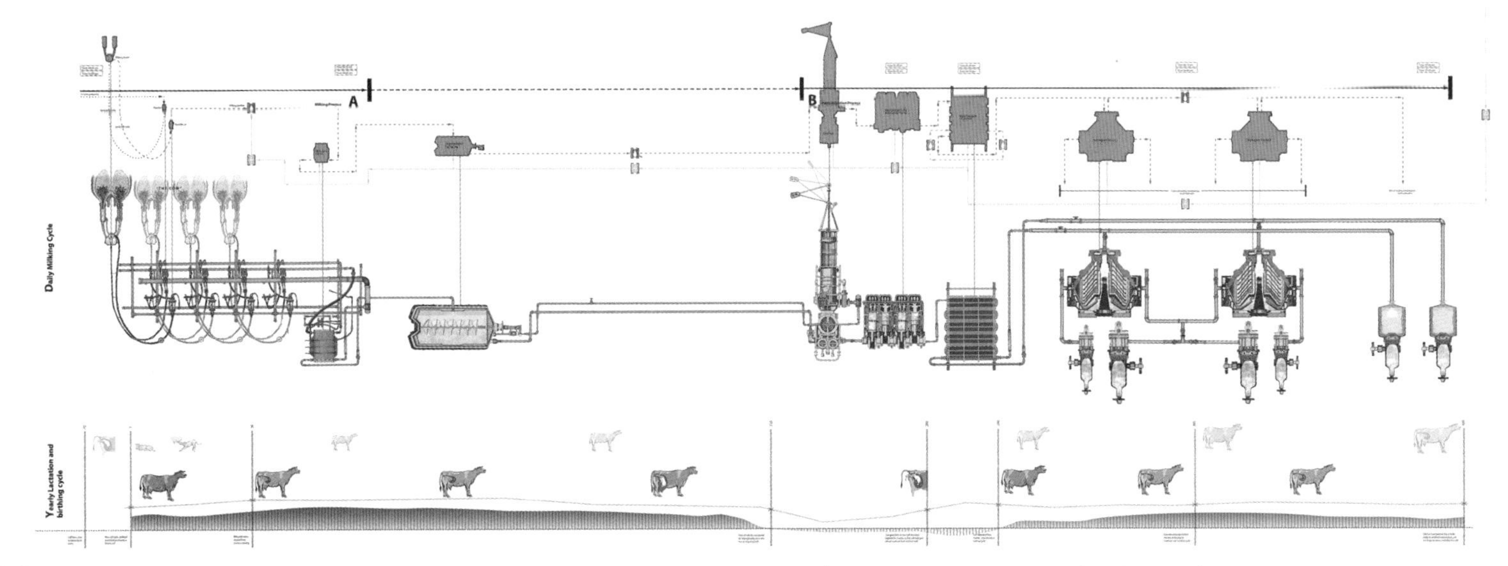

▼ - Long Section 长向剖面图

▲ - Milking & Pastuerisation Process 取奶及高温杀菌工业工程示意图

▼ - Renders and Model 效果图与模型

▼ - Comprehensive Drawing 综合表现图

—— LIVE PROJECTS
—— 实践设计项目篇

097，098

Designing and Constructing Nursery Schools in South Africa

Project Jouberton, Project Limpopo

In 2008, DABE responded to an invitation to collaborate with the Social Architecture programme initiated by Education Africa, a Johannesburg-based charity that has focused on the provision of pre-school nursery school within black township communities. With state schooling starting at the age of seven, the formative education benefits of an early learning programme was being lost to the vast majority of black children in South Africa.

The Social Architecture programme is aimed at Schools of Architecture from around the world to design, fund-raise and construct nursery schools in conjunction with local communities. To date, universities from Austria, Japan, Switzerland, Germany and USA have taken up this challenge, together with the University of Nottingham as the pioneering representative from the UK.

It was clear for the team at Nottingham that projects of this type should be incorporated within the studio programme. This would enable the teaching and learning outcomes to become part of the normal curriculum for the ARB/RIBA validated courses in architecture, as well as the invaluable advantages of working on a live project with real communities in need.

The project opportunities had been introduced to the Year 1 undergraduate students at the end of their first year of studies, to allow individual fundraising initiatives to be undertaken over the summer vacation for travel, accommodation and building materials. Up to fifty students volunteered for each of the two completed schools to date, Project Jouberton and Project Limpopo. These allowed the autumn semester of Year 2 to concentrate on individual site analysis and design proposals, which were developed into teams working on competitive programmes, with the eventual winning schemes being selected by a panel of design tutors and architects within DA+BE before the end of the semester.

For the spring semester, the winning scheme proposals were developed by a smaller team of Year 5 postgraduate students into construction packages with detailed drawings for individual components, based on robust, established and sustainable construction methods related to the availability of materials in South Africa.

With a limited amount of preliminary works undertaken on each of the sites using local labour, the student and staff teams travelled out to South Africa for an intense six or seven week programme of construction over the Easter period to complete as much as possible of the scheme proposals within a rigid time and financial framework. Both projects were then handed over to the respective communities.

南非幼儿园设计建造项目

朱博顿项目，林波波项目

在2008年，应“教育非洲”组织所发起的“社会建筑计划”邀请，建筑与建造环境学院与这家位于约翰内斯堡的慈善团体合作，希望为当地的黑人社区设计建造一所幼儿园。在南非，国家规定的小学入学年龄是7岁，可对于大多数黑人孩子们而言，接受学龄前教育仅仅是一种奢望。

“社会建筑计划”旨在全世界范围内号召建筑院校与南非的土著社区合作，来设计，募捐并建造小幼儿园。在过去几年里，来自奥地利、日本、瑞士、德国与美国的大学已经接受了这一挑战，而诺丁汉大学则是英国在此领域内的先驱。

对于诺丁汉大学而言，很明确这一项目必须与设计课程相联系。因为这样一来，项目的教学成果可以被英国建筑师注册委员会与英国皇家建筑学会所承认，成为正规建筑教育中的一环，此外，为真实的用户设计建造他们所需要的真实建筑则能为学生带来无价的经验。

在第一学年年末，可以参与这一项目的消息被通知给学生们，从而使有意参加的学生可以在暑假期间开始募捐，为将来的旅行，住宿与建设材料筹募资金。距今为止，我们已经完成了两所幼儿园建设，分别位于朱博顿与林波波，每个项目均有大约50名学生志愿报名参加。在第二学年秋季学期里，这些学生起初将集中精力设计各自的方案，之后再通过竞赛选出数个优秀方案，并围绕这几个入选方案组成数个小组，继续深化设计。而在学期末，由学院内教师与建筑师所组成的评审小组则会选出最终的优胜方案。

在春季学期里，最终的优胜方案将由以小组5年级学生继续深化发展为施工详图。这里包括所有的部件大样，并力图基于南非当地的材料，采用坚固，实用且环保的构造设计。

当地工人为这两个项目的前期准备提供了一些非常有限的支持性工作。而我们的学生与教师则利用复活节假期组队前往南非，进行6到7周的建设工作，力图在有限的时间与财力支持下尽可能多地完成项目所要求的工作量。如今两所小幼儿园均已被交付当地社区使用。

▲ - Team of Jouberton 朱伯顿项目组

▲ - Team of Limpopo 林波波项目组

- Jouberton Project 朱伯顿项目

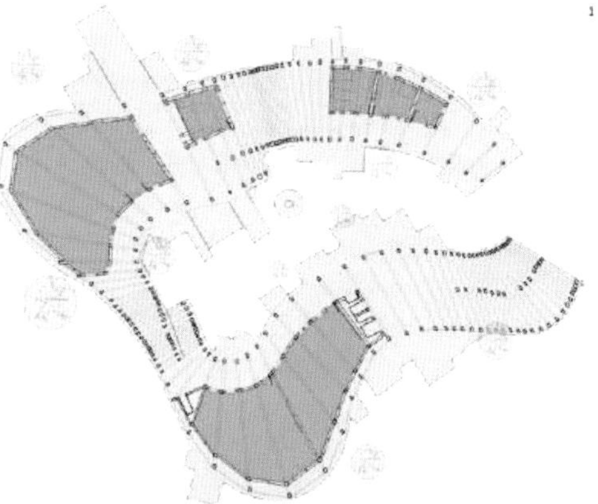

1

2

Limpopo Diary:
-- Studio Design programme

At the start of the academic year in September 2010, the Year 2 studio design unit of fifty students were introduced to the Social Architecture programme and outline brief. Both the studio tutors and project directors, John Edmonds and John Ramsay, are chartered architects with considerable design and construction experience. They had briefly visited the sites for both the completed Jouberton and the chosen Limpopo location to prepare the site information for the students to develop into their scheme designs.

The studio support team was assisted by two further tutors, Dr Wang Qi and Helen Jones. Structural advice and support came from Stephen Fernandez from ARUP. Briefing information came from the Thusanang Trust, a charity based in Limpopo with local education outreach to the scattered and remote rural communities. The proposed nursery (to be known as the Khomatso Crèche) at Calais village in Limpopo was located 50km south of Tzaneen, the nearest regional centre for construction materials and accommodation.

Calais village, located idyllically at the foot of an impressive mountain range, had been established to serve the fertile farmlands producing fruit crops consisting mainly of avocados and mangoes. However, political changes had left many of the farms derelict resulting in high unemployment levels within the village. Ten years previously a large plot had been allocated for the nursery school, but lack of funding had resulted in a totally overgrown site, where there was neither water supply nor electricity.

The student teams in the studio quickly established the key parameters for such a project, with allocation of learning spaces and the essential 'heart' of the crèche. In addition, orientation and environmental aspects to keep cool within a sub-tropical climate, together with the inclusion of an effective sanitary system, were particularly challenging.

The eventual chosen scheme was a combination of two team proposals, with the modularity and construction methods established by the 'Chiefs' in conjunction with the site layout of 'Tshabalala'. These principles were subsequently developed into detailed design packages by the more experienced team of Year 5 postgraduate students in the spring semester of 2011.

Together with the design development, fundraising had been prolific, initiated and administered by Angela Aston and further enhanced by match-funding and gift-aid through the University Development Office to double the student contributions.

林波波日记：
——方案组设计计划

2010年9月，在学年开学伊始之际，我们向选择本项目的50名二年级学生做了有关“社会建筑计划”的简介，并向他们发放了设计纲要。该项目的设计指导教师与项目负责人，约翰•埃德蒙斯与约翰•兰姆西均获得了注册建筑师资格，并拥有丰富的设计与施工经验。在此之前他们已经简要了解并参观了业已竣工的朱博顿项目，并在林波波省为本项目挑选出了合适的场地，收集整理了场地及周边环境信息，以供学生设计方案时参考。

该设计组还拥有另外两名指导教师——王琦博士与海伦•琼斯。来自ARUP的史蒂芬•费尔南德兹为本项目提供结构方面的建议与支持。当地的基础信息则来自于苏珊昂基金会，这是一家位于林波波省的慈善团体，主要与分散在当地偏远乡村中的教育机构合作。计划中的幼儿园（又名库马索-克里奇）位于林波波省加莱斯村，位于查嫩城以南50公里处。而后者则是该区域内最近的中心城市，原材料供应地以及住宿地。

位于山峦脚下的加莱斯村景色宜人，土地肥沃。这里盛产水果，鳄梨与芒果是其主要作物。然而，政治剧变使得许多农场遭到荒废，并同时导致了村子里极高的失业率。大约10年前，村子里的一大块地被规划用来建设一所新的幼儿园，但是资金匮乏使得该计划从来未能实施，而规划用地上则是荒草丛生，既没有水也没有电。

在诺丁汉，学生们很快就为该项目设定了任务书，这其中包括教学空间的位置与“克里奇”核心（围绕一棵大树的庭院）所在。此外，为了在亚热带地区保持室内凉爽宜人，建筑朝向与环境设计得到了重视，厕卫系统也被优先考虑，这些都是该项目带来的特殊挑战。

最终的胜出方案是基于两个设计小组的方案合并而成的。“酋长组”在模式化与构建方法方面拔得头筹，而“沙巴拉拉组”则在场地规划方面十分优秀。这些要素随后在2011年春季学期被一组经验更加丰富的五年级学生细化为成套施工图纸。

在设计方案顺利发展的同时，资金筹募也进展喜人。在安吉拉•阿斯顿女士的具体操作与管理下，项目得到了诺丁汉大学发展办公室的“一镑对一镑”补贴与捐助，从而使学生募款得以成倍翻番。

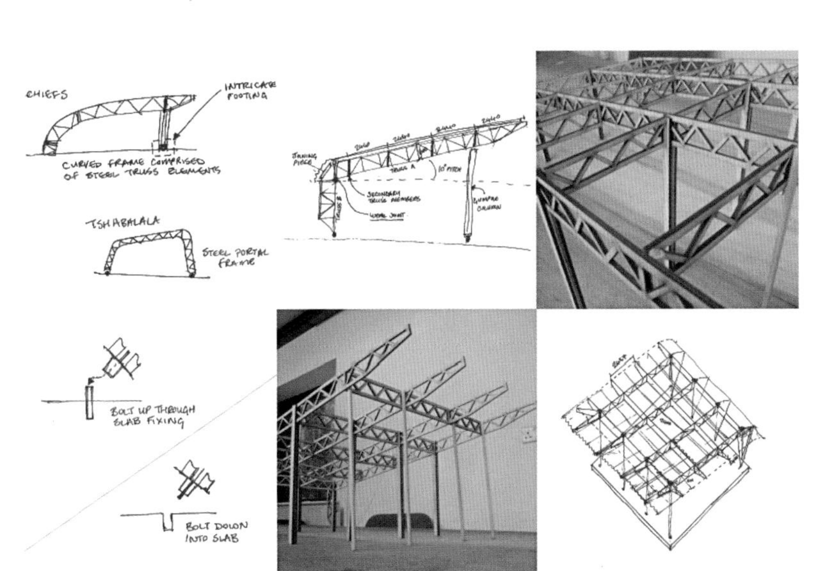

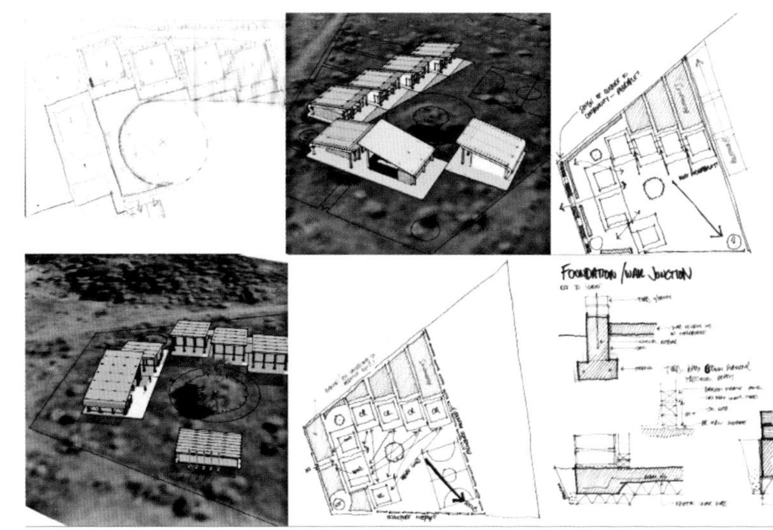

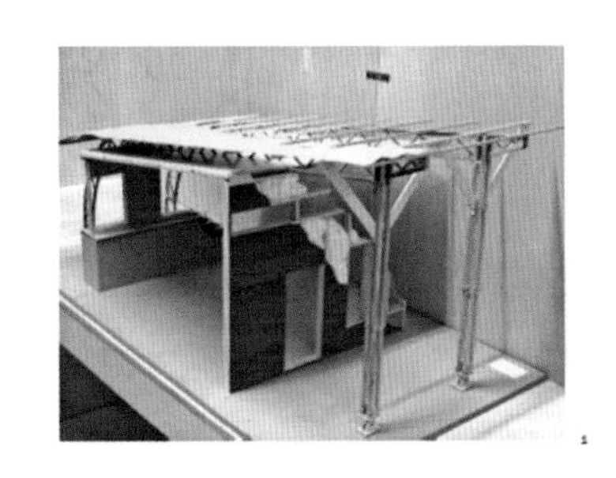

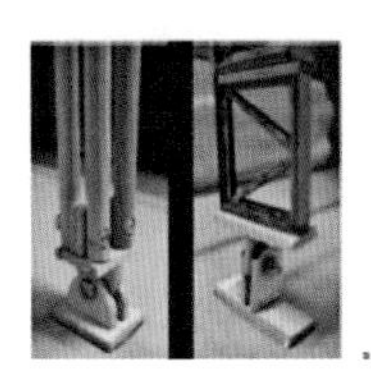

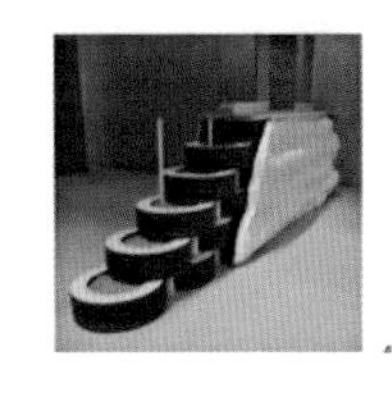

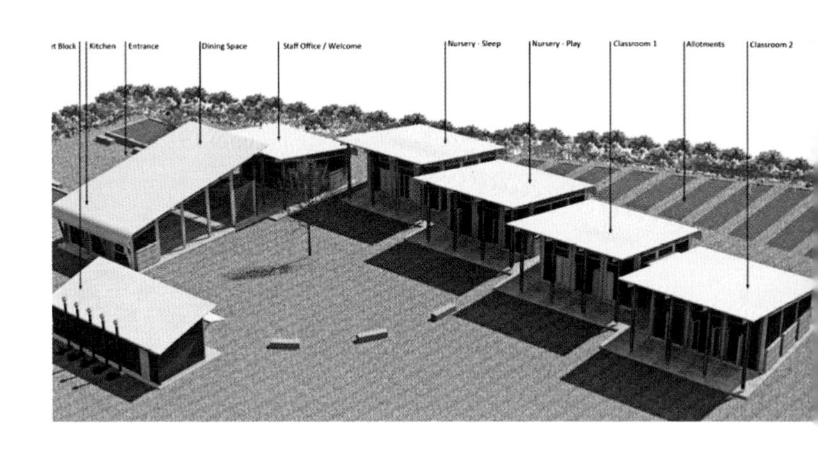

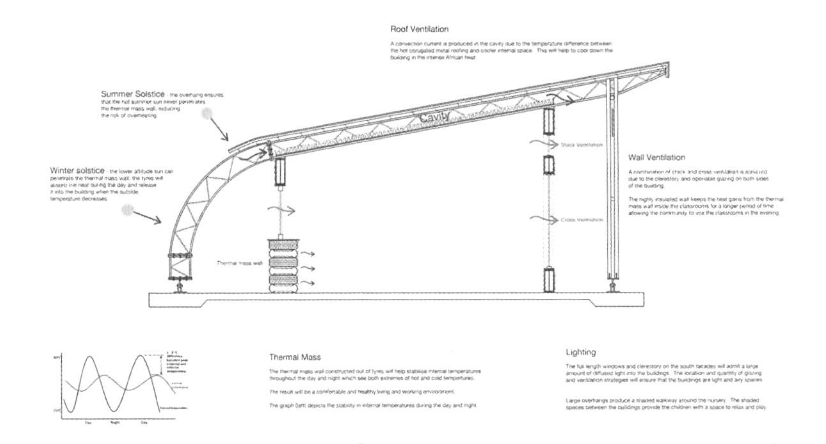

- Scheme of "Chiefs" 酋长队设计方案

- Scheme of "Tshabalala" 沙巴啦啦队设计方案

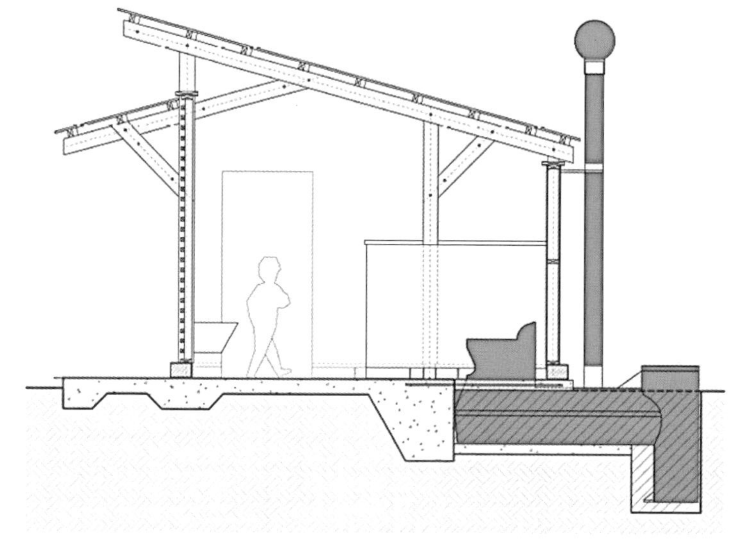

- Detailed Design of Year 5 Students
由五年级学生深化的详细方案设计

Project Limpopo, Phase 1，March 2011

The construction programme at Calais village was split initially into two phases, each of three weeks overlapping the Easter vacation. Phase 1, including twenty-five students and staff, left Nottingham late in March 2011 to Johannesburg, followed by a six-hour coach journey to Tzaneen, where the staff and students were accommodated in a comfortable and reasonably-priced hotel. Locally-sourced minibuses and 4x4 pick-up trucks (bakkies) were main transportation for daily commute between Calais village and Tzaneen.

However, despite prolonged earlier negotiation from the UK with local suppliers in Limpopo, difficulties of logistics were soon experienced with significant delay form local material suppliers together with a critical water shortage at the only concrete plant in the area, which led to strategic decisions to construct temporary structures for a site workshop and external paved working areas whilst awaiting the release of embargoed timber from the local sawmill.

The site workshop was soon equipped with workbenches designed and built by each group of students to mount the power bench tools. Electricity was provided by a petrol generator in the absence of mains supply, with on-site training from Mark Merrills and David Edwards, skilled technicians from the University.

Although due to the serious delays in material supply, the first bolted timber trusses were completed in time for the handover to the Phase 2 students, which was celebrated by the joint raising of the first frame.

林波波项目，阶段一，2011年3月

在加莱斯村的施工过程最初是被分为两个阶段，各包括复活节假期中的三个星期。在阶段一，25名学生与教师于2011年3月末离开诺丁汉。团队首先到达约翰内斯堡，然后再乘坐6小时的大巴抵达最终目的地——查嫩。在这里，学生与教师均住在一处环境舒适且价格公道的宾馆内。而每天我们则需要借助当地的小中巴与4x4越野客货两用车往返于加莱斯村与查嫩城。

然而，尽管与当地供应商的材料供应洽谈早在我们尚在英国时就已经展开，供应方面的困难还是接踵而至。先是木材供应严重滞后，而后当地唯一的混凝土供应站也由于严重缺水而无法工作。面对这些情况我们及时调整了计划，在等待当地锯木厂提供那些“被扣下”（由于没有任何明显的原因，我们对供应商的延误十分不满，于是戏称这些木料被无端“扣”下了）的型材同时，学生们先搭建起了一座临时性工作棚，而后又在室外铺出了一块工作平台。

工作棚中设有学生自己设计装配的工作台，所有的台式电动工具都被固定在台面上。由于没有电网供电，所需电能只能来自一台汽油发电机。两位来自诺丁汉大学的技师——马克•美林斯与大卫•爱德华兹则为学生提供了如何使用工具的现场指导。

尽管材料供应方面有严重的困难，我们还是赶在第二批学生来时完成了对第一榀木桁架的锚固组装。它在所有两批学生手中缓缓竖起，大家在桁架下共同欢庆这一标志性的时刻。

Project Limpopo, Phase 2， April 2011

Using the stock of completed trusses fabricated by the Phase 1 team, the students from Phase 2 soon developed a timber column system to manually erect the trusses by using a simple rope system. The concrete slabs for each classroom were laid and power-floated by Dale Eberhard, a local contractor. As soon as the four frames for each classroom had been erected and cross-braced, the infill wall frames and roof rafters were added to complete the structural skeleton.
In parallel with the classrooms, a toilet block was under construction using a similar timber frame. Drainage was fed to a septic tank system, believed to be the first within the village. A free-standing water tower was constructed to serve this block, but was subject to erratic mains water supply from a nearby reservoir, which resulted in often water barrels in wheelbarrows filled from the local river.
The student teams soon became skilled in various building techniques, sawing, drilling, and bolting timbers, followed by hammer and nails for the infill panelling. External cladding options were reconsidered on site, with the eventual use of galvanised corrugated sheeting laid horizontally for the blank side elevations and the locally- produced concrete blocks for the rear walls. Standard off-the-shelf steel window frames were incorporated within the structural timber frame, shielded from the intense sunlight by the wide overhanging roof.
The Phase 2 team were able to complete the two main classrooms and most of the WC block during weeks 4-6. Despite the supreme efforts from all the students, the roofing was still awaited from the supplier so the school could not be handed over as intended. Contingency plans did allow for further phases.

林波波项目，阶段二，2011年4月

第一阶段学生们留下了很多业已组装好的桁架，而第二阶段的学生们很快组装起了成对的柱子，并利用简单的绳索将一榀榀桁架竖立了起来。当地建筑商——黛尔•艾伯哈德公司为教室浇灌并抹平了混凝土基础。而只要每间教室的四组桁架被竖立到位且固定牢靠，墙体框架与屋顶椽条便可以马上装配到位来完成整个结构骨架。

与教室施工同步进行的是厕所部分。这是一座小型的木架结构体，而厕卫的冲水系统则与一个化粪池相连，这在村里还是头一次见。我们建起了一座独立水塔来为厕所供水，然而由于附近的水库供水十分不稳定，很多时候水塔依然需要用人力小推车和水桶从村边的小河里拉水。

学生们很快就掌握了不同的建造技能——包括锯、钻、锚固木材，以及使用榔头与钉子安装墙面板等等。外立面在建设过程中经历了重新设计。最终我们决定用水平铺装的镀锌瓦楞钢板来装饰实墙部分的外表皮，而村子里自己生产的水泥砖则被用来砌筑后墙。标准钢型窗被装嵌入木构架之中，深远的挑檐则为其遮挡着强烈的阳光。

在第4周到第六周的时间里，阶段二的团队完成了两座教室与厕所的大部分工程。尽管学生们已经竭尽全力，但屋面供应的滞后使得幼儿园最终无法如期交付使用。我们的确需要另一个阶段。

Project Limpopo, Phase 3 August 2011

After budget reviewed back at Nottingham, a smaller task force of twelve students and four staff returned to South Africa during the summer vacation to complete the two classrooms and WC block, as well as the construction of another classroom. The student team was comprised mainly from the Phase 1 participants who had been frustrated by the material supply delays.

With just eleven working days available, the students made incredible progress, assisted by a small team of skilled local villagers who were employed through the contractor. Targets were met, buildings virtually completed during this intense period.

The Khomatso crèche was formally opened in January 2012, with three main classrooms, a WC block and playground in active use for over 60 children already registered. For the headmistress, Sophie, the creation of the nursery school is the culmination of her dreams and will enable so many children to gain from the essential formative years of education.

For the students, the experiences and learning outcomes have been immeasurable. Fund-raising, designing, and constructing their first real examples of architecture at such an early stage in their careers will always remain with them. There are also the intangible benefits of having had the opportunity to work with and significantly help a disadvantaged community, who will also remember Project Limpopo and the students of architecture from the University of Nottingham with pride.

林波波项目，阶段三，2011年8月

在回到诺丁汉后我们重新评估了预算，剩余的资金可供一个包括12名学生与4名教室的小团队在暑期重返南非，去完成两间教室与厕所，并同时建造第三间教室。这一阶段的学生主要来自阶段一的团队，旨在弥补他们由于供应延误而耽搁掉的时间与体验。

该阶段仅有的11个工作日。在一小队由当地建筑商雇佣，手艺娴熟的村民的帮助下，学生们取得了不可思议的进展。既定目标达到了，经过紧张施工，项目最终得以完成。

库马索-克里奇在2012年1月正式开园。60名在册的小朋友现在有了三间主教室、一间厕所，以及一片游戏场地。对于园长索菲而言，这所幼儿园的竣工圆了她长期以来的梦想，从此以后，这里的很多很多孩子将有条件接受必要的正式学前教育啦。

对于我们的学生而言，这一经验与学习成果确是无可比拟的。募集资金、设计，并建造他们的第一座真实的建筑，这一切都在他们的职业生涯中实现得如此之早，也必将使他们受益终身。除此之外，这里面还凝结着与弱势群体一起工作并为他们提供重要帮助而带来的隐性价值——加莱斯的村民们会永远记住林波波项目，以及那一群来自诺丁汉大学建筑系的年轻人。

The Team of Jouberton 朱博顿项目组

Academic Team 教师及辅助团队

Project Director: Adrian Friend; Structural Advisors: John Edmonds, Steve Fernandez; Fundraising & Development Manager: Sarah Newine Moore

Student Team 学生团队

Year 5 Students（五年级学生）:

Jon Hore; Adam Kelly; Matt Wingrove; Mo Gamal; Selina Shah; Simon Gomm; Helen Jones

Year 2 Students（二年级学生）:

Jeannine Moros-Noujaim; Paul Challis; Alex Eagles; Jessica Morrison; Matt Powell; Vicky Fabron; Tom Partridge; Laura Gaskell; William Gowland; Sam Critchlow; Manreshpal Rai; Mike Ramwell; Amy Conneely; Adam Casey; Harriet Pillman; Yuk Lan Wong; Anna Hutnik; Jo Edmunds; Tom Wing-Evans; Heather Brand-Williamson; Hayley Shepherd; Dan Ladyman; Uma Mahendran; Mike Clarke; Emma Harvey; Poppy Trevillion; Cassandra Tsolakis; James Boon

The Team of Limpopo 林波波项目组

Academic Team 教师及辅助团队

John Edmonds; John Ramsay; Wang Qi; Helen Jones; Structural Advisor: Steve Fernandez; Fundraising and Project Planning: Angela Aston; Administration: Lucy Boutlby; Technical support: Mark Merills; David Edwards; Patrick Hodgkinson

Student Team 学生团队

Year 5 Students（五年级学生）:

Matt Vaughan; Dale Muscroft; Ellie Atherton; William Main; Catherine Legg; Sam Smith;

Year 2 Students（二年级学生）:

Team Flamingo: Avril Wheeler; Matthew Palmer; Clara George; Sarah Comfort; Jack Boyns; Trudy McGregor
Team Tshabalala: Daniel Villette; Thomas Dichmont; Samuel Diston; Daniel Parsons; Zoe Watson
Team TIA: Joanna Tatlow; Nathan Craig; Anna Luff; Francesca Attard; Abigail Blumsohn; Philippa Grayson
Team Chiefs: Philip Noone; Nicholas Gaston; Oliver Pedley; William Holley; Joshua Jones
Team Waka Waka: Jyothi Pillay; Alix Blankson; Alfred Roden; Sam Harding; Alana O'Kirwan; Mohammad Mirza; Fatima Ladak;
Team Bou: Christina Agoston-Vas; Taylan Tahir; Geraldine Hallifax; Kate Ottway; Charlotte Page;
Team Ndija Kuthanda: Michael Arnett; Hannah Bass; Samuel Bentley; Stuart Bacon; Tom Bradley; Susanne Bruijnzeels;
Team Elephant: Benjamin Thompson; Laura Sheridan; Haniyyah Rashid; Kristian Bjerre; Becky Gough;

Worded by John Edmonds 撰文：约翰 · 埃德蒙斯

Further Information can be found on: 更多信息：www.projectlimpopo.org www.projectjouberton.com

Nottingham H.O.U.S.E

Team of H.O.U.S.E

The final of the International Solar Decathlon Europe Competition, held in Madrid during June 2010, was the culmination of two year's hard work. The University of Nottingham entry was named 'The Nottingham H.O.U.S.E' (Home Optimising the Use of Solar Energy). The project was designed by Rachel Lee, Ben Hopkins and Chris Dalton, students from the Zero Carbon Architecture Research Studio (ZCARS), a 5th year architectural design studio. The construction of the H.O.U.S.E was then undertaken by a design studio unit comprising 2nd year BArch students and 3rd year MEng students.

Staying cool is obviously a key concern for a house operating in a Madrid summer and good passive design served as the first strategy in achieving this aim. Shading of exposed perimeter openings is provided by a woven mesh, keeping undesirable high altitude summer sun out while still permitting much needed low altitude winter sun to penetrate. Shading was also used to protect the roof, the photovoltaic modules that provide the H.O.U.S.E with its electricity also being used to reduce roof surface temperature. Having worked to keep the sun off the envelope, conduction of heat to the interior was controlled by the high performance envelope with U-values of 0.1 W/m²K for the roof, 0.17 W/m²K for the walls and 0.7 W/m²K for the triple glazed windows. The same strategies that are designed to keep heat out, when coupled with an extremely well sealed envelope that allows close control of ventilation, combine to leave the H.O.U.S.E with a very small winter heating demand for colder climates such as the UK.

Low energy cooling was one of the unique features of the H.O.U.S.E design. Passive downdraught evaporative cooling (PDEC) uses water to maintain comfort temperature during the summer season. Nozzles positioned at the top of the double height space generate a micro spray of water that evaporates in warm external air drawn through the roof light. This generates a plume of cool air that drops into the living space absorbing heat and then exits via the windows.

At night, the nozzles are switched off and cool night air, drawn through the windows, absorbs heat stored in the building fabric resulting in buoyancy-driven natural ventilation that exits via the roof light.

诺丁汉H. O. U. S. E生态住宅项目

H.O.U.S.E设计组

在2010年6月，马德里举办了国际太阳能十项全能欧洲建筑设计竞赛的决赛阶段展示，这一盛事标志着两年的辛苦工作最终开花结果。诺丁汉大学的参赛作品名为“诺丁汉H. O. U. S. E”（这是“太阳能应用最优化住宅”英文名称的缩写）。该项目由来自零碳建筑研究设计组（ZCARS）的学生——瑞秋•李、本•霍普金斯和克里斯•达尔顿设计。而H. O. U. S. E的建造则由一组本科学生负责。他们部分来自建筑设计专业二年级，部分来自建筑工程本硕连读专业三年级。

很明显，在马德里的夏日里，如何保持室内凉爽将是一项住宅设计首要关心的问题。为了达成这一目的，优秀的被动式环境设计至关重要。在建筑外表的所有开口处，我们提供了由细密编织格栅所制成的遮阳板，其角度设计则保证了可将夏日的高入射角阳光遮挡住，且同时允许冬日的低入射角阳光进入室内。房顶设计也考虑到了遮阳问题，那些位于屋面上的太阳能发电板一方面可为H. O. U. S. E提供电能，另一方面又可以保护屋面不至于长时间暴露与烈日之下，降低屋顶的温度。为了使建筑外层维护系统能够减少对太阳热辐射的传导，高热惰性的材料被使用在建筑上，例如，房顶构造的U值为0. 1W/m² K，墙体构造的U值为0. 17W/m² K，而三层玻璃窗的U值为0. 7W/m² K，它们均可以有效控制传导入室内的热量。这一策略可以将外来热能挡在屋外，而当整个外围护系统良好密闭时，自然通风也可得到精确的控制。将这些技术要素合并起来，即使把H. O. U. S. E建在诸如英国这样冬季寒冷的地区，在冬天它也仅仅需要极少量的采暖。

“低能耗制冷”是H. O. U. S. E设计的一项独特之处。“被动式下喷蒸发制冷系统”（PDEC）采用水雾来在夏季保持室内舒适温度。喷口被设置在室内两层通高处的顶部，它们可以喷射出极细的水雾。随后，水雾在降落的过程中随即被透过屋顶天窗弥漫在室内上方的热空气蒸发，从而可即时产生大量冷空气沉降入下方的生活空间中去。这些冷空气则可吸收室内的热能，并重新上升排出室外。

在夜间，喷口停止工作。凉爽的夜晚空气可从窗户进入室内，吸收白天存储在建筑结构中的热量，并由天窗排出室外，从而产生浮力驱动自然通风效应。

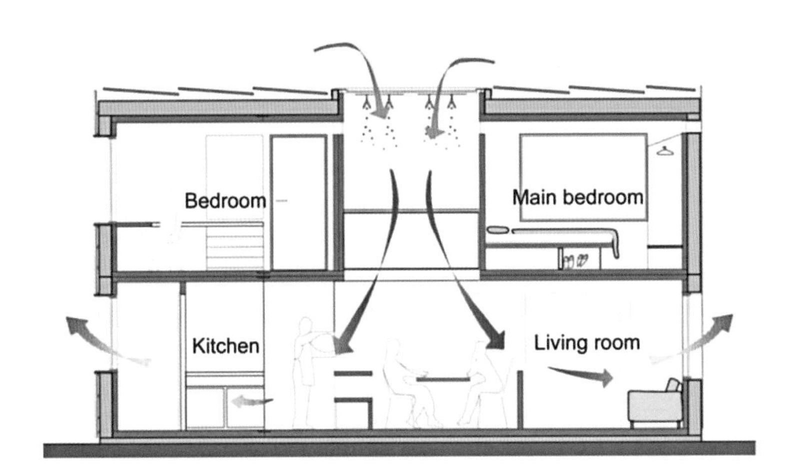

- Summer Day Time Air Flow Path 夏季白天室内通风及制冷示意图

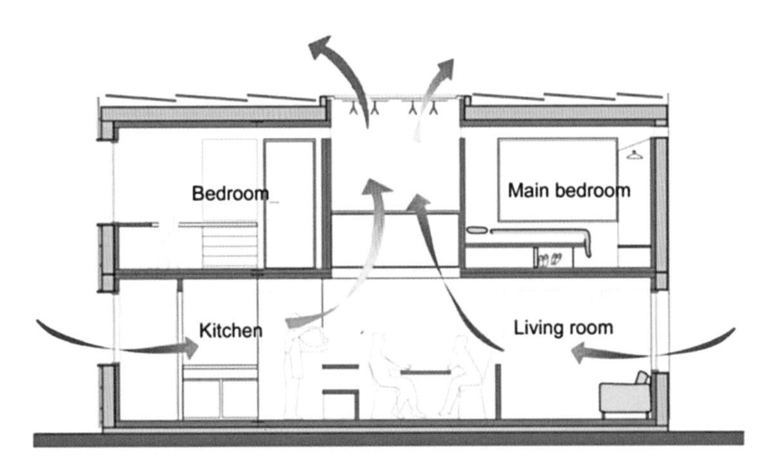

- Summer Night Time Air Flow Path 夏季夜晚室内通风示意图

Measured data from the H.O.U.S.E. collected during the competition week in Madrid (19th - 27th June 2010) demonstrate that the combination of a high performance building envelope, night time convective cooling and PDEC can deliver comfortable conditions throughout the H.O.U.S.E. These results suggest that the absence of an extra high supply tower is no impediment to the successful integration of a PDEC system within a two-storey house. While the data set is limited, these are very important findings as they suggest that PDEC can provide a viable alternative to conventional air conditioning for housing in central and southern Spain and, by implication, to other hot-dry regions of the world.

Architecture students at the University of Nottingham built the H.O.U.S.E in close collaboration with industry and in particular the primary sponsor is Saint-Gobain. This is a principle the Department seeks to follow, having established a Projects Office and keenly using the recent completed prototyping hall on the Jubilee Campus.

The Solar Decathlon competition has always been viewed as the starting point for exploring and developing the H.O.U.S.E concept. The final destination for the 8 modules on their return from Madrid is as part of the 'Creative Energy Homes' project located in 'Green Close' on the University Park campus. The H.O.U.S.E will become the seventh research dwelling and the freedom to explore its behaviour beyond the one week window offered in Madrid will help to provide valuable insight into the technical and non-technical qualities of the design.

In essence the Nottingham H.O.U.S.E is a prototype built of bespoke elements and components, many of which are not yet currently available in the UK. Throughout the project there was a conflict between an essentially experimental construction process and absolute deadlines: the opening dates for Ecobuild and the Solar Decathlon competition were finite and non negotiable. Thus time was of the essence, a constraint that construction eschews for good reason, unless it is essential for the client or a specific architectural typology. The next phase of its life promises to be less frenetic, as the ongoing activity to understand how it performs, provides new opportunities to train researchers within the Department.

在马德里竞赛周期间（6月19—27日），H. O. U. S. E收集了大量实验数据。这些数据表明，将利用高热惰性材料，夜间对流制冷与PDEC系统综合在一起，H. O. U. S. E可以全面提供舒适居住环境。这一结果证明，是否拥有超高中空塔并不会对PDEC系统在两层小住宅内的成功整合产生任何影响。虽然所收集的数据十分有限，但其结论依然是非常重要的发现，因为这证明在西班牙中南部，或其他炎热干燥地区，PDEC完全可以成为传统空调系统的有效替代品。

在建造H. O. U. S. E的过程中，诺丁汉大学的建筑专业学生与实践生产领域密切合作，特别是Saint-Gobain公司为该项目提供了最主要的资助。事实上，这也正是建筑系的一项办学理念。诺丁汉大学建筑与建造环境学院成立了自己的项目办公室，并且正在积极地使用新进完工的，位于朱比利校区的建筑原型建造实验室。

太阳能十项全能竞赛也仅仅是探索与发展H. O. U. S. E概念的一个起始点。在组成整个建筑的八个模块从马德里返回英国后，它将被作为“创新型节能房屋计划”的一部分而永久坐落在大学公园校区那迷人的大草坡上。H. O. U. S. E将是这里的第七栋生态研究住宅，从而将在未来提供远远超过马德里那短短一周时间的长期研究机会，专家们可以在此深入研究它的功效，从而为技术性与非技术性设计提供极有价值的参考。

从本质上来讲，诺丁汉H. O. U. S. E是由一系列定制的组件所构成的建筑原型，这其中的许多构件在英国目前还并没有上市。此外在整个项目设计建造过程中，必要的实验性建构过程与项目既定截止日期之间总是产生矛盾：要知道Ecobuild博览会的开幕时间与太阳能十项全能竞赛的时间是确定且不能更改的。因此，时间就变得非常重要，而与之相应，除了是对使用者而言极为必要和涉及特定建筑形式的方面，建筑的构造体系不得不进行了简化设计。但是H. O. U. S. E“生命”中的下一个阶段就不会这么紧张了，它将成为建筑系的一员，为培训年轻研究人员提供新的机会，使人们能够持续观察并理解它的功效。

▼ - PDEC at Work PDEC系统工作状况

▼ - Nottingham H.O.U.S.E at Solar Decathlon, Mardrid 诺丁汉H.O.U.S.E在太阳能十项全能竞赛现场，马德里

▲ - Nottingham H.O.U.S.E at Ecobuild 诺丁汉H.O.U.S.E在Ecobuild博览会展出

Academic Team 教师及辅助团队

Mark Gillott, Brian Ford, Michael Stacey, Lucelia Rodrigues, Guillermo Guzman John Ramsay, Robin Wilson
Project Manager - Mike Siebert; Technical Support - David Oliver; Office support - Lyn Shaw.

Student Team 学生团队

Chris Dalton, Ben Hopkins, Rachel Lee, Nina Hormazabal, Jose Andre, Jonathan Barron, Timeka Beecham, Peter Blundy, Christian Brailey, Thomas Corbett, Emily Costain, Ashley Evans, Victoria Fillingham, Rebecca Ford, Samuel Fuller-Teed, Philip Gilder, Alisdair Gray, Hannah Griffiths, Teuta Hasani, Benjamin Heede, Shui Hui, Chapman Cheuk Hang Kan, Unjulee Karadia, Simon Kinvig, Frances Kirk, Manya Krishnaswamy, Joe Leach, Frances Lister, Stephen Lloyd, Stephen Lovejoy, Sin Mak, Gareth Marriott, William Marshall, Daniel Mckie, Bhavik Morar, Janet Pang, Trusha Patel, William Slack, Mustafa Tekman, Katherine Tokarski, Elizabeth Vincent, Louise Vitty, Mang Yuan Wang, Long Fei Xiang, Constanza Yanez, Joseph Yates.

Further Information can be found on: 更多信息：www.nottinghamhouse.co.uk

—— THE 100th

—— 第100个方案

DESIGNER-设计人: Kinny Soni

Stories at Nottingham

"One planet one school" – The story begins with an empty green courtyard surrounded by the old and new buildings with a beautiful pink tree. Walking through the gallery seeing glimpses of presentations, I emerge from the old building and see green hills and the Eco House standing still whilst the sun plays with the clouds. Moving down, row of cars take me to my work place where a vintage car makes me curious as to whom it belongs! It meets me every day when I am walking through the dark forest towards the outside world, while the school grows with me, reaching all over the planet from that green and silent courtyard in Nottingham.

诺丁汉故事

"同一个星球，同一个学院"——空旷的庭院绿草莹莹，新老建筑环绕四方，粉色的花朵在枝头烂漫，我的故事从这里开始。穿过展廊，我看到两边贴满了设计作品；踱出古老的学院大门，明媚的阳光与浮云尽情嬉戏，碧绿的山坡与生态小屋跃入眼帘。缓缓下行，沿路停放的一排汽车指向我工作的地方。那里总是停放着一辆漂亮的小跑车，不由的使我好奇不已，想猜猜它的主人到底是谁。每当我穿过那墨绿色的树林走向校外的世界，它总是等在那里与我会面。哦！学院与我一起成长，从那翠色满园，恬静安详的庭院，诺丁汉与整个星球相连。

"My Memories of Nottingham" – I tell about my story that links me to the people and the city of Nottingham. Through this particular work I share my memories and experience of this charming city where I found clouds trying to hide the sun, the landscape with its tales of Robin-hood and his castle. Row of houses with the giant wheel and the sparkling flow, with birds eating my bread becoming my friends brought life in my story.

“我记忆中的诺丁汉”——让我来告诉你一个故事，那与诺丁汉的人们和城市有关。借助画笔，我愿与你共享我的记忆与体验。在这座迷人的城市里，云彩总是想遮住太阳，为罗宾汉的土地与城堡送来阴凉。房子肩并肩站成一排，围绕着中间那高耸入云的摩天轮与银光闪闪的喷泉。飞来与我分享面包的鸟儿是我的朋友，这一切都使我的故事变得灿烂。

STAFF LIST - 教师名单

Professors

Chilton, John
Chair of Architecture & Tectonics
Ford, Brian
Chair in Bioclimatic Architecture
Heath, Tim
Associate Dean for Internationalization and External Relations, Chair in Architecture & Urban Design
Riffat, Saffa
Chair in Sustainable Energy, Head-Institute of Building Technology & Institute of Sustainable Energy
Robinson, Darren
Head of Division for Energy and Sustainability, Chair in Building and Urban Physics
Stacey, Michael
Chair in Architecture, Director of the Institute of Architecture
Yan, Yuying
Professor of Thermofluids Engineering, Head of HVACR & Heat Transfer Research Group

Associate Professors

Altomonte, Sergio
Associate Professor, Director of Diploma in Architecture
Gadi, Mohamed
Associate Professor, MSc Course Director & Admissions Tutor (REA), Institute of Building Technology
Gan, Guohui
Associate Professor, Institute of Building Technology
Gillott, Mark
Co-Director of the Institute of Sustainable Energy Technology
Hale, Jonathan
Associate Professor & Reader in Architectural Theory
Hall, Matthew
Associate Professor in Materials Engineering
Hanks, Laura
Associate Professor, Institute of Architecture
Liu, Hao
Associate Professor, Institute of Sustainable Energy Technology
Omer, Siddig
Lecturer, Course Director and Admissions Tutor, Institute of Sustainable Energy Technology
Rutherford, Peter
Associate Professor, Institute of Architecture
Samant, Swinal
Associate Professor, Course Director, Architectural Studies
Wilson, Robin
Associate Professor, Institute of Sustainable Energy Technology

Lecturers

Agyenim, Francis
Lecturer Sustainable Energy Technologies, Course Director MEng Architecture and Environmental Design
Borsi, Katharina
Lecturer, Institute of Architecture
Boukhanouf, Rabah
Lecturer in Sustainable Energy Technology
Cooper, Ed
Lecturer, Course Director BEng Architectural Environment Engineering
Edmonds, John
Director, Part 3 Professional Practice
Ekici, Didem
Lecturer in Architecture
Guzman Dumont, Guillermo
Lecturer, Institute of Architecture
Lau, Benson
Course Director, MArch Environmental Design. Architect/Lecturer, Institute of Architecture
Lu, Sean
Lecturer, Institute of Architecture
Morgan, John
Lecturer in Architecture
Niblock, Chantelle
Course Director, MArch Technology. Lecturer, Institute of Architecture
Oldfield, Philip
Lecturer, Institute of Architecture
Porter, Nicole
Lecturer
Ramsay, John
Lecturer in Architecture and Tectonics
Riganti, Patrizia
Lecturer, Institute of Architecture
Rodrigues, Lucelia Taranto
Lecturer in Architecture
Su, Yuehong
Lecturer, Institute of Sustainable Energy Technology
Tang, Amy
Lecturer, Institute of Architecture
Wang, Qi
Lecturer in Architecture, Course Director of March Design
Wu, Shenyi
Lecturer, Institute of Sustainable Energy Technology
Wu, Yupeng
Lecturer, Sustainable Energy Technologies and Building Services
Zhu, Yan
Lecturer, Institute of Architecture
Zhu, Jie
Lecturer, Institute of Sustainable Energy Technology

Other Academic

Bromley-Smith, Elizabeth
Undergraduate Studio Leader, Institute of Architecture
Gerber, Nicola
Undergraduate Studio Leader, Institute of Architecture
Howarth, Andrew
Teaching Ad***ministrator***
Nicholson-Cole, David
University Teacher, Institute of Architecture
Short, David
BArch Course Director and Part 3 Co-Course Director, Institute of Architecture

INDEX - 索引

Built Environment 建造环境 v vii viii ix x 085 127 201 216
Nottingham 诺丁汉 i iii iv v vi vii viii ix x xi xv xvii xviii xix 001 002 034 036 038 040 044 052 054 085 114 116 118 127 136 140 142 144 148 150 162 164 166 169 170 174 188 190 192 194 198 201 204 206 210 214 216 217 220 221
Ning Bo 宁波 vii
South Africa 南非 vii xix 201 210
Solar decathlon 太阳能十项全能 vii 214 216 217
RIBA 英国皇家建筑学会 viii 201
ARB 建筑师注册委员会 viii 201
Part I architect 第一阶段建筑师 viii xvii
Part II architect 第二阶段建筑师 viii xvii
Diploma 学位 viii
Charted Architect 注册建筑师 viii 204
Level-A Architect 一级注册建筑师 viii
Humanities 人文 vi vii viii ix x xviii xx 001 034 128
UK 联合王国（英国） viii ix xvi 064 066 070 124 201 206 213 214 216 218 222
UK Green Building Council 英国绿色建筑委员会 ix
Sustainable Architecture 生态可持续建筑 ix
Zero Carbon 零碳 ix xix 122 124 127 164 166 214
East Midlands（英格兰）中东部 x 172
England 英格兰 ix 026
ZCARS 零碳建筑 x
MARS 材料研究工坊 x
Trent River 特伦特河 x
Threefold Spatiality 三重空间性 xii
De-tooled 去工具化 xv
Thought 思想 xii 127 158 164
Critical Conservationism 批判性保护主义 001
Museum 博物馆 xviii 002 016
Concept 概念 v ix xiii xiv xv xvii 024
Concrete 混凝土 002 038 047 108 127 146 147 172 190 206 208
Entropy 熵 xviii 004
Suffolk 萨福克 006
Village of Covehithe 科夫海斯村 006
Coastline 海岸线 004 006 012 086 128 134
Archaeology 考古学 xviii 006 008 056 074 086 128
Archaeologist 考古学家 006 008 128
Victor Hugo 维克多 · 雨果 008
Ecologist 生态学家 010
Marine Conservation Society 海洋保护协会 010
Chiswell 奇斯韦尔 010
Isle of Portland 波特兰岛 010
Fishermen 渔民 010
Platform 平台 xviii 002 012 100 102 103 110 150 152 206
Jurassic coast 侏罗纪海岸线 012 134
Fissure 地缝 012
Limestone 石灰岩 012
Quarry 采石场 014 062 068 072 073
Peninsula 半岛 016
Prison 监狱 xviii 018 062 072
Rupture 裂缝 018
Reminiscence 回忆录 xviii 004 020
Kings lynn 金斯林 020
Cork 科克 022
Asylum 庇护所 022
Chapel 小教堂 022
Archive 档案室 034
Protocell 原细胞 001 026 068
Sherwood Forest 舍伍德森林 026
Decadence 颓废 xviii 028
QMC 皇后医疗中心 xviii 030 034
Incinerator 焚烧炉 xviii 032
Contemplation 冥想 030 034
Community 社区 xviii 006 010 014 018 020 036 042 058 060 062 066 070 072 092 094 098 106 110 120 124 150 162 164 166 174 192
Exhibition 展览 xviii 001 002 030 032 034 042 064 086 098 170 192
Craft 手工艺 xviii 038 127
Car park 停车场 036 038 092 100
Arcade 商业长廊 xviii 040
Context 文脉 ix x 001 008 017 018 042 044 053 056 062 064 074 076 080 148 164 169
Fragment 碎片 xii vi 001 006 012 014 042 188
Architectural Language 建筑语言 ix x 030 044
Lace 蕾丝花边 xviii 044
Cemetery 墓地 xviii 046 186
Rite 仪式 xviii 001 022 048 082
Threshold 门槛 048 062 114 162
Ruin 遗址 xviii 050 064 086 176 178
Parasite 寄生体 054 055
Castle 城堡 016 064 172 221
Legacy 遗产 xviii 066 134
Reconciliation 调解 xviii 072 158
Salisbury 索尔斯伯里 xviii 074 076 078 080 082
Cathedral 天主教堂 xviii 078 082 172
Wind Farm 风力发电厂 xviii 088
Abu Dhabi 阿布扎比 xviii 090 100 101 108
Courtyard 庭院 xviii 004 008 036 074 078 090 091 108 109 134 135 162 164 204 220
Skybridge 天桥 xviii 092 093 102 154 179
Tall Building 高层 ix 090 092 096 102 108 162 171
Green Roof 绿色屋顶 092 104
Ventilation 通风 ix 085 107 116 117 119 120 123 124 166 214 215
Solar 太阳 vii ix x xviii xix 088 100 101 102 104 120 124 150 166 214 216 217 221
Rotterdam 鹿特丹 xviii 096
Singapore 新加坡 xviii xix 098 110
New York 纽约 xviii 088 102 104 106
Day-lighting 日光 102 106
PassivHaus 被动式建筑 104 106
Skyscraper 摩大楼 ix 104 108 180
Façade 立面表皮 094 112
ETRI 能源科技研究所 xix 114 116
Jubilee Campus 朱比利校区 xix 116 118 216
Lighting 采光 ix 085 106 120 124
Derbyshire 德比郡 xix 085 120 122 124
Sixth Form 第六阶（高三） xix 120 122 124
Prevailing Wind 盛行风 124
Microclimate 微气候 120 124
Tectonic 建构 vi vii ix x xv xvii xix 001 126 127 128 136 156 161 216
Stone 石材 xix 130 138 146 147
Geology 地质学 012 134
Conservation 保护 001 010 016 044 046 066 076 086 124 127 130 134 138 154 160 169 176 177
Channel 运河 078
Old Market Square老市场广场 xix 140 142 144
Material 材料 ixx xiii xiv xvi 001 008 030 033 036 048 072 080 082 087 090 127 130 142 146 156 166 192 196 201 204 206 214 216
Liverpool 利物浦 x xix 152
Hub 枢纽 152
Spab 古建筑保护协会 127 154
Newhaven 纽黑文 xix 154
Digital 数码 xix 054 116 118 130 138 156
Baroque 巴洛克 158
Meadows 梅多斯区 x xix 162 164 166
Emission 排放 ix 085 166
Intervention 介入 ix 030 042 169 174 188 192
Lincoln 林肯 xix 169 172
Framework 框架 xix xiii xv xvi 096 127 152 169 174 196 208
Tang Gu 塘沽 xix 176 178 179 180
Diversity 多样性 127
Identity 可识别性 xii xix 008 014 182
Mumbai 孟买 184
Waterfront 滨水区 177
Lobby 门厅 xviii xix 080 081 190 191
Limpopo 林波波 xix 201 204 206 208 210 213 214
Nursery 幼儿园 xix 201 204 208 210
Jouberton 朱博顿 xix 201 204 213
Calais Village 加莱斯村 204 206
H.O.U.S.E 太阳能应用最优化住宅 xix 214 216 217
Story 故事 x xiv xix 016 032 127 180 220 221
Memory 记忆 xvi 001 020 032 046 048 134 169 178 180 188 221
Corridor 走廊 034 046 049 092 104